Ground Truths

Community-Engaged Research for Environmental Justice

Edited by

Chad Raphael and Martha Matsuoka

UNIVERSITY OF CALIFORNIA PRESS

University of California Press
Oakland, California

Suggested citation: Raphael, C. and Matsuoka, M. *Ground Truths: Community-Engaged Research for Environmental Justice*. Oakland: University of California Press, 2024. DOI: https://doi.org/10.1525/luminos.174

Cataloging-in-Publication Data is on file at the Library of Congress.

ISBN 978-0-520-38433-0 (pbk.)
ISBN 978-0-520-38434-7 (ebook)

33 32 31 30 29 28 27 26 25 24
10 9 8 7 6 5 4 3 2 1

Ground Truths

The publisher and the University of California Press Foundation
gratefully acknowledge the generous support of the Atkinson Family
Foundation Imprint in Higher Education.

For Ann F. Wiener (1933–2018) and her lifelong work
for environmental, social, and educational justice.

For Eiko Yamamoto Matsuoka, whose voice and commitment
to community connects so many across generations and place.

CONTENTS

Introduction

Chad Raphael and Martha Matsuoka

This book is about why and how professional and academic researchers contribute to environmental justice by collaborating with community partners to conduct research. While many kinds of research can make useful contributions to environmental justice (EJ), we argue that community-engaged research (CER) is necessary to advance justice through the research process, not only through its outcomes. EJ is best served when communities exercise power to produce and control their own knowledge to inform and influence decisions affecting residents. Unlike conventional research conducted *on* communities, engaged research conducted *with* communities centers local knowledge, inquiry, and experience at each stage—from designing the agenda to gathering and analyzing data to disseminating results and implementing action in response. This research facilitates and elevates truths situated in community knowledges and perspectives, and builds evidence of that truth from the ground up. This approach shifts researchers' role from extracting data and resources from communities to co-constructing knowledge and sharing research resources with community partners. Thus, the means and ends of CER fulfill EJ by valuing and centering community knowledge, building community and movement capacities to generate new understandings, producing actionable data that can influence decisions, and transforming relationships between researchers and communities to be more equitable, respectful, and mutually beneficial.

People in a wide range of fields and institutions can practice CER for EJ. They collaborate using many approaches described in this book, including participatory action research, community-based participatory research, citizen science and community science, Indigenous-led and decolonial research, and more. CER extends beyond academic scholarship to encompass many other kinds of research that can be expressed in myriad genres, from the gray literature to white papers to blueprints, from policies to posters to plays, from maps to marketing campaigns, and much more. The EJ movement has inspired many partnerships among academics and other professional researchers, students, community organizers and policy advocates, government staff and officials, members of religious organizations, development and conservation workers, educators, engineers, designers, artists, and others.

While building research partnerships with communities is the subject of several excellent books, many of them are specific to a particular discipline or CER approach, or include EJ as one of many issues. Drawing on the academic and professional literature of many fields, this book offers a critical synthesis of a wide swathe of engaged research on EJ, describes the major research methods used, suggests agendas for future research, outlines the main steps for conducting engaged research projects, and addresses overcoming institutional barriers to this kind of scholarship in academia. Throughout the book, we employ an original framework that shows how EJ and CER address common dimensions of justice and that links research on the many topics treated in the chapters. We illustrate this discussion with multiple examples and case studies—involving either outside researchers collaborating with EJ organizations or these organizations conducting research on their own. We intend to provide promising concepts, practices, and examples for improving the theory and practice of CER for EJ, not to speak on behalf of the organizations involved in these research projects, or to derive a narrow set of best practices and imply that they can be mechanically replicated elsewhere. While the book is aimed at researchers and students, we point them toward briefer guides and trainings to share with community partners, which are designed to address their needs and perspectives in CER.

The CER approach suggests that anyone can contribute to research if properly trained, but that none of us can produce it alone. It recognizes that knowledge is situated in our individual perspectives and experience, and also produced through our relations with others. Both insights motivated us to start this book project and guided our choices about who to involve in writing it. As we conceived the book, we realized that the principles of CER and EJ, and the dynamism of the field, demanded additional perspectives beyond our own, and more collective authorship, and we therefore sought out co-authors for most chapters. After identifying the chapter topics, we recruited researchers with disciplinary expertise that complemented our own, all of whom were early-career scholars who were Black, Indigenous, or people of color, and who had published work that used or drew on CER to help build the future of EJ research in their fields. We co-developed detailed chapter outlines together and invited our co-authors to serve as lead authors, who developed the arguments and drafted the majority of each chapter, then edited collaboratively to strengthen each other's work and give the book a consistent structure and shared focus. Our collaborators' insights and experiences of CER and EJ produced a far richer and more expansive book than we could have written alone.

STRUCTURE OF THE BOOK

Part 1 establishes the foundations for the book by defining and describing the development of EJ and CER, showing why they are especially well suited to each other and offering a current summary of the major literature on both topics. Chapters trace the expanding definitions, movements, and body of research for EJ in the U.S. and globally since the 1980s, as well as the development of CER and how a CER paradigm can increase the reach, rigor, relevance, and reflexivity of EJ research. We conclude by introducing the unifying framework used throughout the book, focused on how researcher-community collaborations can advance four dimensions of justice common to CER and EJ.

Part 2 addresses how community-engaged researchers co-construct knowledge about EJ with community partners. Chapter 3 summarizes the preparatory work and self-analysis that outside researchers must do before partnering with EJ communities to address power imbalances and bridge differences. Chapter 4 describes how researchers and community partners address power relations in each of the major stages and elements of CER, from sharing funding to co-disseminating findings and taking actions. Chapter 5 addresses the barriers to conducting this kind of research within higher education—related to control of funding and data, research ethics and evaluation, and recognition of community and Indigenous knowledge and interests—and how these barriers might be dismantled through restorative justice.

Part 3 explores the ways that community-based research has been applied to EJ across disciplines. This part begins with a chapter summarizing the range of research methods and methodologies most often employed in CER for EJ. Subsequent chapters explore how this research has contributed to EJ in law, policy, regulation, and public participation; community economic development; public health; food justice and food sovereignty; urban and regional planning; and conservation and restoration. Each chapter presents an overview of CER in this domain, some exemplary case studies, and some directions for future research.

These chapters are organized according to the kinds of work that EJ organizations and their research collaborators seek to inform and influence, such as influencing policies, strengthening local economies, and planning cities. Although some of these applications overlap with academic fields, the chapters are not primarily organized to present each discipline's contributions to the topic, or around specific environmental media (water, air, land) or environmental threats (such as hazardous waste sites, individual chemicals, and so on). While valuable, these ways of defining research foci are more aligned with specialized academic training and priorities than with how communities experience EJ issues as a collection of cumulative and historical harms that are environmental, social, and economic in origin. Similarly, CER typically prioritizes practical and holistic goals to improve communities and build their power, not simply aiming to build disciplinary understanding, advance basic research on the health effects of individual substances, or even improve applied knowledge of how to protect air or water quality in general. EJ and CER goals typically demand research that crosses disciplinary lines, for example to support a community to organize, conduct toxicological and epidemiological studies, bring litigation, and promote policy change.

Our focus on the nexus of CER for EJ means that some important or emerging EJ issues and events are treated briefly or not at all because CER has not yet played a role in them. For example, CER did not contribute significantly to protests against the Dakota Access Pipeline led by the Standing Rock Sioux tribe beginning in 2016, which galvanized global attention to EJ and Indigenous resistance to extractive industries. Conversely, we do not discuss many trailblazing CER studies that do not explore issues of EJ. In addition, while we cite many examples of the growing research on EJ around the world, our and our co-authors' understanding of the topic and selection of cases are shaped mainly by the U.S. experience and by literature published in English. Nonetheless, we hope the book will be valuable to readers elsewhere, and we look forward to learning more from them about how CER can advance EJ around the world.

TERMINOLOGY AND TONE

Because there are many terms used to describe the key concepts mentioned in this book, we chose terminology by considering current and historical meanings. For example, *environmental justice* emerged from the social movement and research in the U.S. that started in the 1980s, in part to link *environmental racism* to additional axes of environmental oppression, such as class, Indigeneity, and gender. In chapter 1, we discuss some of the many other names used to describe elements of EJ around the world to connect these important movements and ideas across borders.

We use *community-engaged research* as an umbrella term for many kinds of research approaches—participatory action research, community-based participatory research, citizen and community science, and others—which have converged in some ways but retain significant differences. Rather than attempting to impose single definitions of other significant terms that have been defined differently, such as *decolonizing knowledge*, we try to provide brief definitions when they are used initially in the context of each chapter.

Terminology for race, ethnicity, and Indigeneity continues to evolve rapidly and consensus is elusive. Throughout, we and our co-authors aim to use terms appropriate to context and used by movement participants themselves today, while recognizing that no term is universally embraced. We name groups in the most specific terms that fit the context: Standing Rock Sioux, youth residents of the East Yards, and so on. When discussing the U.S., we sometimes refer collectively to Black, Indigenous, and people of color (BIPOC) to reflect their similar positioning and frequent solidarity in EJ movements, as well as their distinct identities.

We recognize that language is historically contextual and use the specific terminology used by authors in their earlier works. However, referring only to BIPOC can slight the important, distinctive, and place-based environmental injustices faced by communities of many other races and ethnicities, such as Latinx, Chicanx, Asian, Pacific Islanders, white working-class people, and the hundreds of Indigenous nations and tribes in the U.S. In some cases, we refer collectively to people who face such inequities by the terms *environmental justice communities* or *frontline and fenceline communities.*

Few collective terms for race and ethnicity used in the U.S. describe people struggling against environmental injustices outside the U.S. precisely or adequately, or are recognized by people in those communities. When we cannot be more specific, we sometimes refer to them as residents of the Global South. When discussing Indigenous nations or tribes in CER, we typically adopt the terms they use in their published research. Elsewhere, we try to use specific tribal names when they describe the community relevant to the study, broader regional names for referring to larger collectivities (such as *Native American* or *American Indian*), and *Indigenous peoples* for the highest level of generality about peoples original to their places who face similar struggles for recognition from states (following Gilio-Whitaker 2019).

We recognize that gender and other aspects of identity are often defined at the individual and personal level with terms used by different people in different ways for different purposes and allow for different expressions throughout the book. These identities are important sites of solidarity and social movements that are defined, negotiated, and assumed through discussion and action. We embrace principles of CER that require honest recognition of privilege and power in relationships, and therefore aim to recognize communities' collective self-naming as an important act of empowerment (Kirk and Okazawa-Rey 2019).

Another important challenge of writing about EJ is to find a balance between evoking appropriate outrage and inspiring careful analysis (Lockie 2018). Writing the book during the COVID-19 pandemic and during the widespread reckoning with anti-Black police violence in the U.S. was a constant reminder of how environmental injustices wreak violence on people's bodies, minds, and communities. Environmental contaminants kill fetuses and cause birth defects, cut down young people exposed to toxics in the workplace, and cause or worsen diseases that immiserate and shorten lives. Seizure and contamination of lands displace and destroy communities and cultures. These harms, and the consciousness of them, heap additional trauma on individuals and communities coping with racism, exploitation, colonialism, patriarchy, and other oppressions. Institutions that perpetrate this violence do not simply do so unintentionally or from ignorance, but often consciously designate some communities as sacrifice zones to amass wealth, accrue power, and protect environmentally privileged people from facing the same harms. The fact that much of this is "slow violence" (Nixon 2011), unfolding over years or centuries, makes it even more important to make this violence manifest in scholarly writing.

At the same time, EJ research needs to offer a clear-eyed analysis of the causes, consequences, and potential remedies for environmental injustices. Therefore, we and our co-authors discuss examples of environmental harms as precisely as possible, clarifying who is affected by them and how, and consider potential strategies and solutions critically. We also aim to address multiple kinds and degrees of community involvement in research, rather than dismissing some kinds of CER as irredeemably extractive or authoritarian. We take inspiration from movements and researchers pursuing justice by questioning, and engaging in dialogue about, these matters, which can extend to seemingly small decisions about how to write. For example, while there are some scholars who find it inherently trivializing to abbreviate *environmental justice* as *EJ*, we are less troubled by doing so. Like the choice of terms for racial and ethnic groups, we think that what is written *about* and *around* these terms makes the most significant contribution to illuminating EJ and stoking the fires of change to enact it.

PART 1

Foundations

1

Environmental Justice

Martha Matsuoka and Chad Raphael

Community-engaged researchers who want to contribute to environmental justice (EJ) need a full understanding of the concept of EJ, the movement they want to collaborate with, and the main developments in EJ research. This chapter begins by tracing the expanding definitions of the dimensions of environmental justice, and summarizes the growth of EJ movements in the U.S. and globally since the 1980s to set the stage for more detailed exploration of community-engaged research (CER) for EJ in later chapters of this book.

To ground researchers in history as well as current issues and debates, we devote equal attention to the growth of EJ as a concept, a movement, and a body of research. Yet this is not to suggest that they have played equal roles in the development of EJ around the world. On the contrary, movement thinking, organizing, and demands for change have incubated and motivated much of the theory and research. One of the main arguments of this book is that researchers should deepen their collaboration with EJ movements. To do this, researchers need to consider the multiple dimensions of EJ at stake in any study, how to work with community partners to craft research questions of mutual interest and benefit to EJ communities and movements, and how to employ and improve prior theory and findings on environmental injustices and their potential remedies.

DEFINING ENVIRONMENTAL JUSTICE

Environmental justice is a dynamic and evolving concept because it may be used descriptively or normatively, and because it is a traveling concept that has accrued many meanings as it has spread across diverse political, cultural, and scientific domains around the world (Holifield, Chakraborty, and Walker 2018). EJ is also a concept that continues to grow as activists and researchers confront new

TABLE 1.1. Dimensions of Environmental Justice

Dimension of Justice	In Environmental Justice
Distribution *Who ought to get what?*	Reducing environmental burdens, and increasing environmental benefits and capabilities, for EJ communities and the earth
Procedure *Who ought to decide?*	Participation and influence in environmental decision making by historically excluded groups, particularly in frontline communities Protection of individual and group rights through law, regulation, enforcement, and informed consent
Recognition *Who ought to be respected and valued?*	Respect for EJ communities' diverse environmental cultures and knowledges, and for the interests of future generations and non-human nature
Transformation *What ought to change, and how?*	Restoration of nature and reparation of damages to EJ communities from colonialism, racism, economic exploitation, and other systems of oppression Systemic and structural transitions to create just power relations, regenerative economies, and reciprocal relations with nature

developments in the world. EJ is often defined in universal terms as "the principle that all people have the right to be protected from environmental threats and to benefit from living in a clean and healthy environment" (Davies and Mah 2020a, 4). Yet EJ is principally "an affirmation of an unequal present and yearning for a better future" (4), in which people of color and of low income, Indigenous peoples, women, future generations, and all species can thrive, rather than having their lands, homes, cultures, and lives poisoned or stolen.

As we define EJ more fully below, we distill previous thinking and diverse terminology into four dimensions of justice that have traditionally defined EJ scholarship (see table 1.1). We adopt David Schlosberg's (2009) influential framework, which identifies distributive, procedural, and recognition dimensions of EJ, and we add the emerging dimension of transformational justice. We treat capabilities justice, also discussed by Schlosberg, as an element that cuts across several dimensions. We ground each dimension in the main principles of the movement in the U.S., as they were stated in its constitutional document, the *Principles of Environmental Justice*, adopted at the First National People of Color Environmental Leadership Summit (1991), which articulated the values of grassroots leaders in the nascent movement, identified the distinct and common environmental threats they faced, and developed a shared analysis of and vision for EJ. We also mention some of the typical challenges that arise when applying each dimension of EJ to real-world conflicts, which provoke debate among activists, researchers, and policy makers. We see these four dimensions as interrelated elements of the holistic concept of EJ. This is because it seems both impossible and undesirable to arrive at a fair agreement about how to share environmental benefits and burdens (distributional justice) without involving EJ communities meaningfully in making this decision (procedural justice) and respecting their diverse cultural understandings of the environment (recognition justice), which current institutions are incapable of doing without radical change (transformational justice).

Distribution

The distributive dimension of EJ refers to the fair apportioning of environmental burdens and benefits, and ensuring that environments allow all people to exercise their capabilities. These kinds of issues arise whenever there are disproportionate or intolerably intense harms and deprivations, regardless of whether they are caused by intentional discrimination on the part of specific actors (Kaswan 2021). Distributive concerns formed the initial core of the U.S. EJ movement as it documented and opposed environmental *burdens* on frontline communities, which faced the greatest environmental threats. Protestors fought against contamination from hazardous waste landfills, trash incinerators, oil refineries, chemical plants, mines, and other polluting facilities in majority Black, Latino, and Asian American residential areas; Native American reservations; and rural white working-class communities (Cole and Foster 2001). Groundbreaking research on environmental racism and justice documented the disproportionate exposure of communities of color to hazardous production and waste facilities (Bullard 1990; Commission for Racial Justice 1987).

However, distributive EJ is concerned not simply with *comparative* well-being among groups, but also with the *absolute* well-being of humans and nature. For example, when advocates of waste incinerators accused early EJ activists in the U.S. of being selfish "Not in My Backyarders" (NIMBYs) for resisting polluting facilities, activists replied that contamination did not belong in anyone's backyard and characterized themselves as "Not on Planet Earthers" (NOPEs) (Pellow 2007, 96). Similarly, the *Principles of Environmental Justice* asserted rights to "universal protection from nuclear testing, extraction, production and disposal of toxic/hazardous wastes and poisons and nuclear testing that threaten the fundamental right to clean air, land, water, and food" (para. 5), as well as a universal responsibility to "challenge and reprioritize our lifestyles to ensure the health of the natural world for present and future generations" (para. 18).

In addition, the movement demanded policies for improving EJ communities' access to environmental *benefits*, such as access to clean water and energy, transportation infrastructure, urban gardens and greenspaces, and green jobs (Agyeman, Bullard, and Evans 2003). For example, the *Principles* asserted the "right of all workers to a safe and healthy work environment without being forced to choose between an unsafe livelihood and unemployment" (para. 9), and demanded "ethical, balanced and responsible uses of land and renewable resources in the interest of a sustainable planet for humans and other living things" (para. 4).

The *capabilities* approach to global human rights and development (Nussbaum 2011; Sen 2010) also illuminates distributive aspects of EJ. In this view, justice involves the fair distribution of people's capabilities to function and flourish by realizing their own life choices. Those who apply this lens to global development typically include environmental and physical health as basic capabilities, which are equally important as, and inseparable from, traditional measures of economic well-being, such as income or wealth (Holland 2021). Moreover, individuals' ability to realize their capabilities depends in part on personal and external circumstances. Thus, this approach can help justify equity-based EJ policies, such as adopting stricter exposure limits to hazardous materials to protect people who are most vulnerable to harm (people with compromised immune systems, children, etc.). Capabilities theory has informed measures of collective well-being of humans and nature, such as the United Nations Development Programme's (2018) human development indicators and indices, which now include country-level measures of mortality from air and water pollution, and risk of extinction across groups of species.

However, resolving issues of distributive EJ poses several typical challenges. First, competing principles of distributive justice can lead to different conclusions about how to address unequal benefits and burdens. Should these inequalities be remedied by maximizing overall social welfare (utilitarianism), or by striving for equal distribution of environmental benefits and unavoidable environmental burdens, or by acting in a way to benefit the least environmentally advantaged, or the most historically oppressed, or those in greatest need, or those who deserve greater benefits because they have contributed least to or benefited least from polluting activities, and by other means (Kaswan 2021)? Second, even if we focus not on comparative well-being, but on guaranteeing a common set of capabilities for all, there is still a need for agreement on what those capabilities include and how to resolve potential conflicts among them. Moreover, the theory as a whole has been criticized for conceiving of capabilities solely in individualistic, human, and Western terms that do not reflect other conceptions of fair distribution, especially those of many Indigenous peoples. For example, Watene (2016) points out that capabilities theory conceives of nature instrumentally as a provider of ecosystem services (such as clean air and healthy food) to humans, rather than respecting natural beings as human kin and recognizing that care for their lands is central to many peoples' worldviews and identities. The capabilities view might accept separating Indigenous peoples from their traditional homelands if comparable ecosystem services could be provided to them elsewhere, while the latter view would see this as depriving a people of their existential right and responsibility to maintain their place-based relationships to specific species and sacred sites. As discussed below, conflicts such as this implicate the dimension of justice-as-recognition. For now, it is enough to say that EJ research and activism need to grapple with which principles of distributive justice (and whose) are most appropriate to remedy environmental injustices, and to weigh distributive considerations against other dimensions of justice.

Procedure

Procedural justice concerns "the ability to participate in and influence decision-making processes" (Suiseeya 2021, 38). EJ calls for meaningful participation and influence in environmental decision making by people who are affected by these decisions, especially historically excluded groups in frontline communities, and for consideration of the interests of future generations and non-human nature. This type of justice focuses on whether decision-making processes provide full *access* to information and *inclusion* of participants, whether people and other species are *represented* by those who are authorized to speak for their communities, and whether participants from EJ communities can exercise *power* over outcomes (Bell and Carrick 2018; Suiseeya 2021). Procedural matters also include protection of individual and group environmental rights through law, regulation, enforcement, and requirements for free and prior informed consent by affected communities for decisions and research. Capabilities such as self-determination, control over one's environment, and freedom from discrimination are central to this type of justice (Holland 2021).

Procedural justice has been a central concern of EJ movements, legislation, and treaties. In the *Principles of Environmental Justice*, EJ activists demanded "the right to participate as equal partners at every level of decision-making, including needs assessment, planning, implementation, enforcement and evaluation" (para. 8). The *Principles* also called for "strict enforcement of principles of informed consent, and a halt to the testing of experimental reproductive and medical procedures and vaccinations on people of color" (para. 14). In her keynote address to the 1991 summit where EJ movement leaders drafted the *Principles*, Dana Alston's pronouncement, "we speak for ourselves," claimed knowledge, experience, and voice for the movement in environmental policy making and representation in mainstream environmental organizations (First National People of Color Environmental Leadership Summit 1992).

Formal rights to participate are widely recognized around the world, although participation influences decisions unevenly. At present, over 100 countries have legislated mandatory public involvement in environmental decision making (Suiseeya 2021). In the U.S., legislation such as the National Environmental Policy Act of 1970 triggered reviews of environmental impacts of federally funded projects, as did many counterpart laws enacted by states. The 1998 Aarhus Convention, an international European treaty, establishes some of the strongest public rights of access to environmental information, participation in decision making, and access to the courts. Numerous United Nations conventions and forums—on climate change, biological diversity, parks and protected areas, and illegal trade in endangered species—require Indigenous participation (but typically on a non-voting basis) in international negotiations over environmental and development policy (Suiseeya 2021).

Assessing procedural justice requires careful attention to how power is exercised at each stage of decision making. As Suiseeya explains, "*Whose* problems are identified, *how* problems are defined, and the *salience*, or importance, of particular problems are dependent on *who* constitutes the body of decision-makers and the *relative abilities* of decision-makers to influence the decisions" (2021, 48). Many EJ communities and researchers are skeptical about participatory environmental governance, based on bitter experiences of engaging with state agencies that frame agendas to exclude community concerns, withhold information, refuse to communicate in lay terms and in participants' languages, exclude affected groups from discussion, and treat public participation as an inconvenient bump in the road to ratifying decisions officials have already made.

Recognition

A third dimension of EJ is recognition, including who gets respected and valued. In EJ, recognition entails respect for diverse peoples' environmental cultures (beliefs, values, practices) and knowledge (Schlosberg 2009; Whyte 2018a). This dimension of EJ highlights two broad kinds of injustices (Coolsaet and Néron 2021). One is exclusion of or discrimination against people who deserve equal standing or consideration by relegating them to lesser status because of their identity. Many environmental injustices are rooted in historic and systemic racism and cultural oppression. For example, Pulido's (1996) study of Chicano-led campaigns by farmworkers against pesticide exposure and by small livestock growers for grazing rights reveals how these were not merely struggles over environmental and economic claims, but over "confronting a racist and exclusionary political and cultural system, and establishing an affirmative cultural and ethnic identity" (193). Failure to recognize future generations and non-human nature as worthy of consideration in decisions is also a major violation of justice-as-recognition.

Another kind of misrecognition involves coercive assimilation, which disrespects differences among peoples by imposing dominant cultural and scientific understandings and policy solutions universally. Much Indigenous-led resistance to environmental injustice involves demands for recognition of native peoples' cultural autonomy, self-determination, and land rights, which is "nothing less than a matter of cultural survival" (Schlosberg 2009, 63). For example, when the Standing Rock Sioux protested the Dakota Access Pipeline in 2016, drawing support from around the world to block an oil pipeline that would have crossed the Missouri River on the tribe's reservation, the tribe based its demands on recognition of their kinship with the river and its sacred status, rather than seeking a fairer distribution of the pipeline's environmental risks or protection of proprietary water rights (Estes and Dhillon 2019).

The EJ movement prioritized recognition from the start. The first principle of the *Principles of Environmental Justice* called for recognition of "the sacredness of Mother Earth, ecological unity and the interdependence of all species, and the right to be free from ecological destruction" (para. 2). Additional principles included "demands that public policy be based on mutual respect and justice for all peoples, free from any form of discrimination or bias" (para. 3); an affirmation of "the fundamental right to political, economic, cultural and environmental self-determination of all peoples" (para. 6); and recognition of "a special legal and natural relationship of Native Peoples to the U.S. government through treaties, agreements, compacts, and covenants affirming sovereignty and self-determination" (para. 12). The *Principles* also anticipated efforts to decolonize knowledge by calling for education about "social and environmental issues, based on our experience and an appreciation of our diverse cultural perspectives" (para. 16). Demands for respecting Indigenous knowledge have advanced through the growing influence of traditional ecological knowledge (TEK) in environmental research and regulatory fora (see chapter 2); the adoption of data sovereignty protections for Indigenous peoples' ability to control information gathered about biodiversity and sacred sites on their ancestral lands (see chapters 5 and 12); and the growth of Indigenous-led academic and research institutions in Latin America, North America, New Zealand, and elsewhere.

Addressing conflicts of justice-as-recognition can pose significant challenges, especially because recognition is not always easily integrated with the distributive and procedural dimensions of justice. Some worldviews cannot be reconciled easily, such as the resource view of nature in which a river is a collection of ecosystem services that can be fairly distributed, and a relational view of nature in which a river is a holistic source of life and cultural identity that must be protected because it is sacred. Unequal power in policy and decision making has tended to decide these conflicts in favor of dominant state and economic interests. In other cases, newly recognized rights of nature have granted protection to rivers and landscapes, and assigned Indigenous peoples rights of guardianship to protect these natural features (Akchurin 2015). Procedural solutions also fail to offer a panacea for some conflicts over recognition. Coulthard (2014) highlights the dangers of co-optation and internalized oppression when Indigenous peoples are recognized as partners in decision-making processes but held in a subordinate position. His study of the Canadian government's long-term deliberations with the Dene First Nation over a pipeline project suggests that the process transformed the Dene's relationship to the land, gradually persuading them to think of it in resource-based (proprietary and profit-oriented) terms rather than relational terms, and to accept a pipeline they had initially resisted. These examples point to the importance of considering the quality, extent, and terms of recognition, amidst ongoing pressures of colonization, capitalism, and systemic racism that constrain EJ communities' ability to defend their culture, knowledge, and right to choose their own economic development plans.

Transformation

Transformational justice is an emerging dimension of EJ, which we add here because it is an increasingly important goal for EJ movements. Transformational justice draws on and extends traditions of restorative and transitional justice. Restorative justice, which emerged from criminal justice reform, seeks to engage offenders in dialogue with victims about how they have been affected by a crime, and to have them decide jointly on steps to repair the harm, with the goals of healing their relationship and healing the community (Capeheart and Milovanovic 2020). Transitional justice was developed to guide national transitions from authoritarianism to democracy and from war to peace, typically by organizing official commissions to seek truth about past abuses, establish accountability by responsible parties, offer reparations to victims, and recommend measures to avoid repetition of harms (Killean and Dempster 2021).

Each kind of justice can be applied to abuses of EJ, for example by deciding on reparations for past contamination of and harms to communities of color, or preparing transitions to full-state recognition of Indigenous peoples' land rights, or guiding climate change policy that recognizes rights of workers. The *Principles of Environmental Justice* appealed to restorative justice in affirming "the right of victims of environmental injustice to receive full compensation and reparations for damages as well as quality health care" (para. 10) and demanding that "all past and current producers be held strictly accountable to the people for detoxification and the containment at the point of production" (para. 7). Restorative claims can also include reparations for future adverse impacts, such as anticipated job losses in the fossil fuel industries as part of a just transition to cleaner energy sources (McCauley and Heffron 2018). Harms to individuals, groups, or nature may require reparations that involve redistribution (such as money damages to pollution victims from legal settlements), procedures (such as the inclusion of new groups in the policy-making process), or recognition (of the sovereignty of Indigenous groups over their traditional homelands, or the rights of nature, for example). The *Principles* also called for a transition to reciprocal relations among humans and nature, urging "urban and rural ecological policies to clean up and rebuild our cities and rural areas in balance with nature, honoring the cultural integrity of all our communities, and providing fair access for all to the full range of resources" (para. 13).

However, current models of restorative and transitional justice can be too narrow to advance EJ. Both typically involve government-led, short-term processes focused on a limited scope of issues, and do not question fundamental relations of state power or economic control, which risks restoring unjust relations or transitioning to new injustices (Killean and Dempster 2021; Nagy 2022). Some EJ activists and researchers seek to enlarge these two types of justice to support deeper transformation of societies and their relation to their environments. Transformative approaches typically call for long-term processes led by movements of grassroots organizations that radically redesign structures of power, economic relationships, and dominant cultural narratives (Movement Generation Justice and Ecology Project, n.d.; Nagy 2022). A drive for transformative justice fueled some of the most prominent EJ campaigns in the 2010s, such as efforts to shift from an extractive to a regenerative economy while ensuring a just transition for workers and communities, implement environmentally just recoveries that "build back better" from disasters such as floods and earthquakes, enact rights of nature and return lands to Indigenous peoples to manage, advance alternatives to dominant plans for sustainable development, implement local examples of environmentally just production (of food, energy, and consumer goods), and dismantle racist systems of policing and prisons that create hostile and life-threatening environments.

Assessing transformative justice also poses a variety of challenges. Some of them relate to difficulties of weighing restorative justice. With regard to reparations, what kinds are owed, how much, to whom, from whom, who should decide, and how? Which criteria should be used to decide whether landscapes are restored or repaired (especially if some damages, such as species extinction, cannot be undone), much less human cultures, which are internally diverse and always evolving? Who decides? Some dilemmas are characteristic of transitional justice, such as how to resolve competing truth claims about abuses, and attribute personal and collective responsibility (especially to states and corporations). Some challenges are unique to transformational justice. How much change, and for whom, constitutes structural transformation rather than mere reformism? In addition, because this kind of justice involves an integrated vision of EJ, how should we assess uneven changes that involve improvements in some aspects but not others of economic and environmental equity, democratic decision making, and respect for cultures and nature?

MOVEMENTS FOR ENVIRONMENTAL JUSTICE

The expanding concept of environmental justice is important primarily because EJ movements have made it salient to communities, policy makers, and researchers around the world. Community-engaged researchers who work on EJ do not only enter an ongoing conversation among scholars in their fields, but also enter into high-stakes discussions within EJ movements about their communities' health and survival. Therefore, researchers must be familiar with the broad contours of these movements. Below, we sketch their history, including their diverse origins, their redefinition of mainstream environmentalism and sustainable development, and their characteristic structures and strategies.

While movement leaders developed the initial terminology and organizing for EJ in the U.S. in the 1980s, they addressed a complex of issues rooted in the global history of colonialism, capitalism, patriarchy, the slave trade, and other systems of racial oppression, which seized, exploited, and destroyed lands and peoples for centuries and which continue to shape people's environments and relations with nature today. Many local examples of resistance to these oppressions comprise what has been called the "long Environmental Justice movement" (Pellow 2018, 9).

EJ Movements in the United States

The contemporary EJ movement emerged in the U.S., as Black, Indigenous, and people of color (BIPOC) communities confronted immediate environmental threats to their neighborhoods, workplaces, and health. Local campaigns against environmental racism broadened into a movement for environmental justice as activists identified and opposed common sources of harm, especially from waste dumping and incineration, mining, industrial and agricultural chemicals, energy production, military toxics, and dispossession from ancestral lands. Given the internal diversity of the movement, the 1991 *Principles of Environmental Justice* discussed above were a major step toward building solidarity and networks for organizing in the growing movement.

The movement drew inspiration and activists primarily from movements for the civil, economic, and cultural rights of Black, Latinx, and Asian Americans and Indigenous movements for self-determination or sovereignty, but also from women's movements for health and reproductive justice, from the labor movement (especially farmworkers' campaigns against pesticides and manufacturing workers' occupational safety and health committees), and from grassroots campaigns against toxic contamination in white working-class communities (Bullard 1990; Cole and Foster 2001; Gaard 2018; LaDuke 1999; Peña 1998; Pulido 1996; Sze 2004; Taylor 1997, 2000). As the vision of EJ grew to encompass urban health and its many determinants, organizers and advocates drew inspiration from movements for public health, social work, and urban planning, which reach back to the 1800s (Corburn 2009; Gottlieb 2005; Taylor 2009).

The modern EJ movement reframed Americans' understanding of the environment and environmentalism. Whereas the traditional environmental movement had focused attention on protecting and managing wildlands and waters, the EJ movement redefined the environment to include people's everyday physical and cultural surroundings: homes, neighborhoods, schools, sacred sites, workplaces, and more (Čapek 1993). The EJ movement also forced a reckoning with racism in the mainstream environmental movement. Led by white, economically privileged males, 20th-century U.S. environmentalism had contributed to forced removal of Indigenous peoples from their lands in the interest of forestry and wilderness preservation (see chapter 12), advanced policies that excluded BIPOC residents from white neighborhoods (see chapter 11), promoted nativist movements to exclude immigrants of color from the country as perceived threats to racial and environmental purity (Taylor 2016), and supported coercive sterilization programs targeting people of color in the name of population control (Hartmann 1995). In the 1970s and 1980s, the largest national environmental organizations routinely employed litigation and policy strategies that ignored the interests of EJ communities, or cut deals with polluters and state agencies that undercut local EJ organizers' demands. In response, EJ activists called for a more inclusive environmental movement with increased staffing and leadership by people of color who could reverse the movement's historic racism and hold it accountable to EJ communities (Southwest Organizing Project 1990).

The EJ movement also departed from the structure and strategies of mainstream environmentalism, which was controlled by a handful of large organizations led by professional staff headquartered in Washington, D.C. In contrast, the EJ movement comprised local organizations linked by regionally and ethnically defined networks that provided grassroots organizations with technical, legal, and financial support, and helped them build a wider base of support through organizing (Córdova 2002; Córdova et al. 2000; Schlosberg 1999). While these networks formed the initial glue of the EJ movement, they employed a translocal model of organizing that fostered cooperation between local organizations to build common knowledge and power, while remaining accountable to diverse grassroots constituencies. In contrast to the traditional environmental movement, people of color, especially women of color, formed the majority of the leadership of the EJ movement (Taylor 1997).

While mainstream environmental organizations prioritized national litigation and policy advocacy, EJ activists' initial strategies prioritized community organizing, using tactics of nonviolent protest and direct action to open negotiations with state and corporate actors over influencing facilities-siting decisions, legislation, and regulation (Cole and Foster 2001). The EJ movement also employed a community lawyering strategy, in which attorneys integrated litigation into larger organizing campaigns led by grassroots leaders (see chapter 7), as well as cultural organizing to strengthen members' collective identities based on shared identities, connections to place, and relations to nature and the environment (see chapter 6).

EJ Movements around the World

While the term *environmental justice* is not as widely used outside the U.S., EJ has become a global concern, although it is articulated differently around the world (Martinez-Alier et al. 2016). For Indigenous peoples, including those on lands in what is now called the U.S., EJ is a fundamental dimension of self-determination, protection and return of their traditional homelands, and the right to maintain native cultures and spirituality (Whyte 2018b). In Europe, EJ has often been seen more through the lenses of class and ethnicity than race (Walker 2012), and as an extension of human rights, as in the Aarhus Convention's protections for rights to information, participation, and adjudication of environmental issues. In the Global South, EJ issues are more often framed as matters of decolonization, climate justice and other ecological debts owed by polluters, resistance to multinational corporations, participatory and sustainable development and conservation, food and energy sovereignty, or the environmentalism of the poor (Carmin and Agyeman 2011; Carruthers 2008; Martinez-Alier 2002; Shiva 2016b; Walker 2012). Nonetheless, environmental justice is now a collective action frame that communities around the world use to interpret harms, identify their causes, and mobilize people to act (Sicotte and Brulle 2018). A coherent global discourse of EJ has helped to coordinate and guide policy and action among diverse organizations, coalitions, and governments by providing a common repertoire of concepts, analyses, evidence, and solutions (Agyeman et al. 2016; Walker 2012).

As in the U.S., movements addressing EJ issues elsewhere often arise in reaction to immediate threats to people's surroundings (Sicotte and Brulle 2018). Awareness of these issues has grown worldwide, especially in response to intensified globalization of the extractive economy; relocation of toxic and energy-intensive industrial production from the Global North to the Global South; growing exports of consumer goods to the North and waste to the South; migration of peoples fleeing environmental, economic, military, and political violence; development and conservation projects that displace and disrupt Indigenous cultures and economies; privatized ownership of natural resources and the commons; the globalization of unsustainable agriculture and food systems; existential threats to communities from drought, fire, flooding, and inundation posed by climate change; and the rise of social movements that link environmental rights to economic, social, and political rights (Bickerstaff 2018; Chu, Anguelovski, and Carmin 2016; Martinez-Alier et al. 2016; Peña 1997; Pellow 2018; Shiva 2016a, 2016c; Temper 2018). Notable examples of EJ movements around the world include Kenya's Green Belt Movement, which began by organizing women to plant trees and eventually helped uproot a dictatorial national government (Hunt 2014); the Ogoni people's resistance to oil extraction on their lands in Nigeria (Stephenson Jr. and Schweitzer 2011); and Brazilian rubber tappers' defense of the Amazon rainforest against logging (Keck 1995).

EJ movements increasingly reached across political and economic borders, blurring traditional boundaries of governance and institutions (Pellow 2011; Sikor and Newell 2014). EJ advocates in the U.S. began forming translocal and transnational ties from the 1990s onward, coordinating campaigns and litigation to confront globalized industries where they operated in multiple locales (Ciplet, Roberts, and Khan 2015; Claudio 2007). Movements focused on food sovereignty, biofuels, land and water confiscation, and other issues simultaneously addressed multiple sectors, such as agriculture, energy, mining, trade, and financial markets. Campaigns, such as those against hazardous waste dumping in the Global South, addressed policy and regulation at multiple levels of government around the world (Pellow 2007; Smith, Sonnenfeld, and Pellow 2006). Coalitions organized simultaneous worldwide demonstrations for climate justice, such as the People's Climate March of 2014, which mobilized people in 166 countries with the slogan "To Change Everything, We Need Everyone" (Giacomini and Turner 2015). Global

EJ advocates convened regularly to strategize and promote common visions of an alternative economy and environment at meetings of the World Social Forum and actions linked to the annual United Nations Climate Change Conferences, as well as UN processes on biodiversity and conservation. In doing so, organizers began setting local struggles in larger historical and global contexts, and building solidarity across borders. These strategies reflected the need for transnational alliances rooted in local organizing to address transborder issues, in which economic and political decisions made in distant locations profoundly shape local environments (Mendez 2020; Pellow 2018).

EJ movements have also challenged mainstream environmental thinking at the global level, especially in regard to sustainable development and climate justice. Intergovernmental programs for sustainable development have been faulted for prioritizing market-based economic growth over environmental protection and social equity (Agyeman, Bullard, and Evans 2003; Atapattu, Gonzalez, and Seck 2021). In response, activists have promoted alternative visions of sustainability, including the ideals of Buen Vivir (in Latin America), degrowth (in Europe and North America), Ubuntu (in Southern Africa), Ecological Swaraj (in India), and others (see chapter 8). In addition, by emphasizing the disproportionate impacts of climate change on people of color and people in poverty, EJ movements have reframed the issue as one of climate *justice* (Schlosberg and Collins 2014). They have gone beyond demands for developed countries, which are primarily responsible for historic greenhouse gas emissions, to transfer funds and technologies to help governments in the Global South cope with climate change (Chu, Anguelovski, and Carmin 2016). EJ movements have added demands for their communities to be recognized and to participate as full partners in designing and benefiting from climate resiliency plans, as well as a just transition to an equitable and sustainable economy for workers (see box 1.1).

BOX 1.1. FRAMEWORK FOR JUST TRANSITION

The *Strategic Framework for a Just Transition*, produced by Movement Generation Justice and Ecology Project (n.d.), developed with input from many organizations in the environmental and labor justice movements, offers one snapshot of the breadth of vision among contemporary movements that address EJ (see figure 1.1). The framework lays out pathways for a global transition from an extractive economy devoted to the "accumulation, concentration and enclosure of wealth and power" (7) to a regenerative economy of "ecological restoration, community resilience, and social equity" (15). Its "values filter" reflects demands for distributive justice (by democratizing wealth and promoting racial justice

(*Continued*)

BOX 1.1. (*CONTINUED*)

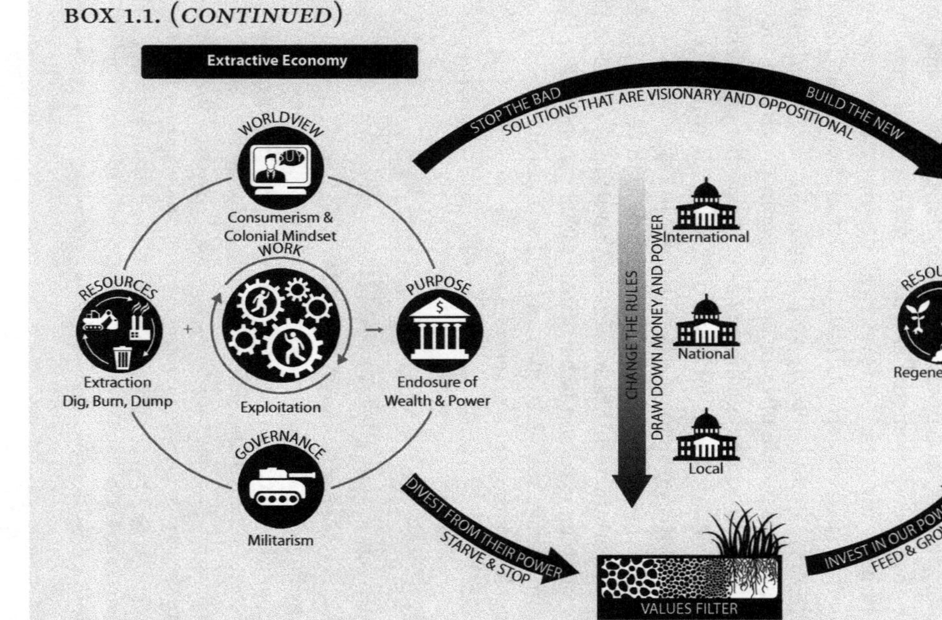

FIGURE 1.1. Strategic framework for a just transition.

SOURCE: Movement Generation Justice and Ecology Project (n.d.).

(*Continued*)

BOX 1.1. (*CONTINUED*)

and social equity), procedural justice (democratizing the workplace and transferring economic control to communities), and recognition justice (advancing ecological restoration, and retaining and restoring cultures and traditions). The framework envisions transformational justice via multiple pathways from extractive to regenerative worldviews, re-envisioning relationships to natural resources, ways of organizing work, means of governance, and purposes of the economy.

This expansive vision has informed the drafting of more detailed policy frameworks by frontline and allied organizations in the EJ movement to improve the proposed Green New Deal legislation in the U.S. (United Frontline Table 2020). Movement strategies for implementing this framework include multiple points of intervention: rewriting dominant narratives in public discourse and education, community organizing and base building to strengthen local power, involvement in policy development and implementation, electoral work to ensure responsive representation, and direct action through grassroots organizations and movements accountable to communities. The framework also informs the campaigns of major umbrella organizations working for environmental, economic, and racial justice, such as the Climate Justice Alliance (climatejusticealliance.org) and People's Action (peoplesaction.org).

The *Framework* and strategies for adopting it present a strong contrast to dominant discourses of sustainable development and mainstream climate policy. The latter embody top-down frameworks in which the most legitimate agents of change are states and intergovernmental organizations, which are informed by economic and technical experts and influenced by multinational corporations and the largest global environmental NGOs (Agyeman, Bullard, and Evans 2003; Atapattu, Gonzalez, and Seck 2021).

ENVIRONMENTAL JUSTICE RESEARCH

In addition to understanding the dimensions of EJ and the growth of EJ movements, community-engaged researchers need to be familiar with how EJ research has expanded over time, including research that does not employ a community-engaged approach. Directions within EJ movements have shaped many of the changes in the research agenda, but the growth of this expansive and pluralistic body of work has also been driven by its own dynamics. As Davies and Mah (2020a) observe, EJ research has spread *conceptually* to include additional aspects of justice (and, we would add, disciplines and methodologies); *horizontally* to additional topics, places, and peoples; *vertically* from consideration of local to global scales; and *temporally* to consider longer time periods and future generations. Because this research is voluminous, and much of it is addressed in later chapters of this book, we limit our citations here to a handful of pioneering studies and recent summaries that provide gateways into broad areas of research.

Conceptual Expansion

Tracking the expanding definition of EJ, research has broadened from an initial focus on fair distribution to include questions of procedural, recognition, and transformational justice, which required additional disciplinary and methodological approaches. Spatial analyses of the socioeconomic *distribution* of facilities and exposure to pollution formed the core of early EJ research. Foundational studies in the U.S. provided systematic evidence that hazardous waste was disproportionately sited in BIPOC and low-income communities (Bullard 1983; U.S. General Accounting Office 1983). A major national study by the United Church of Christ Commission for Racial Justice (1987) established that race was a more powerful predictor of proximity to waste facilities than socioeconomic characteristics. The study's findings and recommendations helped to legitimate the EJ movement and set its initial policy agenda (Agyeman et al. 2016).

Spatial-distributional analysis also set the research agenda for many years. In response to skeptics' challenges to these early studies, researchers applied more fine-grained measures of distance and emerging technologies for mapping, supplemented the plotting of polluting facilities with measures of emissions and exposures to residents and workers, and moved from studying single sources of pollution or individual pollutants to studying populations' cumulative exposure to environmental and social threats (Chakraborty 2018). This research continued to confirm disparities in exposure to toxics and other hazards by race, class, or both (Agyeman et al. 2016). Longitudinal studies addressed debates over the underlying causes and dynamics of these inequities, including discriminatory siting decisions, local land use regulations, and housing policies (Bullard 1990; Kaswan 2021). Exemplifying many of these advancements, a major follow-up study conducted 20 years after the United Church of Christ report demonstrated ongoing disparities from the clustering of multiple environmental hazards in communities of color (Bullard et al. 2008). The study also found that in most cases it was not that people of color moved into polluted areas in search of cheaper housing, but that polluters targeted existing minority neighborhoods for siting hazardous facilities.

The sociologists and geographers who produced these early studies were soon joined by urban planning researchers, who documented inequitable access to transportation, housing, parks, and other amenities (Anguelovski et al. 2018; Karner et al. 2018). Research in the health sciences began to study urban residents' disproportionate exposure to air, water, and noise pollution; agricultural workers' and communities' exposure to pesticides, chemical runoff, and noxious fumes; and workers' and fenceline communities' exposure to industrial chemicals and other hazards (Brown, de la Rosa, and Cordner 2020). Health researchers increasingly integrated methods of exposure monitoring, mapping, toxicology, and epidemiology to analyze the distribution and impacts of cumulative environmental and social stressors in EJ communities and workplaces (Solomon et al. 2016). Agricultural and food researchers analyzed inequitable opportunities to grow and consume healthy food, as well as poor labor conditions, in urban and rural settings and across food systems (Alkon 2018; Shiva 2016a). Community economic development and sustainable development researchers around the world applied EJ principles to research the inequitable impacts of the extractive industries, energy production, and urban development (Bickerstaff 2018; Urkidi and Walter 2018). Health, planning, food, and development researchers were especially responsible for introducing CER methodologies to the study of EJ.

While ongoing research on distributive issues is undeniably important for drawing attention to injustices, research has expanded to address other dimensions of EJ. As the EJ movement confronted polluters, it inspired legal, political, and economic analyses and case studies of the *procedural* barriers to participation and influence in the courts, regulatory processes, legislative arenas, and intergovernmental institutions (Foster 2018; Konisky 2015; Suiseeya 2021). This work also contributed to the development and evaluation of EJ policy and law, and included a significant strand of research conducted in collaboration with movements.

Additional disciplines produced studies relevant to the growing demands by Indigenous and other communities for *recognition* of their cultures, identities, and knowledge in environmental policy making and research forums. Anthropology, philosophy, history, as well as Indigenous, ethnic, gender, and environmental studies helped to illuminate diverse peoples' relationships to their environments and to misrecognition and repression by states, and called for decolonizing environmental knowledge (Gilio-Whitaker 2019; Jarratt-Snider and Nielsen 2020; Nelson and Shilling 2018; Rodríguez 2021; Whyte 2021). As chapter 2 describes, community collaborations helped to recover traditional ecological knowledge, providing valuable insights and alternative conceptions to Western environmental science. CER in the health and social sciences also helped frontline communities to develop their own popular epidemiology, environmental monitoring, biomonitoring, and other techniques for contributing local knowledge, which corrected official sources of data and challenged regulatory science's unwillingness to acknowledge the impacts of pollution on health.

Issues of *transformational* justice loom larger in recent EJ research, much of it provoked by, and some of it produced with, movements. This includes research on envisioning and evaluating local experiments in just and sustainable production of food, energy, and consumer goods (Agyeman et al. 2016; Apostolopoulou and Cortes-Vazquez 2018); policing and prison systems as environmental injustices (Pellow 2018); just transitions and community development (Harley and Scandrett 2019; McCauley and Heffron 2018); just recoveries from disasters (Bullard and Wright 2012; Chu, Anguelovski, and Carmin 2016; Howell and Elliott 2019);

and enacting rights of nature and alternatives to top-down conservation, which often involve returning lands and self-determination to Indigenous peoples (Atapattu et al. 2021; Ryder et al. 2021). Much of this work recognizes and strengthens EJ communities' place-based attachments and claims for justice, for example through planning and design that recognizes all residents' right to the city in culturally diverse metropolises, and through conservation plans that preserve Indigenous peoples' access to their ancestral homelands (Agyeman et al. 2016). Research on transformational justice is increasingly transdisciplinary (London, Sze, and Cadenasso 2018), conducted by researchers who cross and transcend the borders of their fields, and develop new ones, such as the conservation sciences, sustainability sciences, environmental studies and sciences, political ecology, development studies, regional studies, environmental communication and psychology, the environmental humanities and arts, and engineering and design sciences, to name a few.

Researchers concerned with transformational justice have debated whether EJ is possible without radical shifts away from extractive and racial capitalism, and settler colonialism, and how EJ movements should take part in legislative, regulatory, judicial, and consultative efforts. Should activists continue engaging in state-led processes or withdraw from them and challenge their legitimacy, while pursuing mutual aid strategies and creating alternative institutions (Pellow 2018; Pulido, Kohl, and Cotton 2016)? To what extent is EJ possible without efforts aimed at working both against *and* within states, with the aim of radically transforming them to wield their power for EJ, especially as a counterweight to corporate power (Purucker 2021)? How does an anti-state strategy square with the fact that some Indigenous peoples are themselves governments, which demand colonialist states' recognition and engagement in state-to-state relations as equals (Nagy 2022)? The conviction that EJ is not possible without radical change has also drawn attention to alternative economic visions (see chapter 9). Research has assessed attempts to enact these visions and others through prefigurative politics and community resilience strategies—from Central American and African American farmer networks, to urban agriculture, local energy cooperatives, and many other efforts to model how communities can build power to provide for their own needs (Scurr and Bowden 2021; White 2018).

Topical, Geographic, and Intersectional Expansion

The horizontal spread of EJ research means that it is now applied to a broad range of issues, places, and peoples. Benford (2005) identified 52 EJ issues in the literature, not including climate change. By 2021, the online EJ Atlas (https://ejatlas.org/) organized around 3500 case studies under ten broad categories developed by researchers and activists around the world, and by more than 60 different commodities.

EJ research has also broadened its geographic scope. An initial focus on the U.S. reflected the origins of EJ movements and the establishment of research infrastructure in U.S. research and funding institutions. From the 1990s onward, scholars in the U.S. who were allied with the movement created centers and programs on EJ, especially at historically Black colleges and universities, schools of public health and medicine, agriculture, and environmental sciences and studies. Some of these programs formed larger consortiums with each other, with movement organizations, and with independent research centers to conduct collaborative research. Federal funding from the National Institutes of Health, the National Institute of Environmental Health Sciences, and other sources began to support EJ research, much of it involving CER. New journals devoted to EJ appeared, such as *Environmental Justice* and *Local Environment*.

Research on EJ also began to expand globally. In 2009, of scholarly articles published with the keyword *environmental justice*, almost half were authored by researchers based in the U.S., 20 percent were written by authors in the U.K., and 60 percent exclusively addressed U.S. cases (Reed and George 2011). While this distribution likely reflected global scholars' preferences for different terms for EJ issues, it also signaled the need to extend the research community beyond dominant academic institutions and terminology to address EJ around the globe. Academic calls for "seeing from the South" pushed researchers to recognize more diverse perspectives and expand parochial theoretical assumptions (Roy 2011).

In response, new networks and institutions that fostered EJ research developed outside the U.S. Indigenous-led research institutions and universities expanded in North America, Australia, New Zealand, and Latin America, many of which nurtured CER on EJ and other concerns (Díaz Ríos, Dion, and Leonard 2020; Rodríguez 2021). In the 2010s, the European Commission funded the Environmental Justice Organisations, Liabilities and Trade (EJOLT) project, a multinational and multiyear effort linking researchers at universities and EJ organizations in Europe, Africa, Latin America, and Asia. The project helped launch the EJ Atlas, which features case studies written and edited by researchers and activists around the globe, with especially broad coverage of Latin America, Africa, and Asia.

While most EJ research continues to center analyses of injustice based on race, Indigeneity, class, and gender, research increasingly reveals how environmental and health burdens are also unevenly distributed based on ethnicity, nationality, immigration and citizenship status, sexual orientation, age, physical ability, and the intersections among these categories (Chakraborty, Collins, and Grineski 2016; Gaard 2018). Aligning with movements that embrace broad-based organizing on economic, social, and environmental issues—such as Black Lives Matter, #NoDAPL (to stop the Dakota Access Pipeline), and climate justice—researchers are also taking an intersectional approach to analyzing power and how different axes of identity can compound oppression. This research offers more complex accounts of why environmental injustices continue, how they affect groups differently, opportunities for solidarity and allyship, and how to evaluate the justice of environmental solutions for multiple populations (Di Chiro 2021; Estes and Dhillon 2019; Malin and Ryder 2018; Pellow 2018).

Scalar Expansion

Early EJ research focused on documenting and resisting local inequities caused by single-point sources of pollution and exploitation at a moment in time. Today, organizers and researchers are more likely to consider how local injustices are situated within national and global networks of governance, investment, trade, transportation, and pollution. This multiscalar approach is better able to reveal how decisions and hazards generated in one place exert complex effects on people and ecosystems in other places, especially by externalizing harm and resulting EJ conflicts from environmentally privileged to environmentally burdened places and peoples (Agyeman et al. 2016). Examples include the dire threats from climate change to vulnerable communities around the world that have prospered least from climate-altering industrialization and consumption (Chu, Anguelovski, and Carmin 2016), the disproportionate burdens of air and noise pollution (and therefore of asthma, cancers, and stress) borne by communities near major ports and freight corridors for global trade (De Lara 2018; Hricko 2008; Matsuoka et al. 2011), and how workers in the global electronics industry suffer outsized risks of occupational cancers and miscarriages to produce and recycle products that few of these workers can afford to buy (Smith, Sonnenfeld, and Pellow 2006; Smith and Raphael 2015). As Sze and London (2008) write, "research that weaves together multi-leveled, multi-scalar, and multi-method analyses of historical, spatial, political, economic, and ecological factors" can best explain how environmental inequalities arise, why they endure, and what could be done to address them (1344).

Temporal Expansion

A multiscalar approach also drives researchers to examine how environmental injustices unfold over longer time periods through complex chains of causation and within enduring but dynamic structures and systems of oppression, such as colonialism, capitalism, and racism. Much of the research on transformational justice discussed above takes the long view by imagining an environmentally just *future* and considering questions of intergenerational EJ for ancestors, descendants, species threatened with extinction, and sites vulnerable to destruction. Yet much of this research also aims to recover the *past*, employing historical or longitudinal analysis as a necessary basis for understanding present conditions and how to change them.

This work makes several important contributions. One is the tracing of the "slow violence" (Nixon 2011) of attritional harms that unfold over human lifetimes or longer—for example, cancers due to long-term exposure to workplace chemicals or air pollution, the gradual poisoning and destruction of fenceline communities around mines and hazardous waste sites, and creeping threats to lives and cultures from deforestation and desertification. Research on slow violence identifies its historic causes in structures of oppression and the decisions of powerful actors such as corporations and regulators, and reveals these harms as acts of violence rather than normal features of the natural or social landscape (Cahill and Pain 2019). These studies typically draw on multiple methods to uncover the deep roots and complex causation of continuing environmental injustices. Sandlos and Keeling (2016), for example, draw on historical records, observations at public meetings, and CER to show how 50 years of arsenic contamination from the Giant Mine, perpetrated by two mining companies and abetted by federal minerals policy, gradually deprived the Yellowknives Dene First Nation in Canada of safe drinking water, traditional foods, and medicinal plants, acting "as a historical agent of colonial dispossession that alienated an Indigenous group from their traditional territory" (7).

This historically grounded research also illuminates how environmental traumas affect the well-being of people in EJ communities over time. It traces physical and mental effects over human lifespans, such as post-traumatic stress and the cumulative physiological damage from chronic environmental and psychosocial stressors (Solomon et al. 2016). It documents intergenerational traumas, such as depression and anxiety, caused by disasters and compounded by survivors' distrust of authorities. For example, Ezell and his colleagues (2021) summarize studies of the mental and physical harms to BIPOC survivors of the lead contamination crisis in Flint, Michigan, as well as Hurricanes Katrina and Maria, and how these traumas were exacerbated by distrust of the healthcare system in BIPOC communities. This research also recognizes and examines cultural traumas from the splintering of communities and erasure of cultures by dispossessions and dislocations caused by colonization, conservation, climate change, disinvestment, urban redevelopment, gentrification, and wartime destruction (Anguelovski 2013; Chalupka, Anderko, and Pennea 2020; Draus et al. 2019). Other studies, such as Howell and Elliot's (2019) longitudinal study of how disasters have worsened income inequality in the U.S., examine economic traumatization.

Yet historical research also helps to recover the past as a resource for envisioning a just future. EJ researchers, often in collaboration with community partners, have produced counter-histories that excavate past cultural practices and knowledge that can help restore environmentally just relations. Research on protecting traditional foods, plants, and farming practices has shown why it is important to protect them from biopiracy, corporate monopolization, and industrial agricultural practices (Shiva 2016a). Research on applying Indigenous traditional ecological knowledge has helped to revitalize management of land and fisheries (see Gilio-Whitaker 2019; Jarratt-Snider and Nielsen 2020; Nelson and Shilling 2018). Historical research on Black farming is a reminder of African Americans' intergenerational knowledge of how to live well with nature, and how farming and urban agriculture can be ongoing sources of Black communities' resistance and resilience to oppression and dispossession (White 2018).

This chapter has told three stories about the development of environmental justice—as a multidimensional concept, a multifarious movement, and a multiplying body of research—with which community-engaged researchers should be familiar. More and better research grounded in and driven by community knowledge and linked to action is needed to document and make visible environmental injustices, strengthen movements, develop innovative and effective policies and practices, reform governance, and remake economic and social institutions to create the conditions for EJ. As the next chapter argues, CER approaches are especially valuable for meeting these challenges.

2

Community-Engaged Research

Chad Raphael and Martha Matsuoka

Research closely linked to organizing and advocacy has played a crucial role in the struggle for environmental justice (EJ). Consider some of the most influential studies that helped give birth to the modern EJ movement in the U.S., which did not include community-engaged research (CER) as we will define it below, but did help set the stage for it by demonstrating the value of research that responds directly to community priorities. Research by Robert Bullard (1983) for a 1979 civil rights lawsuit in Houston, TX, provided the first systematic evidence that hazardous waste sites were disproportionately located in neighborhoods of color. Later, organizing against toxic contamination in primarily African American communities inspired the Congressional Black Caucus to order the first federal government study of racial and income disparities in hazardous waste siting (U.S. General Accounting Office 1983). A larger study by the United Church of Christ's Commission for Racial Justice (1987) established these linkages more clearly, and found that race predicted proximity to hazardous waste facilities more powerfully than income, property values, or closeness to waste production. Over the next six years the federal government began to adopt many of the report's recommendations.

Responding directly to calls by grassroots leaders and EJ advocates to document environmental racism, these studies influenced public discourse and policy significantly because they were connected to organizing, litigation, advocacy, and regulation to address the emerging issue of environmental justice. The fact that the researchers who conducted these studies were affiliated with academia, a government agency, and a civil society organization demonstrates that EJ research can emerge from diverse institutions. Recalling the early days of this movement in the U.S., activist Vernice Miller Travis said:

> We gave birth to a conversation that people would recognize as their own. We gave it a language, we gave it words, we gave it a science base, we gave it a public policy base, and we gave it a base that was rooted in the power and mobilization of people on the ground so it couldn't be denied. (U.S. Environmental Protection Agency 2014)

By integrating their studies into a current political discussion driven by a growing movement, researchers supported activists and advocates to develop the language, science, and policy of EJ.

In the years that followed, EJ researchers incorporated CER approaches by involving community members themselves in the research process to develop local capacities for public participation and to accomplish more and better research. CER has contributed to the EJ movement in several important ways (Cole and Foster 2001). CER has documented disproportionate threats from environmental dangers to EJ communities, inspiring campaigns to block the siting of additional hazards. CER also helped to provide the evidentiary basis for demands for investment in healthier and safer facilities, more protective regulations, and more effective enforcement. CER aided EJ leaders in understanding how local problems were part of larger systemic patterns of injustice rooted in historic racial, economic, and political oppression. CER also helped to justify policy changes, suggest organizing and legal strategies, and identify promising policy instruments. While research using a traditional approach has contributed to each of these goals as well, CER did so by partnering with community organizations to build their capacities to conduct research with and without academic and other professional researchers and strengthen their influence over the research agenda. Thus, CER contributed not only to the analysis of causes, solutions, and strategies for change, but also to the development of grassroots leadership that has been crucial for building EJ knowledge and the movement.

This chapter prepares researchers to contribute to this body of work by providing a definition of CER as a research paradigm and introducing its main goals and evaluative criteria. We go on to describe some of the major types of and influences on CER that emerged in the Global North and South, and Indigenous research traditions. While we value the large body of EJ research that has not employed CER, including the foundational studies mentioned above, we argue that CER can make a unique contribution by building research partnerships, practices, and knowledge about EJ that strengthen grassroots leadership of the EJ movement, and that produce research with greater reach, rigor, relevance, and reflexivity. To show how CER makes a distinctive contribution to enacting justice in the research process, we employ the dimensions of EJ defined in chapter 1 to introduce a justice framework that relates CER and EJ, which is used to examine CER practices in the chapters that follow.

DEFINING COMMUNITY-ENGAGED RESEARCH

CER as a Paradigm

Community-engaged research is an umbrella term for a paradigm—an overarching theoretical framework of beliefs and understandings that guide research practice—used by professional researchers (in academia, government, and independent research institutes), students, and community partners to co-create knowledge. As a paradigm, CER is not defined by a specific choice of methods for gathering or analyzing data—such as surveys, ethnography, or geographic information systems (GIS)—but by the fact that "participation on the part of those whose lives or work is the subject of the study fundamentally affects all aspects of the research" (International Collaboration for Participatory Health Research 2013, 5). CER is also defined by its beliefs that knowledge is inherently social and action oriented, that it is co-produced by researchers and communities, and that these partnerships must address power relations inherent in knowledge production, respect local cultures and assets, be of practical benefit to communities, and advance liberation and equity (Israel et al. 2013b; Wallerstein and Duran 2017).

Like other paradigms, CER can embrace a broad range of disciplines, theories, and research methods. CER has been applied across the social and natural sciences, arts and humanities, and professional and applied fields (Chevalier and Buckles 2019; Lepczyk et al. 2020; Wallerstein et al. 2017). Similarly, CER researchers employ many theories, especially critical race, feminist, and decolonial theories (Deeb-Sossa 2019; Smith 2021). CER embraces diverse qualitative and quantitative methodologies of different origins, such as community-based participatory research (emerging especially from the U.S. health sciences), participatory action research (from the Global South), community-based research (in Canada), collaborative action research (especially in Australia), and participatory appraisal (in development research). Additional CER methodologies include collaborative inquiry, reflexive practice, feminist participatory research, tribal participatory research, research justice, street science, citizen science, community science, and many others (for summaries, see Davis and Ramírez-Andreotta 2021; Israel et al. 2013b; Wallerstein and Duran 2017).

CER is also enabled by multiple institutional relationships, such as individual projects, long-term collaborations with community partners, and community-university partnerships to improve local capacities and conditions over decades (Raphael 2019b; Welch 2016). As discussed in chapter 4, the kinds of community partners and the degree of their engagement in a CER project can vary considerably. In addition, this research is conducted not only by academics, but also by researchers in community-based organizations, coalitions, and network organizations; in independent research institutes and government agencies; and by advocates, lawyers, and others. (Therefore, we use the term *community-engaged research* to refer to this work as a whole, and reserve *community-engaged scholarship* for studies involving researchers in academic contexts.) Researchers who do CER for EJ often collaborate with grassroots organizations of people who live on the frontlines and fencelines of environmental injustices, coalitions and national networks of community-based organizations, intermediary research and policy organizations, large national and international environmental organizations, tribal governments, or other government agencies (Davies and Mah 2020b).

Despite its internal diversity, CER is a coherent paradigm because it includes a common set of philosophical assumptions about reality (ontology), knowledge (epistemology), and values (axiology) that inform the purposes and conduct of research (DeCarlo, Cummings, and Agnelli 2021). Because of its unique set of assumptions, CER draws upon but does not fit exclusively within any of the other research paradigms that are most frequently mentioned in methods textbooks, including qualitative (or constructivist or interpretative), quantitative (or positivist), critical (or emancipatory), or postcolonial (or Indigenous) research traditions (DeCarlo, Cummings, and Agnelli 2021; Denzin, Lincoln, and Smith 2008; Pabel, Pryce, and Anderson 2021). CER is least aligned with positivism, which aims to produce objective, value-neutral, quantifiable, and generalizable knowledge. However, as chapter 6 shows, CER can employ both qualitative and quantitative methods or a combination of these kinds of methods. In addition, the purposes of CER are often aligned with critical and postcolonial research. However, unlike any of these approaches, CER understands knowledge as co-produced by professional researchers and community partners, and CER is evaluated largely by whether it shares power with and benefits all parties to the research by creating a web of reciprocity and mutual benefit (described further below). For example, only CER *requires* researchers to co-develop the research agenda with community partners and to involve them in the research process to the extent that community partners desire.

Therefore, the most relevant framework for understanding CER is one that contrasts it with expert-oriented approaches to research of all kinds (quantitative, qualitative, or critical) (Saltmarsh 2010). Like CER, an expert-oriented approach is defined by its assumptions about the relationship of researchers and communities, and about knowledge and power, not by whether researchers employ a particular research method, such as surveys, ethnographies, or ideological critique (see table 2.1).

The main ontological differences between expert-oriented and CER approaches concern where and with whom real knowledge resides. An expert orientation assumes that authentic knowledge originates in research institutions (academic, government, or independent), where it is governed by disciplinary and methodological expectations, and produced by credentialed professional researchers (O'Meara and Rice 2005). Knowledge travels outside these institutions when policy makers or the public consume it as a good or service, or when researchers

TABLE 2.1. Expert-Centered and Community-Engaged Research Paradigms

	Expert-Centered Research	Community-Engaged Research
Ontology *What is real?*	Research institutions are primary seats of knowledge	Research institutions are collaborators in a network of knowledge production
	Disciplines are primary governors of knowledge	Authority over knowledge is shared with relevant communities
	Researchers are expert producers of knowledge, which is consumed by or applied to communities	Knowledge is co-created with communities and inherently action oriented
	Research is *on* or *for* the community	Research is *with* the community
Epistemology *How we know?*	Knowledge emerges from researcher expertise	Knowledge emerges from researcher and community expertise in facilitating co-production of knowledge
	Knowledge is vetted by professional peer review	Knowledge is vetted by professional and community peer review, where it is applied and tested for relevancy and action
	Applied knowledge is spread by replicating best practices	Relational, contextual, local, and experiential knowledge is spread by adapting promising practices from one community to others, while respecting their differences
	Knowledge flows unidirectionally from experts to communities	Knowledge flows multidirectionally among experts and communities
Axiology *What is valued?*	Dominant knowledge systems, even if hegemonic and colonizing	Recognition of diverse knowledges, knowledge as power, and Indigenous and decolonizing knowledge
	Community engagement to advance researchers' goals	Community partnerships, participation, and control of research, and outcomes that advance liberation, equity, cultural recognition
	Extractive partnerships	Reciprocal, mutually beneficial partnerships
	Researchers' assets cure community deficits	Community assets strengthen capacity for just practices with and within the community
	Technocracy and vanguardism	Grassroots leadership by those most affected

apply it to communities as a remedy or design, much as a doctor prescribes an approved treatment or an engineer applies calculations to design stable structures. In contrast, CER sees research institutions as one node in a larger web of knowledge production and circulation. Within this network, researchers co-produce knowledge with equally authoritative community actors, who do not simply

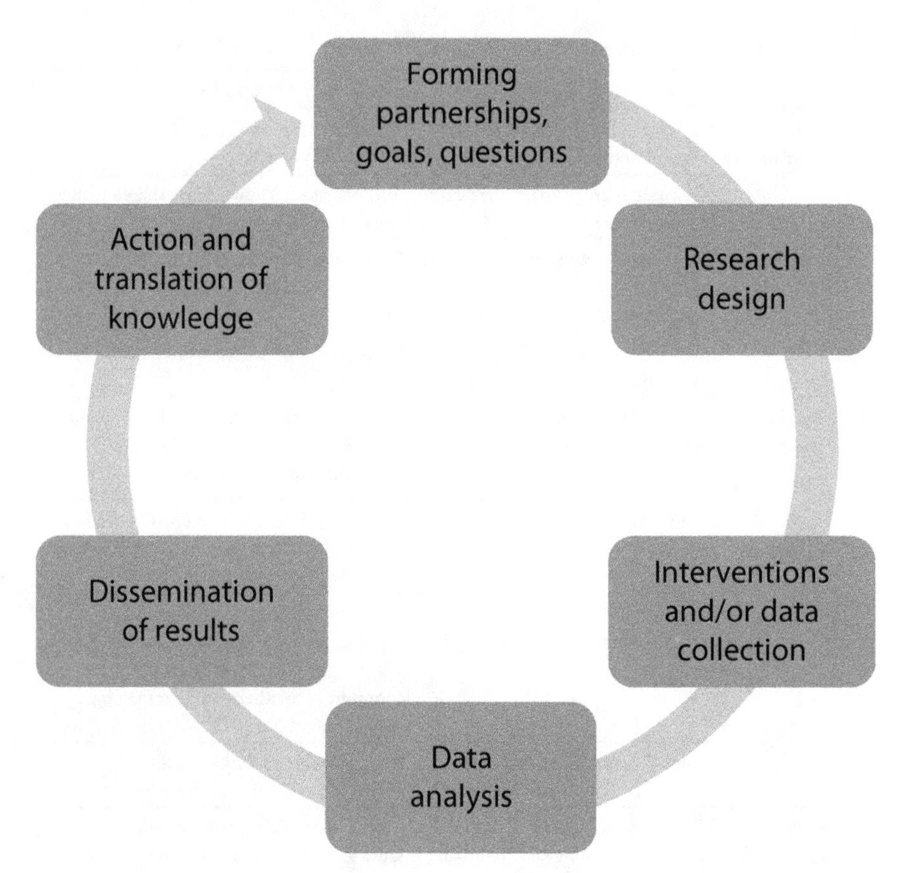

FIGURE 2.1. The CER process.
Adapted from Bacon et. al. (2013).

provide raw data or access to research sites, but contribute to ways of understanding local conditions and experiences in diverse contexts, and to the goals that motivate research and endow it with meaning. The participatory research process is itself a form of joint action in a community—building local capacities and leadership, for example—and informs further actions to change policies, practices, and power relations, including power relations between researchers and communities (Kindon, Pain, and Kesby 2007). Thus, CER understands research as a relationship and process that unfolds *with* communities, rather than on them (as a passive object) or for them (as a client or ward). Ideally, this is a cyclical process of shared inquiry and collaboration to design and conduct studies, and implement actions based on the findings, which leads to new questions and interventions for future research partnerships, deepening the relationship between communities and researchers (see figure 2.1).

CER also rests on different epistemological assumptions than the expert-driven approach. In the latter, knowledge springs from researchers employing their expertise—whether it is to produce quantifiable data and generalizable conclusions (as in positivism), or empathetic and insightful interpretations of informants' experience of the world (as in constructivism), or trenchant analyses of oppression and emancipation (as in critical research). In CER, knowledge emerges from the interactions between researchers' and community partners' expertise. This expertise includes capacities for facilitating the co-production of knowledge, such as building relationships and mediating conflict among research partners; engaging community residents in research; translating among different languages, cultures, and disciplines; generating relevant findings and disseminating them to diverse groups; and brokering and implementing action plans in response to research results (Karvonen and Brand 2014). These kinds of expertise require intentional relationship building between researchers and community partners, and explicit training and personal transformation to develop researchers' cultural competencies and cultural humility to work less ethnocentrically and more respectfully with community partners (see chapter 3). In the expert-centered paradigm, professional peer review evaluates the production of new knowledge to ensure its quality, while CER also includes community review, which adds criteria related to the quality of community participation and potential for practical improvements (discussed below). Rather than disseminate practical knowledge by replicating successful models in new contexts, CER sees practical knowledge as inherently rooted in and bound by the context in which it is created, which can be carefully adapted to other places and peoples, but not mechanically transplanted. Like the entire process of knowledge creation, translating knowledge involves a mutually beneficial relationship among researchers and communities, rather than a one-way flow of enlightenment from experts.

CER's values (axiology) challenge experts' tendency to ignore or accept how they wield power by applying dominant knowledge systems and cultural assumptions. For example, mainstream Western scientific, economic, and cultural conceptions of nature often present it as a warehouse of "natural resources" or a suite of "ecosystem services" for humans to manage and use, rather than seeing nature as humans' kin or as bearers of rights, as many Indigenous and other peoples do. CER leads researchers to recognize diverse contextual concepts and systems of knowledge, to question hierarchies of power, and to challenge knowledge and cultural biases rooted in colonialism, racism, and the exploitation and destruction of people and nature. This requires researchers not simply to conduct studies that advance understanding of how to challenge oppression in the world (as in other critical research), but to collaborate with community partners in ways that practice equitable power relations and pursue mutual benefits. If traditional experts collaborate with community organizations, they tend to do it extractively—to enable data gathering, assemble an audience for a study, or derive credibility in the eyes of potential funding agencies. Instead, CER strives to create reciprocal relationships, in which community partners are co-equal participants in designing the research and identifying desired benefits to the community.

Conventional experts often see communities as sources of deficits—deprived of sufficient resources or critical consciousness—which experts can fix by mining raw data about community problems and pathologies, and producing analyses that light the way to solutions. The implicit theory of change can be technocratic (leadership by the best-trained and most expert) or vanguardist (leadership by the most critically conscious). Community-engaged researchers understand that even the most stressed and oppressed communities possess valuable assets, such as schools, churches, nonprofit organizations, health and social services, informal social ties, and mutual aid networks, which can also incubate critique of oppression. CER aims to build on this infrastructure of existing strengths, sources of resilience and resistance, and latent potentials to conduct research and plan responsive actions (Sharpe et al. 2000). In a CER theory of change, the research process makes as important contributions as research outcomes. By sharing authority over knowledge and developing communities' capacities to know and transform themselves, CER strives to strengthen grassroots leadership and power.

CER TRADITIONS AND INFLUENCES

CER has developed from diverse influences and traditions in the Global North and South, and from Indigenous research. We present each of them to ground researchers in how these different strands of CER arose from specific historical and institutional conditions and struggles, and to familiarize researchers with the most widely influential kinds of CER practiced today. Much CER draws on more than one of these traditions, which are not mutually exclusive. Knowing how and why these approaches emerged is important preparation for CER.

The Global North

In North America and Europe, inspirations for CER emerged from efforts to apply research in communities to improve agriculture, planning and development, public health, social services, and democracy (Wallerstein and Duran 2017). While not practicing the extent and kinds of community participation that CER does today, these precursors sowed seeds for some of the ideas and practices of CER. For example, the U.S. Department of Agriculture's Cooperative Extension, created in 1914 and run through the nation's land grant universities, co-developed research and educational programming with local farming communities (see chapter 10). In its early years, the program reflected rural reformers' views that farming communities, not just academics, could contribute important local knowledge, and that agricultural modernization depended in part on strengthening local democracy and civic life (Shaffer 2017). Sociologists such as those associated with the Chicago School in the 1920s used ethnographic methods to draw on local knowledge, producing social science that aimed to intervene in, not simply describe, social problems of urbanization and industrialization (Munck 2014). At the same time, the philosopher John Dewey (1916, 1934) provided an influential rationale for efforts to develop community-based learning and research to address practical problems and social issues, by urging schools to model the life of democratic communities, make learning an experiential and collaborative experience among teachers and students, and connect formal education and research with tackling social problems in field settings. Dewey's thinking was deeply influenced by his observations of Jane Addams's Hull House in Chicago and the urban settlement house movement, which integrated civic education, community organizing, and social services for immigrant youth and adults (Saltmarsh 2008).

Institutional efforts to develop CER in the U.S. accelerated in the 1990s. Proponents aimed to reverse the post–World War II specialization of academic knowledge, its retreat into a stance of value neutrality and objectivity, and the reduction of universities' purposes to producing knowledge and employees for the market (Boyer 1996; Post et al. 2016). Interest in CER also emerged to address academia's growing need to demonstrate its extramural contributions in response to cuts in public funding for higher education and state pressure to justify universities' tax-exempt status (Doberneck and Schweitzer 2017). Some universities launched place-based learning initiatives and anchor programs in their communities, which sparked new CER partnerships. These collaborations pursued two main goals: to provide opportunities for civic learning and research across the curriculum; and to strengthen community capacities to improve local education, health, services, and economic development (Hodges and Dubb 2012). Three CER approaches have become especially influential today.

Action Research. Initiated by social and organizational psychologist Kurt Lewin in the 1940s, action research challenged positivist assumptions that researchers could study objective social phenomena that existed separately from meanings created by researchers and participants as they acted in the world, and that theory could be separated from practice and applied universally across social contexts. Instead, Lewin and his followers developed research that aimed to solve social problems through an iterative cycle of planning practical interventions in a particular community, taking action, studying the results, and adjusting interventions accordingly. Thus, the concept of action referred both to the importance of studying social behavior in diverse real-world settings and to the goal of research improving social action (Lewin 1946). Lewin's own action research, focused on reducing racism in public housing projects, inspired other social scientists to apply this approach in fields as diverse as education, rural development, community studies, public health and social work, and organizational studies, among others (Bradbury 2015; Kindon et al. 2007). Many action researchers aim to engage communities in researching their own problems and potential solutions as a contribution to a more democratic culture, workplace, and community.

Community-Based Participatory Research (CBPR). From the 1990s onward, health science researchers increasingly saw health inequities as rooted less in disparities of healthcare, lifestyles, or genes, and more in differences among the social, economic, and physical conditions in which people live (Corburn 2009). Researchers developed CBPR largely to engage communities most affected by the underlying causes of health inequities in studying and acting to address these disparities, which are rooted in issues of environmental and social justice (Shepard et al. 2013; Wallerstein et al. 2017; Wilson, Kenny, and Dickson-Swift 2018). CBPR has also strengthened translational research to speed the dissemination of applied findings from healthcare trials, campaigns, and therapies into wider practice among underserved communities and constituencies (Cyril et al. 2015; De las Nueces et al. 2012). National health institutes in the U.S. and other countries began to fund CBPR extensively in the early 2000s. By 2013, U.S. Surgeon General Regina Benjamin wrote that CBPR "has become the preferred model for conducting [health] research in communities" (quoted in Blumenthal, Hopkins, and Yancey 2013, xii). CBPR has influenced community-based research in public and environmental health, and many other fields, by developing frameworks for integrating CER into community organizing and policy advocacy to build community capacities for exchanging knowledge, organizational collaboration, and improving care (Cacari-Stone et al. 2014; Drahota et al. 2016; Minkler and Wakimoto 2022; and see chapter 9).

Citizen Science and Community Science. Practiced in the natural and social sciences, citizen science refers to "the scientific activities in which non-professional scientists volunteer to participate in data collection, analysis and dissemination of a scientific project" (Haklay 2013, 106). Community participation varies considerably in these projects (Cooper et al. 2021). In most citizen science, the community's role is limited to gathering crowdsourced data, while professional scientists exercise control over funding, the research agenda, and data analysis. The primary goals are improving data sensing, democratizing access to scientific resources, and increasing the public's literacy and interest in science. However, this mainstream version of citizen science has failed to engage many residents of EJ communities, instead attracting participation mostly from white, college-educated adults with above-average incomes (Pandya and Dibner 2018; Pateman, Dyke, and West 2021).

At the same time, mainstream citizen science has been successful at institutionalizing public participation in research. Citizen science is widely used around the world to study environmental health and quality (Haklay and Francis 2018; Lepczyk, Boyle, and Vargo 2020; Pandya and Dibner 2018; U.S. Environmental Protection Agency 2015). Citizen science can produce data admissible in legal and policy processes (Emmett Environmental Law and Policy Clinic 2017; Mueller and Tippins 2015). In the 2010s, the U.S. Environmental Protection Agency (2016) and European Union launched new funding programs to support citizen science tools and programs, and the United Nations recognized citizen science methods as legitimate for measuring progress toward the UN's global Sustainable Development Goals (De Filippo et al. 2018).

A variant of citizen science, increasingly called *community science*, avoids the language of citizenship, which can be both exclusive (i.e., of immigrants) and coercive (particularly of Indigenous peoples who feel stronger affiliations to tribal governments and natural kin than to the nation-states in which their lands currently reside) (Cooper et al. 2021). Most importantly, community science goes beyond crowdsourcing data gathering to engage community organizations in co-designing research questions, grants and other research resources, and each aspect of conducting and disseminating scientific research.

Community science draws inspiration from action research and CBPR methods (Cooper et al. 2021). This approach also has roots in the radical science movement of the 1960s onward, which sought to democratize scientific research, and from science and technology studies, which revealed how mainstream scientific institutions and constructions of expertise reinforce epistemic, economic, and political domination (Davies and Mah 2020a). In response, community scientists advocate for open data (ensuring that anyone can freely share and use data) and open science (ensuring research is accessible) as important components of power sharing in the research process. EJ groups have employed this grassroots-led science in urban street science (Corburn 2005) and popular epidemiology (Brown and Mikkelsen 1997), which engage residents in research to inform community organizing on issues such as air pollution, toxic contamination, transportation planning, and access to healthy food (Davies and Mah 2020b).

The Global South

In South America, Africa, and Asia, CER arose from the 1960s onward amidst decolonization and other struggles against structural underdevelopment and authoritarian rule. Compared with CER in the North, the Southern tradition showed greater concern for emancipating knowledge and research from control by foreign and local elites, and supporting communities to empower themselves to create broader social transformation (Hall, Tandon, and Tremblay 2015). Two research traditions are especially relevant to EJ.

Participatory Action Research (PAR). The influential work of Brazilian educator Paulo Freire (1970, 1982) and Columbian sociologist Orlando Fals Borda

(1987, 2006) emphasized the role of education and research in liberating oppressed peoples to develop critical understanding of their conditions and develop their own transformative solutions. Freire advocated collaborative research as part of popular education that helped people living in poverty and oppression to develop a critical consciousness of the structural causes of their conditions of poverty, and transform them. For Freire, learning began with reflection on participants' own knowledge and experiences, helped them develop broader explanations and critiques of their conditions, and fostered their strategies and plans for social action. Fals Borda developed a parallel set of guidelines for PAR researchers, including respect for community knowledge and cultures, skepticism about elitist visions of history and science, and commitment to demystifying the research process for nonspecialists. PAR emphasized marginalized peoples' agency, liberation as the goal of research, and local and experiential knowledge as a source of resistance and change (Chevalier and Buckles 2019). PAR has informed CER for EJ on issues such as urban air pollution (González et al. 2007), climate justice activism (Reitan and Gibson 2012), and recycling co-ops (Gutberlet 2008).

Participatory Development and Conservation Research. From the 1960s onward, a host of CER approaches arose from demands to shift from top-down to bottom-up economic development and resource conservation planning (Cernea 1985; Chambers 1997; Hirschman 1984). Participatory approaches offered grassroots communities one way to resist extractive and exploitive economic and agricultural plans, and "fortress conservation" schemes that banished local residents from protected lands to preserve biodiversity, imposed by national governments, multinational agencies such as the World Bank, and global NGOs. For example, participatory appraisal and planning (Chambers 1997) promoted collective and grassroots identification and framing of problems, participatory research and education, and experimentation with community-generated solutions based on local experience and knowledge. Other researchers inspired by similar aims employed action research and PAR to work directly with rural land reform movements and urban neighborhood organizations, eliciting Indigenous and local knowledge and experience to address issues of environmental and economic justice (Keahey 2021).

By the late 1990s, participatory strategies became co-opted and depoliticized by some governments and multinational NGOs, which failed to share substantive control over development and conservation policies and projects (Cooke and Kothari 2001). Nonetheless, researchers continue to find ways to integrate CER into authentic participation and to inform organizing to transfer power and resources to community-led conservation and development plans (Hickey and Mohan 2005, and see chapters 8 and 12).

Indigenous Decolonial Research

Across the Global North and South, Indigenous peoples' struggles for self-determination and the preservation of their ancestral lands, languages, and cultures after centuries of colonization have exerted growing influence on CER. Indigenous peoples' distinct worldviews and experiences of environmental injustice highlight the importance of incorporating respect for cultural and epistemological differences into research. CER can contribute to decolonization by elevating recognition of Indigenous knowledge, and by adopting research agendas and processes that restore Indigenous peoples' access to land and power over managing it (Neale et al. 2019).

Research by and with Indigenous peoples often adopts different conceptions of community, measures of environmental injustice, and definitions of health than are used in other EJ research (Gilio-Whitaker 2019; Vickery and Hunter 2016). For example, Native Americans may be defined by blood quantum levels, citizenship in a federally recognized tribe, residency on a reservation, or self-identification on census forms, and the method chosen can dramatically affect health statistics and policy responses. In addition, traditional EJ indicators, such as proximity of homes to industrial facilities, do not reflect Indigenous communities' broader connections to the land, which include needs for access to sacred sites, ceremonial plants, subsistence hunting and fishing, and sovereignty over their traditional lands. These criteria, which are part of the definition of public health and well-being for many Native communities, are not captured in typical health risk assessments (Arquette et al. 2002). CER has helped to integrate these culturally specific elements into research, including on EJ issues, although not without tensions with mainstream health science methodologies (see chapter 9).

Much Indigenous-led research and CER applies traditional ecological knowledge (TEK) to improve health, planning, natural resource management, climate mitigation, and biodiversity conservation. *TEK* is an umbrella term for the diverse and expansive knowledges that Indigenous peoples have accumulated over millennia and continue to develop about their homelands around the world. TEK encompasses "adaptations for the generation, accumulation, and transmission of knowledge; the use of local institutions to provide leaders/stewards and rules for social regulation; mechanisms for cultural internalization of traditional practices; and the development of appropriate world views and cultural values" (Berkes, Colding, and Folke 2000, 1251). This knowledge is recovered and passed down to new generations through ceremony, storytelling and oral history, music, arts and crafts, gathering of plants to make medicines, preparation of traditional foods, and increasingly through mapping, digital media, and formal CER and Indigenous-led research.

While some researchers use both Western scientific and TEK concepts and methods, TEK is not simply an input into mainstream science. Instead, TEK presents alternative conceptual frameworks and ways of knowing that can ground environmental, biomedical, and social research in a more holistic understanding of just and sustainable relationships among humans and other nature (Finn, Herne, and Castille 2017; Smith 2021). Some aspects of these epistemologies are not easily translated from Indigenous languages or assimilated into Western conceptions of space, time, subjectivity, and gender relations (Smith 2021; Whyte 2018b). CER need not exoticize Indigenous peoples or romanticize their relationship to nature to recognize and respect these differences. For example, in contrast to dominant notions of scholarly independence, objectivity, or devotion to discovering abstract truth, in many Indigenous research methodologies what is most "important and meaningful is fulfilling a role and obligations in the research relationship—that is, being accountable to your relations," which include nature (Wilson 2008, 77). Indigenous researchers also stress TEK's importance for self-determination of Indigenous nations, including economic independence and spiritual renewal, regardless of TEK's value to mainstream science and to other peoples (Whyte 2018b).

CER in Indigenous communities has also focused new attention on research ethics. As respect for Indigenous knowledge has grown among non-Indigenous researchers, so has the importance of exchanging it in more ethical and respectful ways than researchers have approached communities in the past, which typically involved a one-way extraction and archiving of Indigenous knowledge and artifacts. Changes to tribal laws and the development of tribal institutional review boards to vet research proposals on Indigenous lands have required outsiders to conduct research more collaboratively with Native communities, protected TEK from commercial exploitation and appropriation as intellectual property, and shielded knowledge of sacred sites and natural resources from those who would abuse them (Finn et al. 2017; Whyte 2018b). Principles of data sovereignty such as those of the Global Indigenous Data Alliance—*c*ollective benefit, *a*uthority to control, *r*esponsibility, and *e*thics (CARE) principles—provide guidance to CER researchers on how to comply with expectations for Indigenous data governance (see chapters 5 and 12).

Reciprocal Learning and Practice

Indigenous, Southern, and Northern traditions of CER increasingly engage with and learn from one another. Starting in the mid-1970s, Southern and Northern researchers began to interact as academic and community-based researchers forged institutional ties to strengthen CER. The Highlander Research and Education Center in Tennessee, which had trained organizers in the labor and African American civil rights movements, joined with counterparts in the Global South in emancipatory participatory research, adult education, and community organizing (Horton and Freire 1990). Additional ties were forged by networks such as the International Participatory Research Network (with centers in Canada,

India, Tanzania, the Netherlands, and Venezuela), Australia's Collaborative Action Research Group, the Action Research Network of the Americas, and the United Nations Educational, Scientific and Cultural Organization's Knowledge for Change Consortium. Contemporary volumes on CER reflect the mutual influence of Northern and Southern theories and practices (Bradbury 2015; Davies and Mah 2020b; Munck et al. 2014; Wallerstein et al. 2017), and of Indigenous methodologies and CER (Atalay 2012; Denzin, Lincoln, and Smith 2008; McGregor, Restoule, and Johnston 2018; Smith 2021; Wilson 2008; Windchief and San Pedro 2019).

WHY CER FOR EJ? WHY NOW?

A core argument of this book is that CER is necessary for research to advance and achieve EJ. In what follows, we present two kinds of supporting arguments. One is that CER can make unique contributions to the *quality* of EJ research, which we illustrate with a brief case study. A second argument is that CER employs research practices that align especially well with principles of EJ. We illustrate this argument by presenting a framework that summarizes how CER fulfills the four dimensions of EJ that were introduced in chapter 1. Taken together, the two arguments point to the importance of community knowledge, and reciprocal and mutually beneficial research, for contributing to EJ. We conclude with some thoughts about why a CER approach to EJ research is especially urgent in the current political context.

CER and Quality

Researchers have turned to community-engaged approaches because they make unique contributions to the quality of EJ research by strengthening its relevance, rigor, and reach (Morello-Frosch et al. 2011), as well as its reflexivity (Lockie 2018; Raphael 2019a; Hale 2008).

Relevance is about whether researchers are asking questions that matter to others. In response to professional reward structures and disciplinary demands, many academic researchers are "talking to ever smaller and narrower academic audiences, using a language that educated readers do not understand, publishing in journals they don't read, and asking questions they don't care about" (Hoffman 2015, A48). When research agendas respond to external cues, they mostly come from major funding institutions and government agencies, which rarely include representatives of EJ communities and often demand an expert-centered research approach. CER can ground the selection of research topics in community concerns and maintain this relevance throughout the research process as community organizations participate in all phases of the work.

CER can also strengthen the *rigor* of research by improving study design, data collection, and data analysis. Many EJ communities' mistrust of research institutions presents a major barrier to research that depends on community participation of any kind. Enlisting community organizations as co-researchers can help to identify appropriate research sites and populations, and build the trust necessary to earn access to them by promoting deeper community understanding of the research process and confidence in its goals (Minkler, Salvatore, and Chang 2018). CER can therefore increase sample sizes, survey and interview response rates, and participation in interventions and treatments. Community members correct and enrich data analysis by providing contextual explanatory knowledge. Engaged partnerships can also unlock new sources of funding needed to conduct complex EJ studies.

Engaged research can also *reach* new audiences in ways that inform practice. Community partners bring valuable capacities to disseminate knowledge to diverse audiences and translate it into useful tools for practice, policy, and organizing (Cacari-Stone et al. 2014; Minkler et al. 2018). Researchers and partners express their research in many forms, from journal articles to policy briefings, white papers, fact sheets, opinion articles, testimony in regulatory forums, community activities and meetings, and so on. Community partners play a crucial role in building an active audience for this work, translating it into local languages and lay terms, promoting and applying its findings, and implementing or demanding responses from decision makers. Rather than publishing studies and hoping they have some effect, researchers build relationships and dialogue with their audiences throughout the course of their studies, increasing their reach and influence (Chen et al. 2010).

Participating in engaged research is also uniquely effective for teaching students about EJ. Environmental educators have long recognized the value of place-based learning and community-based learning for deepening students' understanding of abstract concepts and how global problems affect the local level (D'Amore et al. 2016). These active and collaborative pedagogies can also spark the personal commitments to places and communities that inspire students to study and act on environmental problems (Haywood, Parrish, and Dolliver 2016). Research collaborations with EJ communities align well with these ways of learning, and can enrich students' understanding of how social and economic inequities shape environmental conditions (Dittmer et al. 2018). CER projects can help environmental education to expand its scope beyond "pristine nature" to the places where people in EJ communities live, work, play, pray, and learn (Cachelin, Rose, and Rumore 2016).

CER also helps researchers to practice greater *reflexivity* about the nature and purposes of research, power relations within research teams, and whose interests the research serves. Reflexivity emerges from common CER practices of organizing community review boards to craft research agendas and vet project proposals, drafting detailed memoranda of understanding among partners that define their goals and roles throughout projects, holding co-learning workshops to explore the meaning of rigor and validity from researcher and community points of view, and ongoing dialogue and conflict resolution at each stage of the work (Minkler,

Salvatore, and Chang 2018; Wallerstein et al. 2019). These collaborative processes require in-depth consideration of research agendas and methodologies from multiple perspectives. While much academic research begins by asking what scholars in a discipline need to do to improve the field's understanding and influence, CER proceeds from the question of what the world needs from all professional researchers. This reflexivity pushes researchers to worry less about whether they are distinguishing themselves from other fields and more about whether they are collaborating well across disciplines and with community partners to address the most important concerns of EJ communities. While non-CER studies conducted by government agencies and independent institutes may have practical purposes and intended benefits, they still tend to be defined by the interests of officials, professional staff members, and donors, few of whom live in EJ communities.

Subsequent phases of CER projects also demand greater reflexivity. Co-designing research manifests the positional and situated character of all research (Muhammad et al. 2015). Researchers and community partners cannot avoid addressing their differences of power and privilege (Muhammad et al. 2017). When collaborators bridge their diverse perspectives, assumptions, and experiences, they can generate richer and less distorted knowledge about EJ communities than expert-driven studies do (Lockie 2018). This depends on continuous interaction at each stage of the research, not simply sensitizing researchers to different points of view at the start. It involves instructive conflict. Tensions between maximizing the rigor of research instruments and including community-driven research questions (by changing validated scales, for example) require all participants to clarify trade-offs between the internal and external validity of research (Minkler, Salvatore, and Chang 2018). CER draws overdue attention to research ethics controversies over collective consent to research and ownership of data (described above), and individualized reporting of data to participants in health studies (see box 2.1). Conflicts over disseminating research raise important questions about

BOX 2.1. THE NORTHERN CALIFORNIA HOUSEHOLD EXPOSURE STUDY

The Northern California Household Exposure Study (HES) of indoor air pollution around the Chevron oil refinery in the city of Richmond, CA, exemplifies CER's ability to increase the relevance, rigor, reach, and reflexivity of EJ research. The study was co-designed by academics at two institutions (Brown University and the University of California, Berkeley), an independent research institute (Silent Spring Institute), and a statewide EJ organizing and advocacy group (Communities for a Better Environment) (Balazs and Morello-Frosch 2013; Morello-Frosch et al. 2011). Communities for a Better Environment (CBE) offered invaluable local

(*Continued*)

BOX 2.1. *(CONTINUED)*

knowledge about methods of recruiting participants and choosing sampling sites, suggesting a control site that did not have significant air emissions from transportation or industry. The Silent Spring Institute contributed specialized knowledge of chemicals associated with oil combustion to analyze in the study, and supplemented the academic partners' capacity to teach CBE organizers how to conduct air monitoring, dust sampling, and interviewing. The partners' combined efforts helped the HES to document disproportionate exposure to indoor air pollution in Richmond compared with a control community without a refinery, and, more surprisingly, higher levels of multiple pollutants inside homes than outdoors.

CBE and Silent Spring then asked the academic researchers to communicate individual exposure results to all study participants who wanted to know this information, using a protocol that the research institute had used in a prior study. Given the lack of conclusive research on the health impacts of many chemicals, academic health researchers typically have not reported back to participants their personal exposure levels or tried to communicate the risks associated with them. The HES team collaborated to navigate the scientific and ethical challenges associated with this innovative kind of reporting. The research team co-designed materials in Spanish and English, including visual displays of collective and individual results, scientific uncertainties, and strategies for reducing exposure. CBE organizers met individually with households in the study to explain their exposures and the implications. Follow-up research found this strategy increased participants' knowledge of risks, provoked changes in behavior, and supported an organizing campaign to reduce emissions from the refinery (Adams et al. 2011).

In this example, the nonacademic partners boosted the study's *relevance* by involving fenceline residents in the study and inspiring a shift in research practice to include personal exposure reporting. Residents were highly motivated to act on this information, individually and collectively, because they had invested their time in the study and learned about potential risks. Personalized reporting demanded greater *reflexivity* from researchers about the purposes and impacts of their study as they grappled with how to report individual-level risks ethically and accurately to participants. The collaboration among academics, CBE, and Silent Spring strengthened the *rigor* of the study design by pooling different kinds of expertise, adding a control community, and prompting development of a new protocol for communicating findings responsibly. By presenting the findings in community organizing meetings and regulatory testimony, the partners also increased the study's *reach* beyond the academic literature, drawing on their experience and authority as researchers and policy advocates. The HES approach helped inspire other biomonitoring studies to report personal exposures, including a major study in 17 European countries (Exley et al. 2015). Silent Spring, a leading source of research on environmental contributors to breast cancer, used the study to draw cancer researchers' attention to the need to study the EJ dimensions of breast cancer. The study's findings also bolstered the credibility of CBE's local organizing campaign to persuade regulators to crack down on emissions from the Chevron refinery.

TABLE 2.2. Framework for CER for EJ

In EJ	Dimension of Justice	In CER
Reducing environmental burdens, and increasing environmental benefits and capabilities, for EJ communities and the earth	**Distribution** *Who ought to get what?*	Sharing of resources and work among researchers and communities Development of community capacities to conduct their own research and researchers' capacities to collaborate Co-ownership or community ownership of data
Participation and influence in environmental decision making by historically excluded groups, particularly in frontline communities Protection of individual and group rights through law, regulation, enforcement, and informed consent	**Procedure** *Who ought to decide?*	Community participation and influence in the design and conduct of research, including free and prior informed consent, and rights to control data
Respect for EJ communities' diverse environmental cultures and knowledges, and for the interests of future generations and non-human nature	**Recognition** *Who ought to be respected and valued?*	Decolonizing knowledge by recognizing the validity of and differences among local, experiential, and Indigenous knowledges
Restoration of nature and reparation of damages to EJ communities from colonialism, racism, economic exploitation, and other systems of oppression Systemic and structural transitions to create just power relations, regenerative economies, and reciprocal relations with nature	**Transformation** *What ought to change, and how?*	Transformation of academic and government institutions and research to repair their harms to and create just relations with EJ communities and nature

who contributes, who deserves credit, and how partners can speak to lay audiences accessibly without distorting research findings. In sum, CER requires extended social reflection on the most important questions that can be asked about research: why do it, for whom, and how?

CER and Justice

Prioritizing a CER approach to EJ research does not simply improve research outcomes, but advances justice in the research process. Table 2.2 draws on the dimensions of EJ (introduced in chapter 1) to illustrate how CER contributes to justice in the research process. The framework presents descriptors of CER and EJ as a whole, rather than an exhaustive or specific list of criteria for evaluating individual research projects. As in chapter 1, the four principles of justice are presented as distinct dimensions, rather than a linear path that must be followed from one type of justice to another.

CER contributes to *distributive* research justice by encouraging researchers and communities to share tangible resources (such as grant money and labor) and intangible resources (such as authority and credibility with different constituencies), as well as the workload involved in research, on terms that all participants consider fair. Distributive justice can also involve training that builds community organizations' capacities to conduct research in the future, either with new research partners or on their own, which enhances their self-determination. This kind of justice also involves community partners' co-ownership or ownership of data generated by research, which can be a potentially valuable resource that is vulnerable to exploitation by others.

CER advances *procedural* research justice when community partners have both voice and power over each phase of the work, even if they freely choose to participate more fully in some stages than others. This also involves free, prior, and informed consent (FPIC), an international human rights principle that reflects Indigenous demands for self-determination. The principle extends beyond traditional research ethics requirements to encompass community-level consent to the research, with the community uncoerced and fully informed about potential consequences (as discussed in chapter 5). Procedural justice also entails the community's power to control the use of data generated by the study, as required in many Indigenous data sovereignty protocols to protect their sacred sites from vandalism or looting, and to block exploitive or unauthorized uses of biological or ecological specimens.

Recognition in research justice reflects calls to treat community-based sources of knowledge as valid, while respecting their differences from dominant knowledge systems (such as Western science), and striving to represent these knowledges fairly and accurately on their own terms. This may be referred to as epistemic or cognitive justice, or as decolonizing knowledge in contexts involving Indigenous and other formerly colonized peoples.

CER contributes to *transformational* research justice when its collaborative process or the goals of the research help to repair historic harms of omission and commission by research institutions against EJ communities. CER can build trust and address previously neglected needs for research on the most pressing issues confronting frontline and fenceline communities—what some call "undone science," which is undone because it poses a threat to dominant interests (Frickel et al. 2010). CER can also begin to reverse a long history of extractive research practices and conclusions that have justified environmental destruction and other forms of oppression of EJ communities. As chapter 5 discusses in more depth, CER may also aid in larger efforts to enact restorative or corrective justice for the institutional impacts of academic and other research institutions, many of which were built on lands taken from Indigenous peoples, some of them built by conscripted and slave labor, almost all of them funded and run by economic and political elites and, increasingly, run like for-profit corporations. By striving to practice more just relations with EJ communities and with nature, CER can help to prefigure much-needed changes in research and research institutions.

Why Now?

A hard turn toward CER is especially necessary in the current moment. Authoritarian political attacks, many of them made on behalf of extractive industries, increasingly aim to discredit researchers and research institutions because of the inconvenient news they can deliver—about the destruction and injustices caused by fossil-fuel-driven climate change, the industrial food system, racist policing of communities of color, the COVID-19 pandemic, and much more (McCarthy 2019).

We doubt that the authority of research will be enhanced, or that justice will be done, by defending the citadel of traditional science and research without transforming whose science it is and how research is conducted, and how and for what research is utilized. Years of experience have taught many in EJ communities that outside researchers take interviews and specimens but rarely share their findings, that regulatory science ignores evidence of harms by powerful polluters or demands impossibly high levels of certainty, and that when officials or researchers confirm that harm is real, they rarely help to stop it (Cable, Mix, and Hastings 2005; Cole and Foster 2001). In response to many external researchers' historic disrespect for the rights and knowledge of Indigenous communities, research became "one of the dirtiest words" in their vocabularies (Smith 2021, 1).

What would it mean for researchers and research institutions to embrace a research paradigm worthy of EJ communities' trust? We think it would include researchers and their institutions sharing their considerable resources with local partners, collaborating to shape the research agenda, respecting the knowledges that reside in EJ communities as additional sources of expertise, and building relationships aimed at regeneration rather than extraction. Partnerships grounded in reciprocal relationships can bridge gaps of knowledge and trust between community members and researchers who are genuinely committed to EJ, as they work alongside one another to establish common understandings of environmental threats and their causes, and devise just responses. In a research context defined by power, politics, and competing values, strengthening CER partnerships and practices to produce rigorous research is more important than ever. Developing just and effective remedies based on sound research depends on elevating attention to details of systematic data gathering and analysis, while understanding how power is structured and exercised. Community members who have invested themselves in conducting this kind of research and discovering the results for themselves are more likely to believe and act on the findings than if they are asked to passively accept outsiders' findings and recommendations (Balazs and Morello-Frosch 2013; Lewin 1948). Politicians and polluters are less likely to be able to persuade the public to dismiss the evidence and policy prescriptions that result from community-generated studies than from research conducted by experts or advocates alone. In addition, CER that is directly disseminated to the public and policy makers can be harder for centralized authorities to censor, massage, or bury than reports by federal or state government agencies, in which political appointees can interfere in the work of researchers.

CONCLUSION

While we have argued that a CER approach should be at the forefront of EJ research today, this is not to suggest that CER is the *only* legitimate approach to doing EJ research. Some literature reviews, legal analyses, and documentations of environmental injustices that do not involve community partners can make important contributions to advancing EJ. Such studies may be necessary preparatory work to understand issues and evidence, and build credibility with future community partners. In addition, not every situation is ripe for CER. Researchers or their partners may lack full awareness of and commitment to the principles of collaboration. In some cases, involvement in research may pose a risk to the health and safety of community partners. Some communities may be so tired of taking part in studies, or so disappointed by the lack of tangible benefits from past research, that they refuse to participate. Some communities may lack organizations that could legitimately represent their interests, or that care enough about EJ, at the time of the study. In other instances, community organizations may prefer to devote their resources to organizing and to delegate a study to trusted researchers, as long as they remain accountable to serving the community's needs and do no harm. Some community and advocacy groups are quite capable of carrying out sophisticated research without the aid of outside researchers (Pastor, Benner, and Matsuoka 2009).

However, given the value of CER for EJ research, we think that the burden is on researchers to explain why they should *not* collaborate with the community that is the focus of a study, not why they should. The best tests of whether researchers have just reasons for not employing CER may be whether potential community partners can accept these reasons as legitimate or whether EJ is better served by researchers *not* partnering with community collaborators because it would make them more vulnerable to reprisals, or because no local organizations are interested in or supportive of EJ. Even when researchers do not enter into a formal collaboration, making good faith efforts to align a project with community organizations' goals ensures that the research maintains relevance and delivers local benefits, rather than working against community interests and purposes.

We also do not want to suggest that CER is easy. Even researchers and partners who have committed to a full collaboration must wrestle with fulfilling the promise of CER amidst imbalances of resources, expertise, and power. As discussed in chapters 4 and 5, it is challenging to produce research that is simultaneously useful to community partners, recognized as a legitimate contribution to academic scholarship or the professional literature, and in compliance with foundation or government agencies' requirements and priorities. Additionally, many academic and government research institutions continue to raise impediments to CER. Nonetheless, those who conduct CER for EJ embrace these challenges as integral to their missions as engaged researchers and recognize that collaborating with community partners is a uniquely powerful way to integrate the theory and practice of EJ into research. The next chapter describes how researchers can prepare themselves to do that work.

PART 2

Collaborations

3

Preparation for Community-Engaged Research

Floridalma Boj Lopez, Chad Raphael, and Martha Matsuoka

Community-engaged research (CER) for environmental justice (EJ) requires researchers to redefine their traditional roles, which involves unlearning dominant ways of seeing and being as much as learning new knowledge, skills, and dispositions. Knowing oneself in relation to others is a necessary step in co-producing knowledge with communities. Participants need to prepare themselves by examining their own positioning in multiple structures of privilege and oppression. This self-examination is vital for developing the commitment and capacity to redress power imbalances between and among researchers and communities during the research process (Foronda et al. 2016; Tervalon and Murray-Garcia 1998). The goals of this inquiry are to liberate oneself and others from potential abuses of power, but also to move beyond cynicism about the ability of differently situated people to collaborate or paralyzing fear of doing harm, which can prevent researchers from engaging with EJ issues and communities altogether (Lockie 2018). Researchers' examination of themselves in relation to EJ communities is a continuous commitment, not a one-time task, because of the complexity of the work, and ongoing needs to respond to new circumstances and build new relationships.

This chapter lays out the groundwork researchers need to do before building a formal relationship with a community partner to engage in the research process. The chapter presents a framework that researchers can use to examine their positioning in multiple structures of power, including researchers' individual characteristics, disciplines, institutional affiliations, and project-related factors. Doing this groundwork is crucial for anticipating potential barriers between researchers and community partners, and preparing to bridge these obstacles to collaboration.

TABLE 3.1. Preparation for CER for EJ

Dimension of Justice	In Preparation for CER for EJ
Distribution *Who ought to get what?*	Developing an initial understanding of how community members view the root causes and remedies of environmental and social inequities in the community, and defining roles for researchers in helping to build communities' capacities for research
Procedure *Who ought to decide?*	Preparing to share power over the design and conduct of research with community partners, based on a thorough understanding of the community's and potential organizational partners' history, situation, strengths, concerns, and internal diversity
Recognition *Who ought to be respected and valued?*	Recognizing the complex and intersectional nature of privilege and oppression in research relationships Engaging in anti-oppression training and reflection Developing cultural competences and humility to value communities' knowledge Assessing how one's discipline and institution respects or devalues community knowledge
Transformation *What ought to change, and how?*	Transforming researchers' training, traditional roles, disciplines, and institutional practices to prepare the ground for creating trusting and reciprocal relationships with EJ communities

Table 3.1 summarizes these aspects of self-preparation, showing how they relate to the four dimensions of justice common to CER and EJ.

Our approach is grounded in the epistemology of CER, which begins with the idea that what we know is influenced by where we stand and with whom we interact (Young 2000, 136). Intersectional theory draws attention to how an individual's position is crisscrossed by locations in multiple social groups, and how distinct forms of oppression and privilege can be compounded by these multiple identities (Crenshaw 1989). To be a Black woman, for example, is to contend with a mix of environmental racism and sexism that is different from the environmental oppression that Black men or white women experience (Ducre 2018). Yet our perspectives do not automatically determine our opinions, interests, or beliefs. A perspective consists, instead, "in a set of questions, kinds of experience, and assumptions with which reasoning begins, rather than the conclusion drawn" (Young 2000, 137). Residents of EJ communities have diverse perspectives, but they are often distinct from the vantage points of people situated elsewhere, including most credentialed researchers. Thus, community-engaged researchers must grapple with how to build bridges to and among the multiple perspectives within EJ communities.

Based on this epistemology, we present a framework and set of questions that can help guide researchers' inquiry into their positioning, issues of power, and necessary preparation for CER in an EJ community (summarized in table 3.2). We

TABLE 3.2. Researcher Positioning, Power, and Preparation

Positioning	Power Relations	Preparation for Researchers and Research Teams
Individual	**Individual**	**Individual**
Ascribed Characteristics: Race, ethnicity, class, Indigeneity, gender, nationality, citizenship status, sexual orientation, religion, physical or mental ability, etc. *Achieved Characteristics*: Education, job, social position, languages, etc. Which ascribed and achieved characteristics do you share with members of the community with whom you want to collaborate? Which do you not share? Which are most important to how you perceive yourself? Which do you think will be most relevant to how community members will perceive you?	Which of your identity characteristics are sources of privilege (dominance)? Which are sources of oppression (subordination)? How might your privileges and oppressions influence your interactions with community partners? With other community members?	What research and training in anti-oppression, cultural competences, cultural humility, and conflict management would be most necessary for you to work with your community partners? How could you build empathy with community partners and earn their trust from the start? Which kinds of identity characteristics would be most valuable for researchers who want to work with your community partners? Which role(s) in the research process ("uses of self") would be most appropriate for you? How can you and your community partners benefit from a diversity of identities, skills, and viewpoints in the research? Which roles for researchers do you most need to learn and unlearn to share power with your community partners?
Disciplinary	**Disciplinary**	**Disciplinary**
What is the historic relationship of your field to the community in which you aim to collaborate? What roles and actions have people in your field typically taken in this community? How is the community likely to view you as a representative of your field?	In what ways do theories and methodologies in your field reflect colonizing, dominant knowledge? In what ways does your field contribute to decolonizing, liberatory knowledge?	What research and training do you need to do on decolonizing, liberatory approaches to knowledge in your field? Are there additional disciplinary perspectives and research capabilities you/your team need, to work with your community partners?

(*Continued*)

TABLE 3.2. *(Continued)*

Positioning	Power Relations	Preparation for Researchers and Research Teams
Institutional How does your institution hinder or reward CER? What is the historic relationship of your institution to the community in which you aim to collaborate? What roles and actions have people in your institution typically taken in this community? How is the community likely to view you as a representative of your institution?	**Institutional** How is your institution organized to share power and resources with community partners, or not? How does your institution act as a source of domination or liberation in the community with which you want to work? Which of your institutions' programs or actions may be most important to your community partners?	**Institutional** Which interventions might you need to make in your institution to collaborate with your community partners (educating your IRB, getting approval to share resources, etc.)? Which role(s) in the research process ("uses of institution") would be most appropriate for your institution to play?
Project Related What are the initial topic, purpose, community benefits, time commitment, level of change, and model of change? Which of your plans are you willing or able to negotiate with community partners, and which are you not willing or able to change? What is your initial definition of the community and its boundaries (geographical, social, etc.)?	**Project Related** How might your plans align or conflict with your community partners' purposes, time frame, level of change, and model of change? How do your plans reflect dominating or liberatory understandings of these elements of the project? How do your assumptions about the community and its boundaries reflect dominant or liberatory understandings?	**Project Related** What do you need to learn about the community situation and context? How will you explore multiple ways of defining the community to avoid imposing your assumptions? How will you research the community's internal diversity and reflect it in the composition of your partners and/or perspectives?

draw on insights and tools for considering how researchers' relationships to communities may be structured by the identities of individual researchers (Hyde 2017; Axner, n.d.), the research team (Garzón et al. 2013; Muhammad et al. 2015), and their institution (Collet 2008). We add ways of thinking about how researchers are positioned by their disciplines and by their initial plans for specific research projects, which need to be open to redefinition with community partners in CER.

INDIVIDUAL POSITIONING

Examining Identities

Researchers' identities are formed in part by the characteristics listed in table 3.2, which frequently position people in relations of domination and subordination. In EJ research, different *ascribed* characteristics may be especially relevant in different contexts. In the U.S., environmental injustices and EJ movements have been shaped especially by race, Indigeneity, and class (see chapter 1). Therefore, researchers in the U.S. must especially examine their own positioning within structures of white supremacy, settler colonialism, and racial capitalism. Researchers from the Global North doing transnational fieldwork in the Global South must consider how their positionality and power stems from their nationality and language, and their relation to specific histories of colonization, development, and cultural and economic globalization in the local context (Sultana 2017). All researchers need to consider how their *achieved* characteristics also shape their relationships and power in the relevant community. Education level, access to funding, status as credentialed experts, and exclusive scholarly languages privilege researchers in relation to most members of EJ communities, regardless of whether researchers share other attributes in common with community members.

To recognize privileges and prepare to build trust with community partners, researchers can begin by mapping characteristics they share and do not share with members of the community. Community members may perceive different aspects of researchers' identities as more relevant than researchers themselves do. In particular, people from dominant groups are socialized *not* to perceive themselves as defined by their whiteness, maleness, heterosexuality, and so on, while these may be the most important initial markers of their identity for EJ communities. The assumption that one's race, gender, or other characteristics are normal or unremarkable is a privilege of power. Considering how each of our attributes may be a source of oppression or dominance, and how they may influence relationships with community partners and other members of the community, is critical.

This reflection should be informed by anti-oppression study and training salient to the community with which researchers want to collaborate. Many universities and other institutions offer training in how to practice allyship and solidarity, informed by resources on antiracism (DiAngelo 2018; Kendi 2019), antisexism and sexual violence prevention (Crimmins 2019), decolonizing relations with

Indigenous peoples (McGuire-Adams 2021; Swiftwolfe 2019), creating safe spaces for LGBTQ+ people (Woodford et al. 2014) and undocumented immigrants (Sanchez and So 2015), and intergroup dialogue (Zúñiga, Lopez, and Ford 2014). CER researchers and community partners also provide guidance on how collaborations can address race and ethnicity (Environmental Justice and the Common Good Initiative 2020; Fernandez et al. 2017; Murphy et al. 2013), national origin and immigration status (Collet 2008; Vaughn and Jacquez 2017), and how these intersect with differences of class and expertise (Muhammad et al. 2015, 2017). Eng et al. (2017) and Yonas et al. (2013) specifically address antiracism training for CER.

The most valuable of these resources link the personal and the political. They help researchers examine how to unlearn oppressive language, assumptions, and actions; build relationships based on respect for others' differences; and intervene in everyday interactions to promote liberatory and respectful relations. At the same time, they teach allyship strategies that respect the leadership of people from subordinated groups, rather than attempting to speak for them. These resources also link the study of interpersonal and intergroup relations and communication with the history, laws, and policies that continue to influence domination and subordination. For example, working with a community threatened by deportation of undocumented members requires researchers to familiarize themselves with current immigration policy and work carefully to include undocumented people's participation, while shielding them from risk. The more that researchers take responsibility for learning and acting on histories and ongoing structures of domination, the less likely it is that researchers will impose upon community partners by asking them to provide an education they have little time and less responsibility to give.

Anti-oppression work can also help researchers from marginalized groups address the challenges they face in research institutions and communities. These researchers can draw support from mentoring relationships, study and affinity groups, professional associations, and social movements that address the challenges of operating within dominant institutions and provide alternative communities of practice (see box 3.1). A healing justice approach, which stems from community organizing, can also help researchers cope with stress and trauma from being treated as second-class outsiders within academia and the public sphere, overcome internalized oppression, and avoid horizontal hostility among subordinated groups who are often pitted against one another for resources and recognition (Axner, n.d.; Pyles 2021). This approach directs attention to preparing to heal personal, interpersonal, and institutional harm by caring for our and our partners' physical, mental, and emotional well-being while conducting CER and working for change. Healing justice practices of dialogue, mutual support, and mind-body care can take any form that feels culturally relevant to participants, from celebrations, feasts, and purification ceremonies to yoga, mural painting, storytelling, basketball games, and many other activities.

BOX 3.1. MARINA PANDO SOCIAL JUSTICE RESEARCH COLLABORATIVE

When Kristie Valdez-Guillen and I (Floridalma Boj Lopez) decided to start the Marina Pando Social Justice Research Collaborative, our goal as members of East Yard Communities for Environmental Justice (EYCEJ) who were also pursuing doctoral degrees was to create a space where young, first-generation college students who had been part of EJ youth organizing in Los Angeles could have a positive and welcoming experience while learning to do research. The research itself was not the priority, but rather an avenue through which we could continue fostering relationships to youth members who went off to college and were dealing with their own forms of alienation at universities.

The collaborative also pushed back on academic discourses that critique community efforts and instead used my own community knowledge and research skills in the service of movement building. As a first-generation student, I wrestled with how to connect the countless struggles of my multiple communities with what I was doing in my doctoral program. The collaborative became my humble contribution to blurring the boundaries between research and community. I assumed a facilitator rather than a principal investigator role, given that many of the issues raised by the youth were not what I was trained to research. With the expertise of the EYCEJ staff, my co-facilitator and I pooled our collective knowledge to support the youth to carry out these research projects. The EYCEJ staff were particularly excited because the program was coming from community members like myself with explicit goals for research that would benefit the participants, the organization, and the larger movement.

The collaborative's paid summer research fellowships also became an opportunity to support young people who had left the neighborhood for college to return home to apply their college-level skills to the issues they had already been organizing against as high school students. Students already had a deep knowledge of the issues and relationships with the EYCEJ staff; this would not have been the case had we recruited random college students who were unfamiliar with the community, environmental racism, and the organization. After an intensive week of full-day trainings on the nuts and bolts of collaborative research and data collection, we met with participants weekly to discuss research challenges and guided them through the writing of a research report, creating a research poster, and ultimately presenting their research to the community. While we confronted challenges like the need for more technical guidance, the time crunch of generating research during the summer, and the need to fundraise for stipends, the program produced interesting and accessible research. We culminated the program with student presentations of their research to the community members with whom the students had organized. While the students produced great projects on food apartheid, heavy metal contamination, industrial water runoff, and other issues, the real measure of our success was how many of these young people decided to remain members of the organization. Some have since joined the board of directors or staff of the organization, so the program also helped build new leadership.

Cultural Competence, Humility, and Preparing for Conflict

CER practitioners also need to familiarize themselves with their community partners' values, practices, languages, and other cultural characteristics. Organizations and researchers involved in CER prepare themselves by developing *cultural and linguistic competence*, which means they

- have a defined set of values and principles, and demonstrate behaviors, attitudes, policies, and structures that enable them to work effectively cross-culturally;
- have the capacity to (1) value diversity, (2) conduct self-assessment, (3) manage the dynamics of difference, (4) acquire and institutionalize cultural knowledge, and (5) adapt to diversity and the cultural contexts of the communities they serve;
- incorporate the above in all aspects of policy making, administration, practice, and service delivery and systematically involve consumers, key stakeholders, and communities (National Center for Cultural Competence, n.d.).

Researchers also need to be familiar with the environmental justice movement's values and culture. The "*Jemez Principles* for Democratic Organizing" (www.ejnet.org/ej/jemez.pdf) and the Second People of Color Environmental Leadership Summit's "Principles of Working Together" (www.ejnet.org/ej/workingtogether.pdf) provide foundational principles for forming partnerships with academic institutions and lawyers who recognize community expertise. Both documents help illuminate how movement organizations aim to build respectful relationships, address cultural differences, practice leadership that is accountable to the grassroots, resolve conflicts, and share resources fairly.

Scholars must also develop *cultural humility* that goes beyond acquiring cultural knowledge and communication skills, to respect community perspectives (Tervalon and Murray-Garcia 1998, 120). Humility requires ongoing commitment to personal and social transformation to redress power imbalances between dominant and subordinate groups, and between professional researchers and community members (Foronda et al. 2016). Sensitivity to the complex ways in which cultural power and privilege can affect research relationships is crucial for earning community members' trust, designing more respectful and effective studies, sharing the research appropriately within communities, and applying evidence from one setting to another (Fernandez et al. 2017; Vaughn and Jacquez 2017; Murphy et al. 2013). Researchers should prepare to address relevant issues of culture and power that can arise in partnerships with specific communities by consulting past research conducted with similar communities, such as case studies on doing research with people who are Asian American (Islam et al. 2017), LGBTQ+ (Kano,

Sawyer, and Willging 2017), deaf (Barnett et al. 2017), or HIV positive (Rhodes et al. 2017) and with members of faith-based groups (Kitzman-Ulrich and Holt 2017) and with youth (Arredondo et al. 2013; Mueller and Tippins 2015; Ozer, Piatt, and Willging 2017; Fernández 2021).

Humility also prepares researchers to recognize communities as sources of knowledge and to enact transformative justice for past abuses of power in the research process. Researchers must open themselves to how community partners conceptualize their environment and health, their visions of EJ, and their goals for research. For example, McGreavy et al. (2021) reflect on multiple projects on forest conservation, river restoration, and co-management of fisheries by an interdisciplinary team of Native and white settler scholars with the Penobscot Nation of the Wabanaki Tribal Nations in Maine. The partners faced fundamental tensions between academic and Penobscot researchers' conceptions of science, place, and time. They addressed these tensions by drawing on Wabanaki research methods and these nations' practices of diplomacy to negotiate differences; building trust over time while meeting academic needs to publish by including pilot studies, iterative engagement, and dialogue among partners; slowing the typical research process to adopt rhythms of collaborative work linked to the seasons and Wabanaki culture; and integrating Wabanaki students into leadership roles in the research team. Additional cases examine how humility has inspired researchers to grapple with issues of cultural power in projects on neighborhood health (Ellis and Walton 2012) and environmental indicators (Garzón et al. 2013; Shepard et al. 2013), and to translate CER principles themselves into culturally relevant and accessible language to ensure research participants can give fully informed consent to participate in projects (Burke et al. 2013).

CER practitioners can also prepare for conflict in research projects, which is normal in any relationship. In addition to drafting clear and specific agreements on roles, responsibilities, and resource sharing (see chapter 5), research partners can agree at the outset on procedures and techniques for addressing conflict that are culturally relevant to the community. The Maine research team, for example, learned to employ Wabanaki diplomacy, which involves frequent rounds of dialogue that incorporate multiple voices, not simply relying on leaders to execute a single memorandum of understanding at the project's outset (McGreavy et al. 2021).

Additional training in conflict resolution is helpful. Nonviolent communication techniques (Rosenberg 2015) can identify how conflict stems from participants not having their basic needs met, such as needs for resources, recognition, or fair treatment (Pyles 2021). Nonviolent communication engages people in identifying which needs are not being addressed, and aiming to devise solutions that can meet everyone's needs, making conflicts more tractable and reaffirming mutual respect. Restorative justice approaches can address harms in the research relationship. Devised as an alternative to the criminal justice system's emphasis on punishment, restorative justice puts victims and offenders in dialogue so that they understand how the victims have been harmed, and so both can agree on ways to heal the breach in their relationship and the community (Capeheart and Milovanovic 2020). Transitional justice and conflict transformation approaches focus on reconciliation by investigating past harms, identifying responsible parties and offering reparations to victims, and designing measures to avoid repeated harms (Killean and Dempster 2021). Training in intergroup dialogue can help collaborators move beyond personal attacks and defensive responses to explore how their cultural differences influence their relationships, including their approaches to conflict, and how they can be reconciled (Zúñiga, Lopez, and Ford 2014). Even simple practices, such as agreeing to "call in" collaborators for private conversations about how to change norm-violating behaviors rather than calling out colleagues by publicly denouncing them, can address conflict effectively while preserving relationships (Pyles 2021). While no single approach will work in every situation, especially if there are unresolved power imbalances among participants, the more training researchers have in conflict resolution, the more durable and mutually beneficial a partnership is likely to be.

Roles in Research Teams

Because researchers need many kinds of preparation for CER, they often form research teams who can bring a broader range of experiences, skills, and identities to the work than any individual can—even before expanding the research team to include local partners. As researchers consider their roles, they can examine how their identities map collectively to the community's, and consider the best "use of self" by each team member to form authentic relationships that advance the research.

One set of questions revolves around who is an "insider" and an "outsider" in relation to the community. The CER paradigm rejects the assumption that researchers who study their own communities cannot discover truth because they lack objectivity. The notion that outsiders are more trustworthy stems from positivist assumptions that detached observers should conduct research *on* communities rather than *with* them (see chapter 2). In addition, dominant groups have deployed this idea to reinforce their power over knowledge by reframing a major limitation of outsider-led research—its inability to understand subordinated communities on their own terms—as a purported "strength." At its worst, this distinction has reinforced racism and colonialism, as white scholars tried to discredit research on Black communities by Black scholars (Morris 2017) and attacked researchers of all backgrounds who developed strong empathy with Indigenous communities for "going native" (Kanuha 2000). Researchers from non-dominant groups still must contend with accusations of bias and lack of rigor when studying their own communities, which researchers from dominant groups rarely face when studying their own or others' communities (Serrant-Green 2002).

In contrast, CER practitioners tend to view "insider" status as an asset for researchers, while also questioning the terms of the insider/outsider dichotomy itself. Researchers who share important attributes with community partners—such as race, gender, or tribal affiliation—are often better positioned to earn their trust; to draw on shared experiences of environmental injustices; to gain access to knowledge the community is reluctant to share with others; and to act as cultural knowledge brokers who can translate meanings between communities and research institutions, helping people rooted in each of these contexts to form common understandings (Davis and Ramírez-Andreotta 2021; Kerstetter 2012; Moore de Peralta, Smithwick, and Torres 2020). Thus, researchers who share identity characteristics with communities may be the most appropriate team members to work regularly with community partners, although all project leaders should also expect to make themselves available to demonstrate respect and accountability to community leaders (Muhammad et al. 2015). Researchers who live in EJ communities are especially valuable because they understand local systems of inequality and have more embodied and nuanced expertise. Because researchers and students from non-dominant groups are often made to feel that they are "outsiders within" research institutions, it is important for their colleagues to act in ways that honor these team members' uniquely valuable contributions.

As research teams expand to include community partners, the full team will need to define additional roles. Rivera and Erlich's (1998) thinking about role differentiation in community organizing offers guidance. They suggest that community residents who share multiple ties (e.g., of race, class, and neighborhood residency) are the most appropriate people to serve as grassroots organizers, working personally and intimately with their neighbors. Similarly, local residents can work most closely with other residents to gather data and disseminate findings (but ought not be restricted only to these roles). People who share ties of race or class with residents but do not live in their community may serve best as liaisons to the larger society. In CER projects, these people may be part of a team based in a research institution, who serve as principal investigators or project managers. Sympathetic outsiders who do not share any primary ties with the community can provide technical assistance and resources to build community members' capacities and leadership to do the work on the ground. In CER, these people may be other members of the institutional research team, from principal investigators who raise money, manage the team, and help root the study in prior research to other team members who train residents to design research instruments and analyze data.

However, because CER involves the co-production of knowledge, all members of research teams should be able to participate in designing studies and interpreting data. For the same reasons, all researchers need to be wary of assuming privileged insight simply by virtue of sharing similar markers of identity or living in the community, or of assuming that others are incapable of shared understanding across differences (Lockie 2018). One way to create this kind of equitable research environment is to cultivate "up, down, and peer mentorship," which recognizes that expertise is collective, and moves beyond status hierarchies among professional and community researchers to "create a circular democratic model where contributions from each unique position become the established norm" (Muhammad et al. 2015, 15).

Reframing Researchers' Roles

To develop respectful and reciprocal relations with EJ communities, professional researchers and students must *unlearn* their traditional roles. In an apologetic essay, Sherry Cable (2012) described how she asked the Yellow Creek Concerned Citizens of Bell County, Kentucky, for permission to study their campaign to protect themselves from toxic waste emitted by a local tannery. One member asked what Cable would get out of the study. "If I can pull it off, I'll publish enough articles in academic journals to earn promotion and tenure, instead of losing my job," she responded. Another member asked what the group would gain from her work. Caught unprepared, Cable admitted, "Nothing" (2012, 21). Fortunately for her, the group's leader found her honesty refreshing and let her study the campaign. In her essay, Cable apologizes to him for acting in the traditional role of researcher as *parasite* and explains how the experience motivated her to develop a CER practice that prioritizes benefits for community collaborators.

Community-engaged researchers also need to avoid thinking of themselves as *saviors*, who assume that EJ communities depend on outsiders to improve their conditions rather than collaborating with residents to emancipate all people (including scholars) from relations of domination. Messianic researchers are likely to try to make decisions alone that ought to be made with community partners, disrespecting their knowledge and agency and failing to see that research is one small contribution to the success of complex, dynamic, and vibrant community-led movements. CER practitioners also avoid presenting themselves as *public intellectuals*, who engage in media punditry or explain EJ communities on their behalf without their approval. Nor should CER researchers be what Fine (1994) calls *ventriloquists*, who, without residents' consent, present researchers' own interpretations of a community in an objective third-person voice or selectively curate residents' voices to illustrate the researchers' own conclusions, rather than collaborating with EJ communities to co-create knowledge with them.

While much EJ research documents inequities and injustices, CER researchers should not consider themselves merely as *damage assessors*. Eve Tuck (2009) calls for researchers to move away from framing communities exclusively as injured. As she writes, "[E]ven when communities are broken and conquered, they are so much more than that—so much more that this incomplete story is an act of aggression" (416). CER recognizes that residents survive and continue to create joy, fight back, and practice their own epistemologies. Researchers collaborate to design studies that begin from community strengths and concerns, and aim to co-produce something of value and benefit to community partners, not a catalog of victimhood (Wallerstein et al. 2019).

Instead, community-engaged researchers embrace a variety of other terms for the roles they play, which reflect careful consideration of how their roles depend on specific contexts and relationships to communities. Some researchers think of themselves as short-term *collaborators* on one-time projects. Other researchers identify as long-term *allies*, acknowledging their differences of power and privilege from many residents of EJ communities, while committing to work in solidarity by supporting community members' leadership over multiple projects. This may involve acting as a *power shifter*, who uses power derived from one's access to funding, academic or government positions, and other sources to transfer power and resources to community partners (Wallerstein et al. 2019). Some researchers call themselves *scholar activists*, who try to integrate their long-term professional and personal efforts for EJ by working with community organizations and movements on CER and in other capacities (Hale 2008; Montenegro de Wit et al. 2021).

Researchers also define their roles in relation to their ties to the community. Nina Wallerstein, a non-Indigenous researcher who has led many research teams from the University of New Mexico that have collaborated with Native communities, describes herself as a *guest* in tribal homelands. For Wallerstein, being a guest means recognizing that "the community owns and has authority over its own geographic and cultural territory," that academics must ask permission to enter, and that they should bring "offerings or gifts as a symbol that one accepts guest status and conducts oneself accordingly by recognizing 'house rules,' or social norms of the community one has been invited into" (Muhammad et al. 2015, 9). EJ researchers with closer ties to communities find other ways to define themselves. Lorenda Belone manages her multiple identities as a member of the same University of New Mexico research team, a Native New Mexican, and a woman, by calling herself a *native researcher* (rather than an academic researcher), who reconciles clan and academic obligations in her work (Muhammad et al. 2015). Magdalena Avila, a Chicana member of the same team, sees herself as a *practitioner of a way of life*, in which CER embodies "the principles that guide my life" (8), including working hand in hand with communities in which she is both an insider and outsider, and deconstructing this distinction with her partners in the process.

DISCIPLINARY POSITIONING

Researchers also need to anticipate how their disciplines position them in relation to the community with which they want to collaborate. For example, many Indigenous communities have experienced anthropologists as people who rob ancestors' graves, and educators as people who rob their children of their culture (Estes 2019). Researchers in these fields and others who win communities' trust have studied the historic relationship of their discipline to the community, including its harms and benefits. These researchers are prepared to acknowledge this history, to explain how their actions will differ from harmful predecessors, and to listen carefully to communities' conditions for collaboration. These researchers have learned, for example, which research protocols a potential Indigenous partner requires and are prepared to follow them before initiating contact. CER researchers have also thoroughly examined how theories and methodologies in their field continue to marginalize specific communities' knowledge, and any guidance the field provides on how to decolonize and liberate that knowledge.

INSTITUTIONAL POSITIONING

Researchers' home institutions also position them in relation to communities, requiring researchers to examine their employer's culture of research and its reputation in the eyes of the community. Most academic institutions raise barriers to CER by valuing the number and prestige of publications rather than their value to communities, rewarding individual scholarship more than collaborative research, and failing to trust community members to observe research ethics and co-manage funding (see chapter 5). Public agencies often restrict government researchers from collaborating with partisan political groups. Researchers need to reconcile their institution's demands with obligations to community partners, while working to transform their institutions to be more supportive of CER (as discussed in chapter 5). In particular, academic researchers must plan to publish peer-reviewed research that meets disciplinary standards to maintain their positions if they want to keep doing CER. They also need to reach out early to their institutional review boards to understand how they apply their ethics requirements to community participants, and to institutional finance offices to understand their stipulations for paying out funds to community partners.

In addition, researchers should study their institutions' historical relationships with specific EJ communities to understand how potential collaborators are likely to view the institution. Is the institution valued as a source of community amenities and jobs, mistrusted as a driver of displacement and gentrification, resented as an occupier of Indigenous homelands? Does the institution operate particular programs that are especially respected—such as community-based learning centers, food pantries, museums, clinics and hospitals, or even athletic teams—that might help researchers establish contacts in a community? Which of the institution's actions and programs are potential sources of emancipation in the community, and how might one ally with them to make the most appropriate "uses of institution" in one's CER?

PROJECT-RELATED POSITIONING

Clarifying Initial Assumptions

While CER involves co-designing research with community partners to meet their needs, researchers' interests and capacities may be limited. Before entering into discussions with community collaborators, researchers should estimate the scope of the commitment they expect to make to the project; initial topics, purposes, and anticipated community benefits of the research; the project's intended level of change (from local to global, individual to collective); and the model of change (as driven by grassroots community organizing, coalitions of established community leaders, social service providers, government agencies, etc.) (Barge 2016). Clarifying which of these initial assumptions are open to negotiation, and which are not, should guide researchers to find compatible community partners.

Researchers need to be ready to discuss which resources and how much time they can commit to the community. Will the research be a brief project or one that requires commitment to a longer-term relationship? What are potential levels of funding, and how much of it might be shared with partners? Does the researcher envision the project as limited to a specific location or case? An opening estimate of how much one can commit to a project helps manage partners' expectations and contributes clarity to discussions about collaborative work, building a foundation of transparency for the partnership. It also helps partners avoid diverting community energy to research that would be better spent on other change strategies, such as organized protest or mutual aid. Research partnerships need a clear view of how their joint work relates to enduring and structural injustices, and how their projects can build communities' capacities for change over the long haul.

At the same time, researchers need to be wary of defining issues as narrow problems that are amenable to study using researchers' own highly specific skill sets, while failing to address communities' priorities. EJ researchers especially need to appreciate how community members view the focus of the research in relation to larger patterns of oppression. In North Carolina, for example, academic researchers were able to partner more effectively with local Black-led EJ groups organizing against industrial hog farming than were white-led environmental groups, who saw this struggle narrowly in terms of controlling air and water pollution. In contrast, community leaders saw it as one aspect of a larger struggle against historic and institutionalized racism, which required research to guide and support many kinds of actions. As one local organizer commented, "One of the things we learned in this whole process was that white people want to solve problems and black people want to solve issues" (quoted in Tajik 2012, 137). Highly responsive CER practitioners are willing to redefine their topic significantly to reflect community knowledge and bring in additional expertise if needed (Wallerstein et al. 2019).

Because drawing boundaries around a community is an act of power, researchers will need to collaborate with community partners to define the community—whether geographically or by social groups or shared characteristics (see chapter 4). The way the community is defined will determine which organizations will lead the research partnership, who the project will recruit as participants, and where to turn for funding.

The selection of research partners is often also a choice of a change model. In EJ movements, grassroots capacity building and organizing are the preferred strategies for social change, although this can take multiple forms and may involve strategic alliances with social service providers, government agencies, and small businesses. CER projects especially seek grassroots organizations that have a strong base in the community, or organizations that are directly accountable to such groups. Many successful research partnerships start among a small group of organizations that are accountable to constituencies who are directly affected by the research problem. These budding partnerships then enlist others who can represent additional facets of the community as co-investigators, advisors, and/or staff members, matching individuals with roles according to their availability, skills, resources, and influence in the community (Hancock and Minkler 2012).

Familiarization with the Community

Researchers should also test their assumptions through preparatory study of the community's historical context and contemporary situation, how residents experience place, and the community's ecosystem of organizations and power relations. When studying the community's history, researchers should seek out sources that represent it through the eyes of groups at the center of the proposed project, not simply academic or journalistic accounts. Street murals, oral histories, and community news media and celebrations are valuable windows into how members of the community understand their past and how it has shaped their present.

Because EJ is place-based work, building partnerships depends on a thorough understanding of the places in which community members live, work, and play. For example, persons conducting CER in South Los Angeles should have a sense of why the area is distinct from other areas of Los Angeles, including the impact of redlining, the Great Migration of Black Americans from the South in the 20th century, deindustrialization, and other environmental and social upheavals, organizing, and social movements. Researchers should examine place-based histories with an eye to how larger logics of white supremacy, colonialism, and capitalism became operationalized through local policies that shape a community's spatial experience, and how places develop in relation to each other. Frontline and fenceline communities face environmental injustice because they have been selected as sacrifice zones to serve the needs of more environmentally privileged communities for energy, consumer goods, and other benefits, while enjoying protection from the pollution they generate. Tracing the threads of these relationships can point to important research questions about how to transform these relationships.

Understanding place goes beyond studying the current locations of people or of toxic sites to encompass communities' knowledge and experience of place. As Meyer (2008) writes, "Land is more than just a physical locale; it is a mental one that becomes water on the rock of our being" (219). Residents of EJ communities experience places in distinct ways. For example, as Ducre (2018) writes, "poor Black women create distinct cognitive spatial maps of their environments as a means to survive the structural violence and environmental degradation of their communities" (22). This is also distinct from how Indigenous people understand place collectively in relation to the very formation and survival of their nations (Simpson 2017). At the same time, researchers should seek to understand the intersectional sources of environmental inequities within communities, seeking to "more accurately, more relevantly reflect the differentiated needs and capabilities of individuals across and within multiply marginalized groups," so as to design research that can help them identify a range of solutions (Malin and Ryder 2018, 2).

This requires initial study of the community and its internal diversity. What are the community's assets—such as schools, libraries, churches, and other organizations—that are sources of resilience and might be good research partners? Which of these organizations are addressing EJ issues, even if these groups do not identify them as such? What are organizations' current and long-term priorities, and how could CER help advance them? What are their missions, leadership models, decision-making processes, and organizational capacities? For instance, working with an incorporated nonprofit that has full-time staff will be different from working with an unfunded grassroots collective. This difference can shape the organization's capacity to take on interns and whether shared funding is the form of reciprocity that the organization values. Researchers can become familiar with these organizations through their websites, and by following their social media accounts and any news coverage they have received. Volunteering in the community with local organizations can often be an important way to build relationships and demonstrate a commitment to residents that fosters trust and insight. Outsiders to the community should seek local contacts who can vouch for them and make introductions. Researchers should also ask whether established organizations adequately represent groups who are affected by EJ issues, or if researchers will need to include these groups by other means.

CONCLUSION

CER requires careful preparation before embarking on collaborative study of EJ issues. Researchers can use the framework presented in this chapter to take inventory of the many ways they are positioned in relation to communities by individual characteristics, disciplinary training, institutional affiliation, and project-related factors. This groundwork is important for anticipating how power differences can distort healthy research relationships and for attaining a clear-eyed understanding of EJ communities, so that all participants in CER can develop reciprocal, respectful, and trusting partnerships.

4

The Community-Engaged Research Process

Julie E. Lucero, Erika Marquez, Martha Matsuoka, and Chad Raphael

While chapter 3 addressed how researchers prepare themselves to embark on community-engaged research (CER) for environmental justice (EJ), this chapter describes critical issues that arise in each stage of the research process that must be negotiated between researchers and their community partners. We show how these collaborators can build healthy working relationships by cooperatively addressing power relations, defining the community relevant to the study, managing conflict, forming community advisory boards, building community partners' research capacities, sharing control over funding, drafting formal agreements on roles and responsibilities, implementing actions in response to findings, engaging in project evaluations, and disseminating knowledge in multiple venues. In this process, researchers and their collaborators can address the four dimensions of justice common to CER and environmental justice (see table 4.1).

ADDRESSING POWER

At its root, CER is a relationship between community and academic partners who co-produce knowledge for social action. CER aims to undo the traditional relations of power in research, in which academic and government researchers apply their knowledge to communities, which are seen as lacking expertise, resources, and rights to produce knowledge about themselves (Tajik and Minkler 2006). Instead, CER aims to develop researcher-community relationships that are carefully and deliberately built on co-learning, reciprocity, shared governance, and reflexivity. Figure 4.1 presents a series of questions that can guide researchers and community partners to design partnerships that are conscious of how power

TABLE 4.1. Community-Engaged Research Process for Environmental Justice

Dimension of Justice	In the CER for EJ Research Process
Distribution *Who ought to get what?*	Ensuring fair sharing of resources and work among researchers and communities, and developing agreements, managing conflict, and building community capacities for an equitable partnership
Procedure *Who ought to decide?*	Community participation in and power to design and conduct research, including defining the community, forming community advisory boards, performing participatory evaluation, and establishing roles, responsibilities, and rights to control data
Recognition *Who ought to be respected and valued?*	Recognizing the validity of and differences among local, experiential, and Indigenous knowledges in defining the community and its representatives, and throughout the research process
Transformation *What ought to change, and how?*	Collaboratively disseminating results and implementing actions to repair harms to, and create just relations with, EJ communities and nature

manifests in every aspect of the research process. Discussion of these questions within the partnership can address power imbalances, addressing known conflicts over the roles of professional researchers and community partners. This discussion can transform research relationships from *power over*, or the application of dominance, to *power with*, or the horizontal development of shared values and strategies among different interests for social equity (Eyben et al. 2006). However, because power manifests and is reproduced through processes of socialization, CER collaborators need to return to these questions throughout their partnership, to monitor and maintain equitable power relations at each stage of the research (Lucero, Boursaw, et al. 2020).

DEFINING COMMUNITY AND PARTICIPATION

Defining the community that is the focus of a CER study is one of the most powerful decisions that researchers and community partners make. To say that research is *community based* may mean that (a) the research is conducted primarily in a community setting, (b) community issues or problems are the focus of the research, or (c) a community, rather than individuals, forms the unit of analysis (Israel et al. 2013a). The community may be defined by geography, occupation, race or ethnicity, or many other factors (in Indigenous communities, for example, the community may include plants, animals, and ancestors). Reflexive CER researchers do not assume that a community is a natural, homogeneous, or harmonious entity that a single organization or public agency can represent, but recognize differences of power and interest within communities, and that the least powerful members need a voice in research (Raphael 2019b). Researchers must learn about a community's situation, context, and internal diversity (see chapter 3).

FIGURE 4.1. Research process and questions to address power for mutually beneficial research partnerships.

In the interests of procedural and recognition justice, researchers should consult widely with diverse community organizations about how to define the community in terms that reflect community affiliations, cultures, interests, and needs that are relevant to the research. It can be valuable to have both insiders and outsiders attempt to define the community and compare their definitions to uncover and question power-laden assumptions about who belongs where (Eng et al. 2013). Community advisory boards, discussed below, can also play a primary role in the process of community definition.

Defining the role of researchers and community participants in the study is another foundational decision about how power is allocated in CER. The degree of researcher engagement with community partners can vary considerably, including in breadth, duration, and reciprocal influence. Some research projects may interact with a broadly representative collection of leaders or residents, while other projects engage narrowly with a single organization or a segment of a community (Huntjens et al. 2014). In some cases, nonprofit, advocacy, and service groups or programs are enlisted as intermediaries between researchers and community residents. This role is unique, complex, and even contradictory, which is why defining roles is critical. The relationship of the intermediary to the community is often leveraged for research purposes, and the organization is expected to deliver on the research team's promises to maintain a favorable reputation. Additionally, research tasks are often added to an intermediary's daily job responsibilities, rather than integrated into them (Caldwell et al. 2015).

Commitments and degrees of engagement also vary. Some collaborations may involve short-term projects of several months, while others require long-term commitments that stretch for many years. Partnerships can be transactional, involving mostly one-way outreach from research institutions aimed at affecting communities, or transformative efforts aimed as much at changing the research institution's role in the community and the institution's research priorities (Saltmarsh and Hartley 2011).

Procedural justice depends on researchers and community partners forging agreements on the degree of community participation. Table 4.2 modifies the International Association of Public Participation's (IAP2 2018) widely used spectrum of public participation in decision making to present the degree of community engagement in a range of research approaches, only some of which fully realize CER (adapted from Raphael 2019a). Each approach suggests different roles for community partners in the research process and aims for different outcomes. Envisioning these approaches along a spectrum helps to distinguish them, while avoiding an unnecessarily simplistic or prescriptive definition of the community's role in CER.

EJ research that incorporates some public participation, but that does not fully practice CER, includes research aimed at *informing* the public of risks and enhancing public understanding of science. This typically occurs when the research involves efforts to provide accurate information to communities in response to focus groups, surveys, and other means of gauging residents' needs and interests. Researchers may communicate this information, or may rely on intermediaries to the community, such as service providers, community workers, or advocates. In these approaches, researchers or their intermediaries build brief relationships with communities based on mutual recognition of each other's legitimacy.

Ethnography, and informal research for governmental public consultation, can promote fuller participation by *consulting* community members about their views and experiences up front and confirming researchers' analyses and recommendations through follow-up public engagement. Ethnographers' reports back to participants of interim findings, also called member checks, can be especially effective at comparing researchers' understandings against community interpretations. An iterative and sincere consultation approach can yield valid interpretations of community views and experiences, and responsive conclusions and steps toward action. However, consultation does not fully practice CER if this approach does not enlist community input on framing research questions. In addition, consultation typically ends with researchers exerting final control over drawing and disseminating conclusions, or with government agencies writing final reports and issuing decisions based on them.

CER is realized more fully by *involving* community members themselves in conducting research. This can be accomplished through crowdsourced citizen science projects, in which participants gather data but do not help analyze or disseminate findings. In action research commissioned by government agencies or nonprofits, participants typically take the lead on defining the study's goals

TABLE 4.2. Spectrum of Community Engagement in Research

	The Community Is . . .				
	Informed	Consulted	Involved	Collaborated With	Leading
Process	• Researchers share information with the community, customized to its needs or interests • Mutual recognition • Brief encounter	• Researchers seek community input, views, voices, and feedback on analysis before dissemination • Dialogue • Short-term and medium-term relationships	• Researchers enlist community partners or work for clients, who contribute to study design, data gathering, and/or execution • Cooperation • Short-term and medium-term relationships	• Researchers share resources and control over all stages of study with community • Co-production of knowledge • Long-term partnership	• Community controls resources and has final say over all stages of study • Community-led production of knowledge • Long-term partnership
Outcomes	• Tailored transmission of research findings to community strengthens their relevance and impact	• Community perspectives and information strengthen interpretive validity of research, responsiveness of action steps	• Relationships strengthen research study design, access to data, validity, community and organizational problem solving	• Cooperative learning partnership strengthens research and community problem solving, mobilization, transformation	• Co-ownership strengthens research and community capacities for further research, mobilization, transformation
Approaches	• Risk communication • Public understanding of science	• Ethnography • Community needs assessment • Public consultation (public comments and hearings)	• Crowdsourced citizen science • Action research and professional consultancies for organizations	• Participatory citizen science and community science • Participatory action research • Community-based participatory research	→ → → • Community-owned and managed research

and providing access to data sources, while researchers choose the methodologies and analyze the data. These cooperative partnerships can strengthen the research's quality and practical value but can fall short of fully involving community partners in each phase of the research and engaging a broad swath of community members.

Participatory citizen science and community science, participatory action research, and community-based participatory research typically lend themselves to higher degrees of participation and the fullest expressions of CER. These approaches often include researchers and community organizations *collaborating* to manage funding and other resources, and co-designing and co-conducting each aspect of the research. Local community knowledge often exerts as much epistemological authority as professional and disciplinary expertise. In rare cases, the same approaches extend to community partners *leading* by maintaining final control over, and financial ownership of, all elements of the research. Some partners prefer to call this community-owned and managed research (Wilson, Aber, et al. 2018). Collaboration and community leadership approaches aim to activate community members to mobilize themselves based on the findings, inspire their efforts for community-level change, and develop communities' own abilities to launch future studies.

While useful, a neatly arranged research model such as this can present dangers. It can tempt researchers to substitute choosing the most attractive label for their work for careful negotiation of the most appropriate terms of collaboration for a particular community context and project. While this spectrum of approaches can help researchers and communities clarify their relationships, it does not excuse them from examining the intent and impacts of collaborations during each phase of a partnership by discussing the questions listed in figure 4.1 repeatedly, not simply at the outset. In a world in which terms such as *participatory research, community engagement*, and *shared power* have been widely co-opted by institutions that do not accept substantive community influence (see Cooke and Kothari 2001 and chapter 2), each study must be evaluated based on how fully researchers share power with community partners at all stages of the research.

CONFLICT MANAGEMENT

As partners in CER build their relationships, conflict is to be expected, and partners need to plan to address it throughout their collaboration, rather than reacting to it after it arises. Much of this conflict stems from power differences between research institutions and community organizations that face environmental injustices (Lucero and Wallerstein 2013). Partnerships typically level power imbalances by creating empathy, building trust, and developing cultural understanding among participants (Lucero, Emerson, et al. 2020; Neubauer et al. 2020). Table 4.3 summarizes some well-documented sources of conflict in CER partnerships that

Reasons for Conflict in CBR Partnerships and Descriptions

Reasons for Conflict	Description	Ways of Addressing	Reference
Communication differences	Language differences	Create multilingual spaces with translation	Lucero and Wallerstein 2013
	Direct communication style (saying what one means, with no hidden messages) versus indirect style (couching or hiding meaning to avoid discomfort)	Know your own communication and conflict style	Yonas et al. 2013
	Differences in or lack of knowledge about conversational norms, rituals, or cultural rules (e.g., turn taking, interruptions, gestures)	Design a communication structure for partnership	
Differences in worldviews and ways of knowing	Differing worldviews, or sets of beliefs and values, arising from cultural backgrounds (e.g., the belief that humans rule the natural world versus the belief that humans are part of the natural world)	Understand core differences between worldviews	Loh 2016
	Differing views about privileged knowledge (e.g., lived experiences, formal education, cultural knowledge)	Develop a set of values for the partnership	Yonas et al. 2013
			Tajik and Minkler 2006
Coordinating collaboration	Different timelines and conceptions of time (annual calendar, academic calendar, cultural calendar) that interrupt research activities	Consult community advisory boards about community priorities	Loh 2016
	Positive or negative expectations about researchers, based on community partners' experiences	Draft memoranda of understanding (MOUs) among partners	LeClair, Lim, and Rubin 2018
	Misalignment of priorities (e.g., of which research topics and benefits are important to researchers and community partners)	Conduct formative evaluations during research projects	Mayan and Daum 2016
Balancing research and action	Competing pressures on community partners (e.g., to pursue other organizational priorities besides research, such as providing services)	Draft MOUs among partners and with funding agencies	Loh 2016
	Competing pressures on researchers (e.g., to publish rather than implement action)	Provide research training for community partners	Tajik and Minkler 2006
	Competing expectations of the funding agency (e.g., to disseminate research to academics rather than community members and policy makers)	Collaboratively plan the dissemination of data in accessible and actionable formats (fact sheets, policy briefings, etc.)	Fletcher, Hammer, and Hibbert 2014
			LeClair et al. 2018
Governance	Unclear or inequitable decision-making processes, data sharing, and data usage policies	Draft agreements about governance, resource sharing, and data sharing; develop community leadership of the project	Lucero, Boursaw, et al. 2020
			Mohammed et al. 2012
			Yonas et al. 2012

address EJ and other issues, major examples of these conflicts, steps that partners often take to address these conflicts (all of them discussed later in this chapter), and relevant sources in the literature where readers can learn more.

ADVISORY BOARDS

Within community-academic partnerships, project oversight or guidance structures take many forms, such as community coalitions, steering committees, community action teams, and advisory boards. The most common structure is the *community advisory board* (CAB), comprising community members who share a common identity, history, and culture and are knowledgeable about the research topic and/or priority population (Israel et al. 1994). Selection of CAB membership should be deliberate and based on the goals of the partnership and project (Green 2001). The composition and role of the CAB should also be guided by efforts to include community-based knowledge and expertise, and to ensure that community representatives exercise voice and influence in decisions.

Just as the oversight leadership structure can vary, so can the purpose of the CAB. CAB members serve as research partners and sources of leadership. Leadership can occur for individual projects, the overall partnership, or a combination of projects and partnerships (Newman et al. 2011). A common criticism of CABs is that members are only allowed to offer advice, and researchers have the discretion to integrate advice or not. However, if CAB members are genuine research partners, then the advice they provide will be discussed, negotiated, and reflected in how decisions are implemented. CABs can facilitate ethical research processes by informing research protocols (Strauss et al. 2001) and offering valuable community perspectives on the research topic and design, risk and benefits of research, recruitment strategies, data collection methods, and how to make data actionable (see box 4.1).

BOX 4.1. THE NEVADA MINORITY HEALTH AND EQUITY COALITION (NMHEC) COMMUNITY ADVISORY BOARD

The NMHEC is a statewide coalition that "promote[s] the health and well-being of diverse communities by pursuing research, capacity building, and advocacy that recognizes the unique cultural and linguistic differences of Nevadans" (https://nmhec.org/our-mission). The coalition has addressed issues at the intersection of health and EJ, such as the disproportionate impact of COVID-19 on Nevada's communities of color and low income because of background environmental injustices, such as crowded housing conditions, reliance on public transportation, lack of workplace occupational safety and health protections, and racial stigma. To incorporate the voices of diverse sectors that contribute to health, NMHEC is guided by

(*Continued*)

BOX 4.1. (*CONTINUED*)

an 11-person advisory board who reside across the state and represent for-profit, nonprofit, school, and government organizations. The advisory board was formed to (a) identify community needs; (b) contribute to interdisciplinary research; (c) determine needs and support capacity building in areas of policy, advocacy, research, and grant development; (d) provide input on policies and practices that address social determinants of health; and (e) identify community members to participate on project steering committees, which direct each research, education, or outreach project. Steering committees include coalition members and external members recruited to join the coalition over time to fill emerging gaps in knowledge and expertise (Nevada Minority Health and Equity Coalition 2021). CAB members chose consensus-based decision making as an important value.

CAPACITY BUILDING FOR RESEARCH PARTNERS

To participate fully and influentially in research partnerships, CAB members and other community-based research partners must also have opportunities to build their capacities—a topic not widely addressed in the literature. Much of the relevant peer-reviewed literature comes from training programs, such as the CBPR Partnership Academy (Coombe et al. 2020), Sharing Power with Communities (Pratt 2021), Transformative Co-Learning Model (Loh 2016), Building Equitable Partnerships for Environmental Justice Curriculum (UCDEHSC and UMLEEDC 2018), and Holding Space Toolkit (Lucero, Emerson, et al. 2020). Collectively, these trainings and toolkits are a means for partnerships to develop all members' abilities to guide and conduct research.

The EJ movement also has a long history of providing capacity building to its members. National organizations and networks (such as the Highlander Center and the Environmental and Economic Justice Project) provided popular education and research trainings for the emerging movement. Over time, grassroots EJ organizations combined research trainings with organizing and advocacy to build in-house research capacity. These efforts often provide an understanding of the deep historical and cultural causes of local environmental injustices in structural racism, colonialism, and economic exploitation. Trainings also include engaging in individual and partnership reflection, developing community and institutional leadership for CER, introducing the research process and specific training in how to do CER, training in protection of human subjects in research, initial and refresher training on project topics and outcomes, identifying funding mechanisms, and grant writing. To link CER to organizing and developing community leadership, EJ organizations typically aim to include community members in these trainings, not simply service providers or advocates for the community. Thus, researchers who want to work with EJ organizers should be prepared to engage residents directly and develop their leadership in the research project. Additionally, routine training should create an ongoing passage for new leaders to emerge when turnover occurs among organizers and residents.

PROJECT FUNDING AND BUDGET SHARING

Practicing distributive justice means ensuring that research partners and participants are compensated for their time and knowledge, and that all partners are comfortable with how funding is controlled. Funding agencies often expect or require that the academic partner will submit the grant application and be the principal investigator, who manages the grant money. In some cases, this is appropriate; academic institutions have administrative infrastructure for project reporting, institutional review boards, and the ability to spend funds when a contract is based on invoicing. In other cases, it is appropriate for the community organization to be the primary agency, especially when it takes responsibility for the bulk of the work. Supporting community partners to apply for their own research grants is a valuable contribution to building their capacities over the long run.

Regardless of which organization becomes the primary agency, a realistic review of the funding amount needed to accomplish proposed work should be undertaken. Hoeft et al. (2014) guide readers through the process of understanding costs needed for research activities such as travel, communication, meeting and food, time, research activities, and how to equitably compensate community partners. CER partnerships can consider providing the community partner with funding that is proportional to their scope of work in a memorandum of understanding or other contractual agreement, and/or providing key academic and community research personnel a similar amount of funds to be applied to their salaries. CER partners need to have potentially hard conversations about fair compensation early in the research process.

Ensuring that community partners are compensated equitably can be accomplished through subawards to partner institutions, hiring partners as consultants, or creating new positions for partners. Subawards and consultancies may be appropriate for sharing funds with experienced EJ organizations that have paid researchers on staff. While subawards typically define clear deliverables and due dates, consultancies can allow for more flexibility about how partners contribute to projects. For example, an individual consultant may only be responsible for carrying out a training while a subawardee would be responsible for the training development, implementation, and evaluation. Furthermore, consultants may be less expensive, as they may not require facilities and administrative costs. Creation of new positions for community partners is another approach that can meet project needs while valuing community knowledge. Black et al. (2013) developed a community engagement model that centers community research fellows (CRF). The CRF criteria and position description was a joint endeavor of academic and community partners of the North Carolina Translational and Clinical Sciences

Institute, which funded a variety of health and EJ focused projects. A noteworthy criterion was the "ability to transfer skills to both community and faculty" (265). Similarly, some academic institutions have developed tribal liaison positions that facilitate relationships with local Indigenous communities to promote education, research, and engagement, and demonstrate institutional commitment to decolonization. In these examples, the hiring criteria are as rigorous as in traditional academic positions, yet researchers' skills and knowledge stay in the community to provide capacity for future initiatives.

AGREEMENTS ON ROLES, RESPONSIBILITIES, AND DATA SHARING

Creating agreements on roles and responsibilities is a CER best practice, which may also be required by some universities' institutional review boards or offices of grants and contracts. These agreements are forms of governance that aim to create procedural justice for all partners in CER. The process of drafting agreements can also advance recognition justice as partners learn about each other's goals, experiences, and capacities, rather than simply negotiating with each other in a transactional manner. CER often involves several kinds of agreements. A memorandum of agreement (MOA) is an official agreement and legal contract that outlines roles, responsibilities, and expectations of each party. A memorandum of understanding (MOU), on the other hand, is not legally binding but is formal, carries a degree of seriousness, conveys mutual respect, and addresses expectations, as well as roles and responsibilities of each party. Another joint agreement option is a written collaborative research agreement, such as a project charter that provides details of partnerships. Any formal partnership document names all organizations involved and outlines partnership goals, operating norms, expectations, responsibilities, contingency plans, and ownership of data (Mayan and Daum 2016).

Data-sharing agreements and management plans—the policies, protocols, and procedures related to the handling of data—are extremely important governance tools for community partners (Woodbury et al. 2019). Many Indigenous scholars have taken the lead on this topic by interrogating policies and procedures as they relate to human subject research, including data security and de-identification of participants, and data ownership (see chapter 5 and Harding et al. 2012; Hiratsuka et al. 2017; Marley 2019). This includes concerns that data and biologic specimens that participants contribute for one research purpose are not used for secondary research without their informed consent. Secondary use of data and specimens can also violate the confidentiality and privacy of individuals and communities, risking harm to their reputations, economic viability, and well-being. These risks demand data-sharing agreements and appropriate ongoing forms of consent beyond general permission.

Included within data-sharing agreements are terms of prior review of materials and manuscripts by CABs or other oversight boards like tribal institutional review boards (IRBs), publications, and public dissemination. Data-sharing agreements stipulate conditions under which researchers can collect, share, disseminate, and return data, including specimens (Harding et al. 2012; Lucero, Emerson, et al. 2020; Woodbury et al. 2019). There is a need for dialogue between researchers and community members, and possibly the funding agency, as to whether data is openly shared or shared with restrictions and what those restrictions entail (Harding et al. 2012). While NIH and other funders require a data-sharing plan, they rarely provide specific guidance. Fortunately, groups like the Colorado Clinical and Translational Sciences Institute have shared best practices, recommendations, and step-by-step development guides (Backlund Jarquín 2012). Researchers and their partners also need to anticipate how data sharing and ownership may affect the project's ability to make data actionable to maximize community benefit. For example, Indigenous community partners typically require restrictions on release of sensitive data about their sacred sites, to protect them from looting and vandalism (Ban et al. 2018).

FROM DATA TO ACTION

Advancing knowledge and driving community change are equally important in CER. Partnerships aim to create actionable data that informs how programs, policies, campaigns, and practices are designed and implemented. Actionable data bridges research and practice, and academic and community concerns. Collaborators need to discuss from the outset how community needs and priority issues will guide which data will be collected, how they will be measured and analyzed, and how they will be expressed in a format that can be used effectively for the end purpose of the project (whether it is a legal case, policy proposal, organizing campaign, community mural, and so on). Zakocs and colleagues identify five characteristics that increase the likelihood of acting on data: the data answer questions that are important, are credible, are reported in a concise and understandable manner, are shared before decisions are made, and are available to stakeholders in time for them to reflect on findings, implications, and possible action (Zakocs et al. 2015).

Careful decisions need to be made about what data will be collected and how they will be used. Stephen Luck succinctly summarizes the issue: "You can't manage what you can't measure" and inversely, "you can only manage what you do measure" (quoted in Pine and Liboiron 2015, 3149). In CER, these decisions are acts of power sharing, which include community partners in analyzing and interpreting data—to build community capacities, learn from partners' unique knowledge, and draw on their insights about how to make results actionable for their communities. EJ researchers can learn from collaborative data analysis strategies pioneered by human rights activists to fill in gaps in official data (Alvarado Garcia et al. 2017), from CER that has involved community partners in analyzing data gathered via multiple qualitative and quantitative methods (Cashman et al.

2008), and from practices such as research reflection meetings, data analysis workshops, and consensus-building activities to arrive at shared findings (Godden 2017).

For example, the Nevada Minority Health and Equity Coalition (NMHEC) led the #OneCommunity campaign, a COVID-19–focused community-engaged outreach and education project in communities most impacted by the pandemic. As noted above, environmental injustices such as crowded housing, lack of personal protective equipment in workplaces, and racial violence made communities of color and low income especially vulnerable during the pandemic. Furthermore, the pandemic thrust community members, leaders, and scientists into developing time-sensitive safety and mitigation responses. To inform these responses, the NMHEC worked alongside community leaders to conduct focus groups across the state in seven diverse populations—Hispanic/Latinx, Black, American Indian, Asian, Pacific Islander, LGBTQ+, and Deaf and Hard of Hearing. Partners co-led the development of focus group questions and surveys to best address community needs and facilitated focus group sessions with community members. Over 23 focus groups, exceeding 200 participants, were conducted over six weeks. The data were collaboratively interpreted with community partners to create culturally tailored messaging for each priority population to address unique concerns and misconceptions. Project partners led dissemination efforts into their communities. Most importantly, the importance of COVID-19 to each community facilitated the mobilization of ten funded local partnerships to take action to reduce the disproportionate impact of COVID-19 in their communities.

Data form understanding of an issue, lead to decision making, and provide the cornerstone for action-oriented approaches, such as building capacity among stakeholders, informing diverse audiences, and driving action (Alvarado Garcia et al. 2017; Pine and Liboiron 2015). Data can describe the scale and scope of a problem by describing how a condition, physical or social, can manifest itself in the population. Furthermore, data support the interpretation of community problems and the process of addressing them. Thus, making data actionable to address community concerns requires more than simply collecting data to identify environmental and other disparities. It requires a strategic approach to interpret the data to drive decision making (Alvarado Garcia et al. 2017).

PARTICIPATORY EVALUATION

Given its practical emphasis, much CER for EJ involves evaluation research, which assesses the effectiveness and power-building capacity of community-based programs, interventions, campaigns, or activities. In traditional approaches to evaluation, researchers or funding agencies define the evaluative criteria, "objective" observers from outside the community conduct the evaluation, and data are often restricted to narrow quantitative measures of outcomes. In contrast, *participatory evaluation* is better suited to CER for EJ, because this approach emphasizes

TABLE 4.4. The Evaluation Process

	Formative Stages		Summative Stages	
Evaluation Type	Needs Assessment	Process Evaluatio	Outcome valuation	Impact Evaluation
Occurs	Before program begins	Throughout the program implementation	As immediate and intermediate outcomes occur	As long-term intended effects occur
Question Asked	What is the need? What can be done to address the need?	Is the program or intervention operating as planned?	Is the program achieving its objectives?	What predicted and unpredicted impacts has the program had?

Adapted from https://meera.snre.umich.edu/evaluation-what-it-and-why-do-it.html.

community partners' right to take part in research (participatory justice) and the value of their knowledge (recognition justice). In participatory evaluation, community partners and members collaborate fully and equally with researchers to identify evaluative criteria according to the community's values and priorities, and these co-evaluators may examine a broad range of qualitative and quantitative measures, guided by professional standards and local knowledge (Wiggins et al. 2017). While the goal of traditional, top-down evaluation is often to hold community organizers and service providers accountable for their performance to funding agencies, participatory evaluation aims to strengthen community organizations' capacities to define their own measures of success, and to research how they can best improve residents' living conditions and build power to make change (Neubauer et al. 2020; Wiggins et al. 2017). These measures often go beyond project-specific objectives to include strengthening an organization's capacities for self-governance, community organizing and power building, coalition building and movement building, and other organizational and political goals.

There are two major categories of evaluation—formative and summative—both of which help optimize the success of a project. Table 4.4 shows the two categories of evaluation, subtypes, when each type of evaluation occurs, and what types of questions each evaluation type answers. *Formative evaluation* is an opportunity to engage community partners to establish a need for the project, shape how it is designed, and monitor its progress (Dehar, Casswell, and Duignan 1993). Formative evaluations also include *process evaluations*, which are used to ensure that proposed activities are implemented to reach the targeted audience and achieve the expected outcomes (Saunders, Evans, and Joshi 2005). As an iterative process, process evaluation is conducted throughout the implementation of the research activities and includes feedback mechanisms for improving achievement of short- and medium-term outcomes. *Summative evaluation* measures both immediate and long-term impacts of a program or intervention.

A citizen science partnership led by the Science Museum of Virginia provides an example of how formative and summative assessment can strengthen CER for EJ (Hoffman 2020). The three-year partnership in Richmond, Virginia, aimed to educate and spark action among residents to build climate resilience by reducing temperatures in urban heat islands, which are especially hot areas in cities that are disproportionately located in neighborhoods of color and low income. Formative assessment of the project's initial educational programming, in the museum and online, showed that these traditional methods failed to meet the project's goal of helping residents connect climate change to their own surroundings and take action to promote resilience. In response to these failures, the project launched a citizen science project, mobilizing residents to drive across the city taking temperature measurements throughout the day. Community partners suggested locations to measure based on their local knowledge of hot spots around the city and recruited drivers to the study. Researchers combined the measurements with data on risk factors that increase residents' susceptibility to heat (such as poverty, and rates of cardiovascular and respiratory diseases) to produce heat vulnerability index maps of the city's neighborhoods. The maps became a focal point of the museum's redesigned interactive programming and of new efforts to engage residents through community-based organizations, such as a program with community partner Groundwork RVA to engage public high school students to plant vegetation in especially vulnerable neighborhoods to reduce temperatures and improve air quality. Summative assessments found these new aspects of the project based on CER improved the project's ability to educate and mobilize residents to act for climate resiliency, and the Science Museum's capacity to collaborate with local organizations.

KNOWLEDGE SHARING AND DISSEMINATION

Dissemination of knowledge in CER is guided by two famous phrases: "we speak for ourselves" (popularized by the EJ movement) and "nothing about us without us" (which has been adopted by activists in many oppressed and stigmatized communities). Both phrases assert community partners' procedural rights to communicate research findings and recommendations for action to their communities and other decision makers, rather than relying on outside researchers or advocates to speak on the community's behalf.

Although dissemination is typically done at the end of the research process, it should be considered at the beginning of a study and is a critical step in trust building throughout the research process. As an exchange of learning, research conducted *with* communities rather than *on* them needs to be shared widely with the community and others who can help improve conditions. A well-thought-out communication and dissemination plan can build awareness, create understanding, and drive action (Harmsworth and Turpin 2000) if the plan includes several elements (Carpenter et al. 2005):

- involvement of community members in drafting and implementing the plan
- relevance to the community's self-defined priorities, needs, and interpretations of research

TABLE 4.5. CER for Communication Campaigns

Elements of Campaigns	CER Activities	Example Research Products
Strategizing	Identifying goals and organizing plans Target and power analyses Framing and cutting issues Choosing communication sources (such as organic community leaders), channels, and messages Identifying funding sources	Comprehensive campaign plan: goals, objectives, organizing and communication strategy Fundraising applications to support campaign implementation and evaluation
Implementing	Research to support outreach, mobilization, and/or advocacy Information sharing and coordination with allies	Reports and white papers on problems and solutions Score cards on the performance of targeted industries and government agencies Organizing toolkits, fact sheets, training curricula Development and testing of frames, messages, and tools (apps, databases, etc.) Presentations, testimony, participatory media

Adapted from DataCenter (2015b).

- clearly defined objectives, such as mobilizing the community to change policies, practices, funding allocations, and so on
- understanding of the priority audiences and the contexts in which they live to tailor information and persuasive messages to their values, priorities, and needs in appropriate, accessible language(s)
- tailoring of communication to the plan's goals, including choosing appropriate channels, using messengers trusted by community members, and employing appropriate timing of communication
- evaluation of the impact and reach of messages

In EJ organizing, communication plans often take the form of campaigns to educate, persuade, and mobilize (Raphael 2019a). EJ campaign goals may include promoting individual attitudes and behaviors but typically focus on bolstering community capacities, mobilizing support for policy and legislative change, or directly enacting changes in corporate and government practices.

CER can support campaign strategy and implementation by informing many kinds of communication products (see table 4.5). CER can help organizers identify

TABLE 4.6. Non-digital Tools for Dissemination

Outreach	In-person tabling at community events allows researchers to engage directly with residents. Educational meetings, workshops, and town halls can share findings via multiple media and visual aids. Phone trees can activate groups by efficiently spreading brief messages to many people. Direct mail campaigns can provide targeted information directly to a person's place of residence
Street Organizing	Street theater can use drama to communicate findings and recruit participants for organizing, while demonstrations can gather people to share information and make demands
Broadcast Media	Television and radio ads, announcements, and programs can broadly share a consistent message about findings
Print Materials	Flyers and brochures can share findings to communities in multiple languages. Organizations' newsletters and networks can elicit authentic community engagement with research findings
Billboards	Billboards can inform people in private vehicles and public transportation
Communications to Decision Makers	Organizations and residents can communicate findings to public officials and other leaders via letters and testimony
Academic Publications, White Papers, Policy Briefs	These publications can provide more technical information for diverse stakeholders and decision makers

Adapted from Marquez, Smith, and Perez (2022).

campaign targets, whose agreement is needed to implement the campaign's solutions, and map power relationships that can be used to leverage change (Data-Center 2015a; UCDEHSC and UMLEEDC 2018). Research findings and recommendations can benefit from being cut and framed to reflect the views of specific constituencies (Center for Story-Based Strategy 2017). For example, a campaign to reduce household lead exposures may generate broader participation if presented as an issue of children's health to families, of preserving the habitability of rent-controlled apartments to tenants, and of securing low-interest loans for lead abatement to homeowners and small landlords (Staples 2016).

Translating research into campaigns also depends on choosing credible sources and effective communication channels. In EJ communities—where residents are often people of color and of low income and/or are first-generation immigrants—it is important to take advantage of non-digital dissemination methods that engage people who have limited internet access and computer literacy and speak multiple languages (see table 4.6). Dissemination should also employ digital channels to share information quickly to people in any location, including their homes. Decision makers in government, corporations, and other institutions can be reached by both kinds of channels (see table 4.7). Academic publications and policy papers are also important publication venues. Because the position of the storyteller is

TABLE 4.7. Digital Tools for Dissemination

Social Media	Social media platforms can inform, consult, and involve the community in conversation and organizing through their personal networks
Email, Digital Newsletters, and Texts	Findings can be shared via digital newsletters and emails to community-based list servers that have credibility with residents. Text messaging and messaging apps can share information quickly and broadly, especially as residents forward it to their networks
Websites	Websites can share online brochures, flyers, scorecards, and toolkits with community members
Webinars	Webinars and virtual town halls can educate the community, solicit feedback about plans, and provide insight about common community concerns
Videos	Videos can publicize findings to people with limited literacy by incorporating residents' experiences, telling stories, dramatizing issues, and teaching people how to take practical steps

Adapted from Marquez, Smith, and Perez (2022).

one of power, it is important for academic researchers to include community partners as co-authors on these publications to ensure the story of the research is being told accurately, share credit for the work, and recognize all contributors' expertise (Mulrennan, Mark, and Scott 2012).

CONCLUSION

This chapter has provided an overview of the major issues that arise in the CER research process. We have focused especially on the need for collaborators to address power relations among community partners and researchers at each step. Power threads through choices about defining the community relevant to the research, addressing conflict among partners, creating accountability to community advisory boards, developing the team's research capacities, sharing resources, crafting agreements on roles and responsibilities, mobilizing action in response to findings, integrating evaluation throughout the project, and disseminating knowledge in multiple venues to make change. Paying attention to power is important for understanding how each choice that partners make implicates one or more dimensions of justice, including how partners share resources, exercise voice and influence over the research, respect community knowledge as well as professional research expertise, and transform relationships among researchers and communities to promote EJ together.

5

Transforming Academia for Community-Engaged Research

Felicia M. Mitchell, Celestina Castillo, Chad Raphael, and Martha Matsuoka

Strengthening community-engaged research (CER) for environmental justice (EJ) requires examining the whole of the relationship between an academic institution and the broader community. While chapters 3 and 4 addressed how to prepare for and conduct CER for EJ, this chapter focuses on transforming institutional barriers and creating supports for this kind of research in academia. Researchers need to navigate, and many institutions need to change, tenure and promotion criteria that fail to recognize CER, reluctance to recognize community advisory boards as equal partners, and administrative systems that make it difficult to share resources with community partners. While many researchers have learned to overcome these obstacles, they continue to stifle projects that are most relevant to EJ communities and limit academic institutions' ability to build just relationships with these communities.

Drawing on promising practices, we also offer recommendations for how academic institutions can be more supportive of CER for EJ. The recommendations and examples presented here are milestones on a long journey toward a more just system of knowledge production and education, one that transforms inequitable relationships of wealth and structural racism across society. Table 5.1 summarizes how the four dimensions of justice common to CER and EJ can inform the changes needed in academia that we will discuss in this chapter.

TABLE 5.1. CER for EJ in Academia

Dimension of Justice	In CER for EJ in Academia
Distribution *Who ought to get what?*	Fair sharing of academic funding and research tools with communities, and development of community capacities to conduct their own research
Procedure *Who ought to decide?*	Community participation in and influence over the design and conduct of research through community advisory boards, data ownership and control, and research ethics reforms
Recognition *Who ought to be respected and valued?*	Practicing epistemic justice and decolonizing knowledge in CER, curricula, co-curricula, and campus archives and museums
Transformation *What ought to change, and how?*	Transforming academic research, criteria for evaluating research, impacts on surrounding communities, and composition of the campus community to repair historic harms to and create just relations with EJ communities and nature

DISTRIBUTIVE JUSTICE

In their mission statements, academic institutions typically state their commitment to produce and spread knowledge, but they rarely share the means of producing knowledge with their surrounding communities. This includes resources such as grants, research tools, and data, as well as access to capacity development (such as skills-based training, knowledge-based education, research experience, and research networks), which communities need to participate in CER and to conduct their own research. Making these resources and capabilities more available to communities will require changes to academic policies, procedures, and administrative systems.

Academic institutions are designed to attract and retain research resources rather than to share them with community partners. Even in CER projects, academic researchers typically control the funding, their institutions take a significant portion as overhead, and community collaborators often receive little to no share of the money. While this may be warranted if community partners lack experience in managing complex grants, many partners can do this, or want to learn, and helping spread these capacities should be a long-term goal of CER (Wilson, Aber, et al. 2018). This may involve academic researchers playing the junior partner role—for example, as subawardees on grants to community organizations—which positions community partners in leadership and grant manager roles and contributes to the partners' ability to obtain future research grants. Academic institutions can also develop training programs that build community partners' skills to conceive, fund, conduct, analyze, and disseminate their own research.

When academic researchers are the lead grant managers on CER projects, their institutions' business offices and financial systems need to reduce barriers to compensating community partners. For example, some partners are required to include the institution on their insurance policies or to submit frequent invoices in order to secure funding. Partners that are not incorporated as nonprofit organizations (such as volunteer, collective membership organizations) may lack the employee identification numbers that institutions often demand to set up contracts and distribute funds. Funds to individuals may require social security numbers, which excludes some immigrants from participating as partners. While institutions must protect themselves from financial risk, they will also have to develop systems that facilitate transferring funds to community partners.

Institutions must also find ways to expand access to research tools so that community partners can participate fully in campus-affiliated CER projects and generate independent research. Academic institutions can share subscriptions to research databases, proprietary data sets, and tools for analyzing and representing data. Establishing neighborhood-based science shops, maker spaces, and research centers can help community groups to develop their own research projects. Academia can build on the radical science shop tradition developed in the 1970s to align research with community-defined needs in collaboration with local nonprofit organizations, officials, schools, and others (De Filippo et al. 2018). Academic institutions could develop more open science and maker spaces, and involve faculty and students in helping teach community members how to use them to address community priorities.

PROCEDURAL JUSTICE

Most scholarship conducted in EJ communities is *extractive:* researchers obtain grants to conduct studies of their own design; gather data from communities; analyze it in researchers' own labs, computers, and heads; and publish it in academic journals and books that are inaccessible to community members. While chapters 2 through 4 describe how researchers can embrace CER, their institutions also need to change to allow community partners to participate fully in designing research, sharing ownership and control of data, and practicing research ethics that align with community values.

Design and Control of Research

Community advisory boards (CABs) are often established to ensure community participation in and power over the design and conduct of CER, including obtaining informed consent and managing rights to control data (see chapter 4). CABs may focus on guiding campus-community partnerships for learning and research across an entire university, a school or department, or a specific research project.

However, community participation in CABs can be limited by structural and systemic oppression within communities, which may prevent members from participating as equal partners (Safo et al. 2016; Wallerstein et al. 2019). Formation of the CAB requires careful consideration of composition and recruitment to ensure community representatives bring a mix of perspectives, expertise, and resources necessary for the partnership and the research. CER must prioritize CABs in which community members are co-decision makers, not simply advisors whose input can be rejected, throughout the research process. Their ability to exercise equal influence also depends on how CABs establish operating principles and procedures, balance power, and make decisions. Once established, CAB members must work together to maintain themselves through reflective and evaluative processes and developing a plan for sustainability to ensure empowerment and capacity building (for best practices and examples, see Newman et al. 2011; Symanski et al. 2020; and chapter 4).

Data Ownership and Control

Higher education can learn from Indigenous peoples' efforts to protect their ownership and control of data. While tribal governments' status as sovereign nations gives them a unique status among EJ communities, these governments' well-developed data policies and guidelines can inform academic agreements with other EJ communities. Like many marginalized groups, Indigenous peoples (individually and collectively) have been the subject of research not sanctioned or overseen by their tribes or communities. In such cases, researchers often extract data without consideration of the harms or benefits to the community, the people, and the land (see chapters 2 and 4).

Indigenous data sovereignty is founded on Indigenous groups' inherent and sovereign right to govern their peoples, lands, and resources. Further, it is the right of Indigenous tribal nations to oversee the collection, application, and ownership of data concerning their people and community collectively (www.gida-global.org). These principles are meant to ensure that data for and about Indigenous peoples and lands is used to advance Indigenous priorities for collective and individual well-being. In table 5.2, we list the Native Nations Institute's preliminary recommendations for decolonizing data and indigenizing data governance (Rainie, Rodriguez-Lonebear, and Martinez 2017) and provide more specific guidance for applying them in academia. Although not an exhaustive list, the recommendations can guide academic data policies with regard to many EJ communities. (For additional guidelines specific to conservation research with Indigenous peoples, see chapter 12).

Research Ethics

CER requires significant changes in how academia assesses whether research projects meet ethics requirements. Since the 1970s, research ethics protections in the U.S. have evaluated whether research designs comply with the Belmont principles. These principles include *respect for persons* (participants in research must take part voluntarily, and there must be additional protections for children and others who cannot make their own choices); *beneficence* (research designs must minimize risks and maximize benefits to participants); and *justice* (research must be designed

TABLE 5.2. Applying the Native Nations Institute Recommendations to Academia

Native Nations Institute Recommendations	Applications to Academia
Acknowledge Indigenous data sovereignty as an objective, and incorporate it into tribal, federal, and other entities' data policies	Acknowledge and incorporate Indigenous data sovereignty throughout academic data policies, including IRB and research ethics policies. Engage policy researchers and facilitate collaborative work with tribal leaders and policy experts to create equity-based policies that benefit tribes
Generate resources and build support for Indigenous data governance, including the governance of Indigenous data by others	Establish institutional resources for Indigenous data governance policies and mechanisms that support tribal sovereignty, including governance of how Indigenous data will be handled by academic institutions and researchers, through data-sharing agreements and management plans and university IRBs and ethics committees
Grow tribal data capacities, including establishing their data governance policies and procedures, and recruiting and developing "data warriors" (Indigenous professionals and community members with skills in collecting, creating, and managing data)	Assist tribes to grow their capacities for data governance and to conduct their own research. Recruit, develop, and retain Indigenous students, scholars, and community researchers as data warriors
Establish strong relationships between tribal leaders and data warriors	Establish strong relationships between academic researchers and data warriors to reduce community research burdens and fairly distribute benefits between Indigenous communities and academic researchers
Create intertribal institutions to practice data leadership and build data infrastructure and support for tribes	Develop academic technical assistance programs, policy institutes, and similar structures to support intertribal institutions to do this work
Build connections among Native nations domestically and internationally for the sharing of strategies, resources, and ideas	Provide academic assistance in bridging Indigenous groups domestically and internationally through institutional alliances and financial supports to share research strategies, resources, and ideas

to balance potential risks with benefits to participants). Federal research-funding agencies and academic institutional review boards (IRBs) have applied these principles to build ethics protocols used to decide whether to approve proposed research projects. However, these protocols may omit many of the most significant ethical considerations of CER partnerships.

CER seeks to prevent community harm while also actively benefiting communities by reframing research relationships and goals to align with communities' priorities and needs. Thus, relevant research ethics for CER expand the Belmont principles' traditional concern with the rights of *individuals as research subjects* to include concern for the rights of *communities as research participants and co-producers*. However, IRBs often fail to address community- or population-level protections and assurances, including rights to consent, participate, share control and ownership, ensure cultural appropriateness of research, and benefit from it (Banks and Brydon-Miller 2018; Beans et al. 2019).

When assessing proposed research in EJ communities, IRBs can go beyond minimal requirements for obtaining consent from individual research participants. One model is the notion of collective free, prior, and informed consent (FPIC), an international human rights legal principle for seeking local communities' approval of development projects, inspired by Indigenous demands for self-determination (Suiseeya 2021). FPIC requires affirmative consent (an explicit assertion of approval) that is free (obtained without coercion), prior (obtained before a project is implemented), and informed (given by people who are fully aware of the impacts of their decisions). Consent is not mere consultation, which does not guarantee communities a right of refusal. Consent must also be demonstrated by representatives of the community, including marginalized groups, not simply obtained from individuals. This principle could more fully inform how academics seek approval of research projects in EJ communities, supplementing requirements that are currently limited to obtaining approval from individual research subjects.

IRBs can also consider communities' rights to participate as peer researchers, ensure cultural appropriateness of the research, and own and control data. Some university IRBs have impeded CER proposals because of reluctance to review ethics compliance by partner organizations, especially to ensure that lay members of research teams are trained sufficiently to protect participants' confidentiality and other rights. In these cases, research may be delayed, community members may be restricted from gathering or accessing data, or local partners may be forced to pay for independent IRB oversight (Morello-Frosch, Brown, and Brody 2017). IRBs have also asserted academic institutions' ownership of research findings as intellectual property, which contradicts CER principles of the collective ownership of data and the co-production of knowledge with community members (Su et al. 2018). IRBs need training to reconcile these rights more fairly (see, e.g., Pearson et al. 2014), and community or tribal review and ethics boards can also assess whether research proposals observe these rights (Gachupin and Molina 2019).

Incorporating community rights in decisions about data dissemination can involve trade-offs with traditional scientific principles. For example, some CER projects omit control groups because partners consider it unethical to deny community members potentially beneficial interventions. In addition, researchers and community partners must grapple with whether and how to publicize negative findings about a community that could stigmatize it and dissuade community members from participating in beneficial interventions. These ethical decisions involve weighing the benefits of scientific rigor against advancing goals for improving the community's welfare (Minkler and Baden 2008).

Conversely, community values may call for disseminating data that traditional research ethics would restrict. For example, IRBs have resisted CER projects' desire to report study participants' own individual-level results of exposures to hazardous substances and other health data if there is scientific uncertainty about their impact (Morello-Frosch et al. 2015, 2017). This resistance stems from concern that participants may endure unnecessary stress by getting access to their genetic data or chemical exposure levels when there is uncertainty about their health implications. Yet many EJ researchers and community partners would prefer to report back these data out of respect for community members' right to know. There is evidence that even if these individuals may not be able to eliminate exposures or alter their genes, participants gain important knowledge about environmental health, take precautionary steps, and involve themselves in policy processes to reduce their risks (Morello-Frosch et al. 2017).

To summarize, CER partners, federal funding agencies, and universities can take several steps to reform research ethics practices (Morello-Frosch et al. 2017). These include

- educating funding agencies and IRBs that are unfamiliar with CER about its principles, benefits, and ethical concerns, such as protecting community rights;
- encouraging funding agencies and IRBs to value statements of "community consent," not only of individual consent to participate in studies;
- involving and training community members in review boards to evaluate proposed CER, which can inform IRB decisions;
- reforming the guidelines of major funding institutions, especially federal granting agencies, to offer guidance in handling human subjects concerns specific to CER;
- encouraging IRBs to assess the quality of training of peer researchers and respect data collection methods common to CER, rather than raising unnecessary barriers to community participation, and to devise new criteria for reporting individual health data to study participants;
- encouraging IRBs to require CER ethics programming and population-specific ethics trainings as part of academic researchers' routine certification to do research involving human subjects (see box 5.1);
- fostering respect and knowledge about the importance of Indigenous cultural review and ethics boards, tribal IRBs, and/or forging informal agreements with tribal leadership.

These reforms require ethics to be considered as an ongoing set of issues, dynamics, and relationships throughout the partnership and research process (Glass et al. 2018). CER highlights how all aspects of research projects we discuss in this chapter—sharing resources, making decisions, recognizing marginalized knowledges, and transforming academia's relationships to exploited communities—involve ethical choices because they are matters of justice (Flicker, Guta, and Travers 2017). Academic researchers and IRBs must be trained to assess these broader ethical criteria for research as well.

BOX 5.1. ETHICS TRAINING FOR RESEARCH IN INDIGENOUS COMMUNITIES

Many tribal communities have well-developed research review boards and IRBs with specific criteria rooted in principles of tribal sovereignty (Parker et al. 2019). In other tribal communities, research approval may take other forms, including endorsements from tribal leaders or appointed councils. However, many academic IRBs' policies conflict with tribal research ethics. In particular, academic training based on the Belmont principles often fails to recognize ethically relevant cultural and community aspects of research involving American Indians and Alaska Natives (AI/ANs) (Parker et al. 2019; Pearson et al. 2014). In addition, academic IRBs have sometimes blocked collaborative research approved by their tribal counterparts by imposing stricter protections for the individual rights of participants (Morello-Frosch, Brown, and Brody 2017).

Researchers who want to collaborate with AI/AN communities ought to go beyond the narrow ethics training required by academic institutions to learn how to comply with AI/AN communities' research ethics. To fill this need, Parker and colleagues (2019) developed the research Ethics Training for Health in Indigenous Communities (rETHICS), a module and curriculum that aligns with AI/AN culture, context, and community-level ethical values and principles. It was developed through an extensive process of community consultation and input from three expert panels drawn from a nationally representative list of AI/AN researchers, including a community expert panel, scientific and academic expert panel, and IRB and policy expert panel.

The rETHICS training was based on foundational constructs that "(a) [were] framed within an AI/AN historical context; (b) reflected Indigenous moral values; (c) linked AI/AN cultural considerations to ethical procedures; (d) contributed to growing Indigenous ethics; and (e) provided Indigenous-based ethics tools for decision making" (Parker et al. 2019, 9). The curriculum is freely available (https://redcap.iths.org/surveys/?s=R3EJPAYD4J) and can be adapted for other cultural groups (Parker et al. 2019). University IRBs could add this to the requirements for EJ researchers proposing work with Indigenous groups and establish new trainings using CER that focus on other EJ populations.

RECOGNITION JUSTICE

To advance recognition justice, academic institutions must respect the value of and differences among local, experiential, and Indigenous knowledges in EJ communities. Many leaders of EJ communities mistrust academia, and therefore are reluctant to engage in CER with academics, because of historic institutional disrespect for these communities' knowledge (Cable, Mix, and Hastings 2005; Cole and Foster 2001). Dominant academic traditions have presented knowledge produced from a Western, white, male, economically privileged perspective as "objective" and "universal," and much of this scholarship has ignored, denigrated, or legitimized the oppression of EJ communities (Smith 2021; Whyte 2018a). In contrast, CER is grounded in epistemic justice, which recognizes multiple ways of knowing both inside and outside academia (see chapter 2). Enacting epistemic justice means making space within academia for non-Western and non-dominant thought, practices, and worldviews essential for effective CER, such as Indigenous traditional ecological knowledge (see chapters 2 and 12). This means approaching epistemology as having both intellectual and ethical dimensions, examining its application in CER and its implications for transforming academia to be more equitable and inclusive through increasingly intentional integration of justice and scholarship, as in efforts to decolonize knowledge (see chapters 6 and 12).

Recognition justice involves opening up research, teaching, and service to non-dominant knowers and ways of knowing. Preparing students for CER requires modifying curricula and diversifying representation of pedagogical and epistemological paradigms in core theory, philosophy, and research courses throughout degree programs. Such modifications depend on faculty in all disciplines expanding their understanding of non-Western epistemologies and community-based knowledge—by recognizing, for example, that Indigenous peoples "have always been data creators, data users, and data stewards [and have used] data to interact with each other and the natural world since time immemorial" (Rainie et al. 2017). CER can help build relationships with more organizers, service providers, artists, and leaders from non-dominant communities and compensate them for sharing their knowledge in classrooms, academic museums and archives, cultural programming, and public spaces. Community-based learning placements, which are ripe for developing CER, must shift from employing a service learning approach in which faculty and students too often adopt a white savior mentality to bless communities of color with academic knowledge and skills, or a preprofessional mindset of enhancing multicultural credentials on one's resume (Irwin and Foste 2021). Instead, these programs need to become opportunities for true learning partnerships and exchanges, and should feed longer-term relationships by linking to ongoing CER partnerships that build community capacities and power. Across each of these efforts, academia needs to become more comfortable with acknowledging profound epistemic differences among dominant and marginalized communities' knowledges, and with attempting to bridge them through dialogue rather than papering over their differences or vanquishing some of them through debate.

TRANSFORMATIONAL JUSTICE

Higher education also needs to acknowledge how academic research and institutional actions have contributed directly to colonization, racism, exploitation, and environmental injustices, as described below. Making an institution-wide commitment to a broad program of CER is one important way for academia to engage in transformational justice that repairs harms to and creates just relations with EJ communities and nature. Reorienting academic research will require broader changes in academic culture and reward structures, transforming institutional impacts on surrounding communities and changing the composition of the campus community. To pursue these long-term goals, academia can learn from practices of restorative justice and transitional justice (see chapter 1).

Transforming Research

Spreading a culture of CER in academia can contribute to transformational justice for research practices that have not only failed to recognize the knowledge of people of color and Indigenous communities, but actively harmed them. For centuries, academic research across the disciplines has played a powerful role in advancing colonization, racism, and environmental destruction. Chapters 7 through 12 in this book describe how research in law and policy, development, planning, public health, food and agriculture, and conservation helped legitimate the contamination and destruction of nature and people in EJ communities around the globe. This research was not conducted by fringe theorists, but by leading scholars in their fields. Academic buildings and prizes continue to bear their names, and many scholars continue to draw on their work. As public funding for research and education has waned in many countries, contemporary institutions increasingly rely on private grants, contract research, and monetizing research services and products, making academic research more reliant on support from exploitive and polluting industries, and the foundations and think tanks they fund to influence public discourse (Canaan and Shumar 2008).

Notorious studies that directly traumatized vulnerable participants have especially led many people in EJ communities to distrust academic research. Many Black Americans know how the U.S. Public Health Service–Tuskegee Institute syphilis study concealed from Black male research subjects their diagnoses of syphilis and left their disease untreated so researchers could examine its progression for 40 years, causing preventable deaths among participants and their families (Smith, Ansa, and Blumenthal 2017). Many Native Americans know that Arizona State University researchers convinced Havasupai tribal members to give blood samples for a study on genetic links to diabetes that might improve health remedies for their tribe, and that researchers then used participants' DNA without their consent to publish stigmatizing research about inbreeding and schizophrenia in the tribe, and distressing research about the origins and migration of the Havasupai that conflicted with the tribe's beliefs (Mello and Wolf 2010). Subsequently, many tribal members shunned diagnostic care for diabetes, leaving their conditions untreated until they needed kidney dialysis, because the tribe no longer trusted medical authorities (Pacheco et al. 2013). To obtain justice, survivors of these infamous studies had to sue the responsible institutions, which did not agree to settlements until decades after the harms were committed. The Tuskegee study also prompted the drafting of the Belmont principles and restructuring of research ethics protocols across the U.S.

Transforming Harms from Research

Rather than acting as a fortress against complaints of research injustices, academic institutions should integrate CER into reconciliation with affected communities. Borrowing from transitional justice practices in postwar societies, academic institutions and professional associations can collaborate with representatives of people harmed by past research to establish commissions committed to healing and transformation. Such commissions can engage academics and community members in CER to examine how especially damaging and flawed studies and research programs became vehicles for misinformation, and contributed to oppressive policies and practices. These commissions can establish accountability by responsible parties, offer apologies and retractions of harmful research, and provide reparations and recommend policies to avoid repeating these abuses. As in restorative justice programs, offenders can participate in dialogue with survivors about how they have been impacted by harmful research, and agree on reparative measures to heal the university-community relationship. While Arizona State University did not adopt these practices, box 5.2 describes some of the measures

BOX 5.2. REMEDIES FOR RESEARCH HARMS AT ASU

In its 2010 settlement with Havasupai tribal members, Arizona State University (ASU) agreed to measures that illustrate some of the reparative justice options available to academic institutions (Mello and Wolf 2010). The university agreed to return remaining blood samples and research materials derived from the samples, ban the university's IRB from approving any research using the blood, and provide the tribe a list of all individuals and institutions with whom the samples were shared. ASU agreed to pay $700,000 to 41 tribal members. The university

(*Continued*)

BOX 5.2. (*CONTINUED*)

also adopted a five-year agreement to support education, health and nutrition, and economic development among the Havasupai. The Arizona Board of Regents also instituted a scholarship program across the state's three public universities for Havasupai tribal members and descendants of individual plaintiffs from the tribe who were parties to the settlement agreement.

ASU has also implemented several policies and procedures to ensure the university does not engage in further harmful research activities and develops just relations with Indigenous communities. The university now abides by a tribal consultation policy, which states that regardless of the authorizing body, any project that could potentially affect a tribe's government, their community, or their members or occurs on or near tribal lands should acknowledge and respect the distinct role of tribal governments, sovereignty, and government-to-government relations in the manner in which ASU engages with tribal nations (Arizona Board of Regents 2018). ASU's IRB now requires a cultural review for any research proposed with Indigenous peoples or on or around their lands. The cultural review board is composed of AI/AN scholars from across the university and community. The academic IRB process also requires an official letter of agreement for proposed research from an appointed tribal representative. Resources and training seminars on conducting ethical research with AI/ANs are available university-wide to all academic researchers. ASU established a special advisor to the president on American Indian affairs. This position is held by an Indigenous person who oversees university initiatives that relate to American Indian issues, develops relationships with tribal nations on behalf of the university, and is responsible for advising the university on programming to improve outreach, retention, and graduation rates of AI/AN students. While the actions were a significant recognition by ASU of past unjust practices, restorative justice requires ongoing and vigilant monitoring of these agreements and systems, as well as continued and deepened relationships with tribes and communities to continue shifting institutional power and practices in the future.

the university eventually took to reconcile with the Havasupai tribe as part of a legal settlement, which provide some example remedies.

Transforming the Culture of Research

Expanding CER depends on changing how higher education defines and values research. The neoliberalization of academia since the 1980s has created a market-driven culture of research that prioritizes maximizing external funding rather than community benefits, aligning research with major funders' and donors' agendas rather than community priorities, boosting research productivity (measured narrowly by the number of publications and citations) rather than building communities' research capacities, and cultivating researchers' competitive individualism and self-branding rather than their ability to develop relationships with community partners. To publish as much as possible, many researchers focus on plowing the same furrow in their field ever more deeply rather than engaging in interdisciplinary or applied research, and shun the time-consuming yet important work of building relationships with community partners. For these reasons, some dissertation advisors discourage early-career researchers, including Indigenous and other researchers of color, from doing CER with EJ communities (Mitchell 2018a), socializing scholars to use their time most productively rather than most meaningfully.

Structural problems demand structural change. Studies of faculty members who engage in CER find that institutional incentives are especially powerful (Ulrich 2016). Faculty members are more likely to adopt a CER approach when their institutions signal that CER aligns closely with the institutions' missions; provide a supportive infrastructure, such as offices for community engagement to help build relationships and manage budgets; offer internal funding and help faculty apply for external support for CER; and assess faculty research using criteria that clearly define and value CER. Unfortunately, few institutions around the world have made significant commitments to create these conditions, despite widespread endorsement of university-community collaboration (Appe et al. 2017; Welch 2016).

Structures for linking higher educational institutions and communities developed over the past four decades provide flawed but potentially valuable resources for transforming the research culture. Many institutions have launched place-based learning centers and anchor programs to promote community-based civic learning and research across the disciplines, and to build local capacities to improve public schools, healthcare, social services, and economic development (Democracy Collaborative 2019; Hall, Tandon, and Tremblay 2015; Hodges and Dubb 2012). While some of these programs have been designed to serve privileged, white students better than communities of color, some anchor programs and faculty participants in them are especially committed to transforming longstanding inequities in their communities (Sladek 2019). Centers, science shops, and maker spaces can host participatory research driven by community priorities if academic institutions are willing to share their findings and inventions openly rather than monetizing them as proprietary intellectual property (Munck 2014). If none of these university-community structures is perfect, each is worth struggling to transform because they are potential levers for change.

Certification schemes could also foster change across higher education. For example, the Carnegie Elective Classification for Community Engagement certifies over 300 academic institutions in the U.S. for implementing a broad range of community-engaged educational and scholarly practices (Carnegie Classification of Institutions of Higher Education, n.d.). Strengthening requirements for CER could help certification systems such as this to drive change across the educational sector. These requirements can push institutions to match their rhetorical commitments to community engagement with the necessary institutional support to build permanent and coherent programs, which require adequate staffing, faculty participation, and experience in managing community partnerships. Over time, organizers for change in academia could persuade the most powerful overseers of higher education—accrediting bodies, legislatures, trustees and regents, and major donors—to convert elective classifications into mandatory commitments to invest in community engagement.

To shift from penalizing to rewarding researchers who do CER, academia can revise criteria for evaluating research (see box 5.3 for a case study on enacting

BOX 5.3. ORGANIZING FOR CER IN TENURE AND PROMOTION POLICIES

While a shrinking percentage of faculty members hold tenure-stream positions today, tenure policies are often the strongest indicators of an institution's priorities and values for all researchers. Occidental College (Oxy) provides one example of how faculty used a community-organizing approach that made progress toward transforming tenure and promotion policy, on which future efforts could build.

Attempts to include CER in Oxy's tenure and promotion policy began in 2005 but stalled because of resistance by campus leaders. In fall 2013, the college's Center for Community Based Learning (CCBL) faculty committee reignited this conversation as the college applied for reclassification as a community-engaged campus by the Carnegie Foundation. Because the application asked whether community-engaged teaching and scholarship were included in the tenure and promotion process, the faculty committee took this as a window of opportunity to reintroduce the proposed policy changes.

The committee's organizing began with a power analysis of the campus, paying close attention to who held faculty governance posts, as well as the academic dean and president's positions on community-based learning and research. It was important to understand how the administration believed CER would benefit or harm the college, in order to develop a strategy that would resonate with them. The committee also mapped a critical mass of instructors across disciplines who conducted CER or taught community-based learning courses, and who had received tenure or been promoted. This mapping also found that several academic departments now included visits to the CCBL for all finalists in faculty searches. These developments suggested that community-based learning had become rooted in the culture of the campus. The committee formed a core group of around 15 faculty allies from multiple disciplines, who built consensus and shared leadership for a new proposal, and then fanned out to initiate conversations about it in their departments.

The faculty committee took care to show how CER aligned with the culture of the college and its peer institutions. The committee researched comparable colleges that recognized community-engaged teaching or research in their tenure and promotion policies, yet also embedded the rationale for its proposal in Oxy's mission and marketing. Committee leaders built trust with those who did not often think about CER or were fearful of what recognizing it would mean for

(*Continued*)

BOX 5.3. *(CONTINUED)*

their own research, tenure, and promotion. The core faculty group readied for meetings with the dean of academic affairs and the faculty council by preparing responses to questions about the description of CER and dispelling fears that all faculty members would be required to do it. In spring of 2016, the faculty unanimously approved the proposal.

Winning approval required some strategic compromises. The committee succeeded in including language recognizing community-based work in all three areas of evaluation: teaching, research, and service. However, the final language omitted several changes that were seen as too radical, including an expanded definition of materials that counted as publications and measures of the impact of scholarship, and a broader definition of peer review to include nonacademic reviewers. Thus, there was more to be done to build adequate infrastructure to support strong CER agendas across disciplines and training for faculty to evaluate CER in their colleagues' applications for promotion.

The Oxy case suggests several steps for organizing to transform tenure and promotion policies:

- taking advantage of windows of opportunity provided by accreditations, other external reviews, campus strategic planning processes, or changes in leadership
- convening a team of faculty leaders interested and invested in CER, culture change, and collective leadership to draft a proposal
- conducting a power analysis to identify existing support across campus, who can influence decision makers, and what other initiatives might support or detract from the goal
- rooting proposals in institutional mission statements and marketing materials
- building trust and support by meeting repeatedly with supportive faculty members in each academic department, potential opponents, faculty governance bodies, and administrative leaders
- making strategic revisions to the proposal based on feedback, in order to begin progress toward long-term policy and cultural change
- planning to strengthen the infrastructure to support CER and training faculty evaluators to implement policies recognizing the value of CER.

these institutional reforms). Many disciplines have developed standards specific to engaged scholarship (Doberneck and Schweitzer 2017; International Collaboration for Participatory Health Research 2013; Kastelic et al. 2017; Sandoval et al. 2011). Table 5.3 presents a set of criteria adapted from an influential rubric created by scholars convened by the Community-Campus Partnerships for Health (Jordan et al. 2009). The rubric integrates standard expectations for academic research, such as academic peer review and publication, with additional qualities relevant to CER, such as community peer review and dissemination of research (see chapter 4 for more detail on implementing these criteria).

TABLE 5.3. Characteristics of Quality CER

Characteristic	Definition
Clear academic and community change goals	Well-defined research objectives and/or questions, and realistic goals for community change
Adequate preparation in the content area and engagement with the community	Demonstration of researchers' content knowledge and preparation to conduct meaningful research with community partners
Appropriate methods: rigor and community engagement	Demonstration of how rigor (valid theory, research, and methods) is maintained and/or enhanced by community collaboration
Significant results: impact on the field and the community	Reporting of research results, knowledge created, and actual or potential effects on the community (e.g., on policy, community practices and processes, outcomes, organizational or individual capacities, or leadership development)
Effective dissemination to community and academic audiences	Co-presenting results with community partners through diverse channels for reaching relevant academic and community audiences (e.g., academic journals, community events and meetings, local media, policy briefings)
Critical reflection on the project to improve the research and community engagement	Assessment of the project's impacts and ways to improve the design, conduct, and outcomes of future research, drawing on community and academic feedback
Leadership and personal contribution	Evidence from academic and/or community arenas that the research has helped the research partners to earn recognition for leadership on the subject (e.g., invitations to present at professional, community, or government meetings, or to serve on advisory, policy-making, and other committees)
Consistently ethical behavior: socially responsible conduct of research	Demonstration of mutually beneficial, trusting, and equitable relationships with community partners; compliance with academic institutional review boards and relevant community review processes, cultural norms, knowledge systems, and data control and ownership protocols; sharing credit with community partners when disseminating the research

Transforming Impacts on Communities

Educational institutions shape their surrounding communities not simply through their research but by how they operate and whom they educate. The accumulated wealth of academia—in land, buildings, and endowments—has a history. In settler-colonial states, many academic campuses were founded on lands seized from Indigenous peoples by state order or religious decree (Tachine and Cabrera 2021). Some campuses were built by conscripted Indigenous laborers and enslaved Black workers, and funded by the slave trade (Harris 2020). Many institutions continue to rely on gifts from donors who made their fortunes exploiting and contaminating communities of color. Today, academic institutions play a major role in land and economic development that gentrifies surrounding neighborhoods, pushing out low-income residents of color (Canaan and Shumar 2008; Matsuoka 2017). In countries such as the U.S., higher education reproduces racial and economic inequality by employing a growing precariat of low-wage and part-time teachers and staff, and by disproportionately graduating the children of white, affluent, and highly educated parents, while saddling low- and moderate-income students with mounting student debt (Cahalan et al. 2020).

CER can invite communities affected by these impacts to take academic institutions themselves as objects of study in the interest of transformational justice. Faculty, staff, and students are increasingly researching their own institutions' roles in slavery and colonialism, and how contemporary campuses erase or celebrate this history in their museums, archives, art and monuments, and building names. For example, the Universities Studying Slavery consortium (https://slavery.virginia.edu/universities-studying-slavery), a collaboration of over 80 institutions in the U.S. and U.K. hosted by the University of Virginia, shares practices and principles for conducting truth-telling projects about academic institutions' historic connections to the slave trade and enduring racism in academia. Research such as this is informing some institutional actions to acknowledge, reconcile, and repair these damages. Initial steps include redesigning campus sites to represent this history from the perspectives of enslaved and Indigenous peoples, acknowledging that campuses sit on Indigenous lands, renaming buildings, providing access to or returning lands and artifacts to Indigenous peoples, creating scholarship programs for and paying reparations to descendants of enslaved and conscripted laborers, and contributing to community and economic development in their communities (Harris 2020; Mamtora, Ovaska, and Mathiesen 2021). More of this research that informs campuses' understanding of their past and adoption of reparatory policies could be conducted with representatives of affected communities to ensure that their perspectives and policy preferences are centered (see box 5.4 for an example).

BOX 5.4. TRANSFORMATIONAL JUSTICE AT RYERSON UNIVERSITY

At Ryerson University in Ontario, the university's Standing Strong (Mash Koh Wee Kah Pooh Win) Task Force (2021) reexamined the historical record and contemporary legacy of the university's namesake. Egerton Ryerson was an educator who led the creation of the Ontario public school system, which included racially segregated schools for Black students and residential schools for Indigenous students, where children were separated from their families, endured physical and sexual abuse and neglect, and were forced to assimilate into Christian and Canadian culture. The task force of faculty, staff, students, and alumni—many of whom

(*Continued*)

BOX 5.4. *(CONTINUED)*

were also active members of Ontario's Indigenous and Black communities—conducted historical research on commemoration of colonial history; engaged in "learning and unlearning" about Indigenous cultures with scholars, Elders, and Traditional Knowledge Keepers; and consulted members of the campus and Indigenous communities through surveys, community conversations, and media outreach. The task force explained that its recommendations were not intended to erase the university's history, "but to reflect a more complete understanding of the past, celebrate current values and set aspirations for the future . . . and reflect the kind of ancestors we wish to become for our next seven generations" (11–13).

Their report began by acknowledging that "students, faculty, staff and community activists—particularly Indigenous and Black community members—have completed paid and unpaid research on, and raised awareness about, these topics" for over a decade. It went on to "recognize the harm that has been caused by the university's failure to prioritize historical research and meaningful community engagement about Egerton Ryerson's work and legacy" (11), despite prior efforts to address truth and reconciliation on campus. Among other recommendations, the report called on the university to take these actions:

- Rename the university through a process that engaged community stakeholders and the university community.
- Adopt five principles of commemoration drafted by the task force and a review process to guide future decisions about commemoration.
- Create exhibits about Egerton Ryerson's life and legacy, and the era in which the university was named for him, and make all archival materials about him publicly available.
- Develop plans to integrate learning about Indigenous and Black history, studies, and colonial relations into all academic programs and faculty and staff training.
- Expand scholarships for Indigenous and Black students, and expand hiring and promotion of Black and Indigenous faculty members.
- Develop land acknowledgements, and use public space on campus to install community-based art installations, plant gardens for growing Indigenous medicines accessible to the community, and conduct a healing ceremony at the former site of a statue of Ryerson that had been pulled down by protestors.

While Ryerson's president immediately accepted the recommendations in full, many of them will likely take years to enact meaningfully. Nonetheless, this task force illustrates the role that CER could play on the growing number of campuses that are starting to reckon with their own pasts. This includes CER for rigorous study of the historical record, systematic analysis of its current meanings to diverse campus constituencies and harmed communities, and careful design of policies that enable transformative justice.

EJ research also has a special responsibility to address academia's impact on gentrification. Large institutions that continually expand into surrounding neighborhoods to build new student housing and academic facilities tend to drive up rents that drive out low-income residents and small businesses. Even well-intentioned neighborhood greening initiatives—such as park cleanups, river restoration, tree planting, community gardens, and attracting healthier food stores—can fuel gentrification by making neighborhoods more environmentally desirable (Rigolon and Németh 2020). CER with an EJ lens needs to integrate anti-displacement goals into colleges' and universities' conventional sustainability programs in their communities (Di Chiro and Rigell 2018).

For example, a faculty-led class project at Occidental College in Los Angeles partnered with community residents and local organizers to document neighborhood changes in the rapidly gentrifying neighborhood outside of the campus. Students produced an online map of archival material and set the foundation for continued collaboration between faculty, staff, students, and community residents. In response to the college's purchase of a building in the neighborhood, the collaborative developed "Principles for Occidental College-Community Neighborhood Development" to guide future off-campus real estate acquisitions. These principles sought to strengthen and expand mutually beneficial collaborative relationships between the college and its neighboring community and ensure that the college's actions as an investor, developer, and landlord would reflect not only its financial interest, but also its mission to promote the public good and community-based learning. Occidental's board of trustees deemed the principles too restrictive and instead adopted a set of "Investment Principles" rather than the "Neighborhood Principles" developed by the campus-community coalition (Occidental Magazine 2019). Building on this CER and ongoing collaboration with community partners, faculty and students continue to promote the collaborative's principles and engage in CER to document changes in the surrounding neighborhoods (Matsuoka and Urquiza 2021).

Transforming the Composition of Institutions

Expanding CER for EJ also depends on recruiting and supporting a critical mass of faculty, staff, and students who come from and care about EJ communities and want to engage in research with them. Black, Indigenous, and people of color (BIPOC), women, and LGBTQ+ researchers tend to engage in CER and in EJ research at higher rates than other faculty members (O'Meara et al. 2011; Vogelgesang, Denson, and Jayakumar 2010). While respectful and culturally humble researchers can be effective allies across socioeconomic and racial lines, researchers who share some aspects of community membership with external partners are often especially well positioned to build trust and co-create knowledge with them (see chapter 3). Greater inclusion of BIPOC students, faculty, and staff is vital to sharing the means of production of knowledge about EJ (for distributive justice). It is important for involving members of EJ communities in shaping CER from *inside* the walls of academia, not just from outside (for procedural justice). It is crucial for respecting and valuing the experiences and knowledge of EJ communities (for recognition justice). It is one powerful means of providing reparations for academia's ongoing harms to EJ communities, and its history of exclusion and oppression of BIPOC peoples and knowledges within universities (for transformational justice).

Of course, academia should become more diverse, equitable, and inclusive for more reasons than advancing CER or EJ. In addition, all faculty members should enjoy the academic freedom to choose their methods and contribute to any field, and all students should be educated to take part in CER and to advance EJ. Nonetheless, one of the most powerful ways to increase this kind of research is for institutions to create campus climates in which underrepresented faculty and students are not only recruited but promoted, and not simply included but belong (Pedler, Willis, and Nieuwoudt 2022; O'Meara et al. 2021). Especially important is investment in student scholarships and stable full-time faculty and staff positions, which afford the time and security to develop programs of CER and to conduct EJ research that challenges institutions to live up to their missions and heal injustices in which academia itself is deeply embedded.

CONCLUSION

In this chapter, we have argued that CER is not simply a research methodology, but an alternative vision of academia's role in society, and that higher education is implicated in environmental and social injustices. In this light, academia needs to do more than make a little space for CER as a boutique program that allows researchers to do more relevant research and their institutions to reap goodwill in their communities. Rather, CER is a challenge and an opportunity to rethink higher education's relationship to oppressed peoples and communities. Doing so will require academia to address multiple dimensions of justice, including how higher education shares research and educational resources, who gets to make decisions about research, whose knowledge is recognized in curriculum and research agendas across all disciplines, and how to remake relationships between academic institutions and communities to transform historic injustices and ongoing inequities.

PART 3

Applications

6

Research Methods and Methodologies

Ryan Petteway, Sarah Commodore, Chad Raphael, and Martha Matsuoka

Community-engaged research (CER) for environmental justice (EJ) employs many methods to measure exposures to hazards, document inequities, represent injustices, and tell the stories of EJ communities. This chapter provides a critical review of characteristic methods and methodologies of CER in EJ research, which are detailed more fully in methods textbooks and the technical literature than we can do in a single chapter.[1] We use *method* and *methodology* interchangeably here because the line between discrete methods (e.g., beta attenuation monitoring) and broader methodological approaches (e.g., environmental monitoring) is often blurry in practice. We focus more on methods for data collection than those for analysis, discussing the strengths and weaknesses of each method, potential uses, and how they are employed in example studies. We include citations to relevant sources that offer greater technical and procedural detail on how to use these methods.

Our discussion is grounded in some commonly held insights of antiracist, decolonial, and feminist approaches to methods and knowledge production. These traditions recognize that to choose a set of research methods is also to choose a set of power and property relations—between research teams and participants, and among credentialed researchers and their community partners—in

1 Especially valuable textbooks and handbooks on how to apply community-engaged research methodologies include sources on action research (Bradbury 2015), participatory action research (Chevalier and Buckles 2019; Kindon, Paine, and Kesby 2007), community-based participatory research (Blumenthal et al. 2013; Israel et al. 2013; Minkler and Wakimoto 2022; Wallerstein et al. 2017), and citizen and community science (Lepczyk, Boyle, and Vargo 2020). For more detail on how to apply Indigenous and decolonizing research methods, see Atalay 2012; Denzin, Lincoln, and Smith 2008; McGregor, Restoule, and Johnston 2018; Smith 2021; Wilson 2008; and Windchief and San Pedro 2019.

TABLE 6.1. CER for EJ Methods

Dimension of Justice	In CER Methods for EJ
Distribution *Who ought to get what?*	Choosing research methods that share resources and access to data, and that develop communities' capacities to conduct their own research Prioritizing methods that allow for community members to materially and professionally benefit from their contributions
Procedure *Who ought to decide?*	Centering community voice and influence in the selection of research methods and in data collection, analysis, and dissemination processes Prioritizing data sovereignty
Recognition *Who ought to be respected and valued?*	Practicing epistemic and cognitive justice, and decolonizing knowledge, by choosing methods that recognize the validity of and differences among local, experiential, and Indigenous knowledges Respecting communities' rights to consent and to control data about themselves
Transformation *What ought to change, and how?*	Transformation of relations between research institutions and communities by choosing methods that co-create knowledge rather than extracting data from the community, and that allow communities to speak effectively to power holders

the production of knowledge. This choice is shaped by the social and institutional conditions in which epistemologies and research methodologies developed. Contemporary methodologies continue to bear the influences of capitalism, settler colonialism, white supremacy, and patriarchy. Many methodologies are designed to extract data from communities—from biological specimens to opinions and beliefs—and profit by converting them into research funding, publications, patents, and professorships. Data analysis typically proceeds without communities' participation or consent, according to dominant epistemologies (such as Western science) that exclude, erase, and disrespect community knowledges, cultures, and values. The knowledges produced—and the production process itself—often function to further subjugate oppressed and colonized peoples, and to build researchers' and research institutions' prestige rather than creating equitable and reciprocal relations with and within researched communities (Petteway 2022). In contrast, the CER paradigm strives to center community knowledges by pursuing new methodologies, and by questioning and remixing mainstream methodologies to transform traditional knowledge, power, and property relations within research collaborations to better align with principles of epistemic, procedural, distributive, and, ultimately, research and data justice. Thus, throughout this chapter, we consider how the choice and application of research methods can advance the dimensions of justice common to CER and EJ (summarized in table 6.1). Of course, research justice is not guaranteed simply by choosing the "right" methodology, but also depends on, for example, how it is applied. Moreover, as many of our examples show, CER often combines two or more methods to address multiple research questions and strengthen the relevance, reach, and impact of the research. Thus, considerations of research justice are a matter of continuous, layered, and iterative reflection and researcher reflexivity.

SURVEYS

Surveying poses questions to gather information from people. The resulting data may be quantitative (collected by asking respondents to rate or rank items on numerical scales), qualitative (gathered in respondents' own words as they answer open-ended questions), or a mix of both kinds of data.

Survey research offers many advantages for CER. Community participants can learn to conduct their own surveys with minimal training and gather their own data relatively inexpensively, rather than depending on complex or proprietary research equipment and data sets. Many community members and advisors can participate in co-creating and reviewing survey questions and procedures, ensuring they reflect local needs and values, and building collective consent for the research. Surveys can help a community to compile and validate many kinds of knowledge about itself—beliefs, attitudes, practices, experiences, identity characteristics, environmental and social risks—and explore the relationships among them. Open-ended questions can allow respondents to share multiple cultural perspectives and kinds of knowledge, not only those anticipated by researchers. Survey methods can respect research ethics that matter to EJ communities: conferring anonymity offers respondents some control over their privacy, and surveying can explore the interactions of multiple risk factors without conducting randomized control trials that might expose community members to hazardous substances or withhold remedies from some participants (Korn and Graubard 1991). Survey data can contribute to transformational justice by providing an overview of community problems, identifying critical needs, soliciting potential solutions, and evaluating progress toward collective goals through repeated measures. These methods provide qualitative and quantitative data that can be used to design organizing campaigns and interventions, and support policy arguments.

The Richmond health survey (Cohen et al. 2016) offers an example of how to apply many CER principles. The survey was prompted by local concerns about exposure to multiple sources of pollution (especially from petrochemical facilities) and elevated levels of cancer in a predominantly minority low-income fenceline residential community. The research partners included the community organization Communities for a Better Environment (CBE) and academic researchers from Brown University and the University of California, Berkeley. This research team generated hypotheses and brainstormed survey questions in community meetings, aiming to identify health problems about which residents wanted community-wide data. The research partners then trained community members to recruit participants and administer the survey with study staff, canvassing neighborhoods on foot and contacting participants from previous studies in the area, which helped to increase awareness of the study. The research found an association between residents' poor health and cumulative stress from multiple sources of pollution (Cohen et al. 2012). CBE disseminated this finding to the community and used this evidence of cumulative impacts as a tool in its organizing campaign for increased regulation of local facilities and against the expansion of a local oil refinery.

However, there are also ways in which surveys can fail to align with principles of justice in CER. Many community members who respond to surveys are unlikely to be involved in helping to design them; this raises the possibility of researchers and community organizations extracting data from residents to advance the research partners' agenda, rather than enlisting respondents as co-creators of meaning. Because the most marginalized members of communities are often least likely to respond, this self-selection bias may mute their voices in survey data. Researchers can fail to respect local cultures and ways of knowing by mistranslating questions into local languages, using questions that are not validated through piloting with community respondents, or relying too heavily on closed-ended questions informed by researchers' narrow assumptions and meanings. These limitations can produce results that misrepresent community conditions, perceptions, and priorities, distorting interventions and actions based on the conclusions.

CER can mitigate these problems by involving diverse elements of the community in each stage of the research. Cohen et al. (2018) draw several relevant lessons from their community-based cross-sectional survey in France. They urge CER teams to design questions that allow respondents to discuss household and community issues, and that honor local knowledge. Research partners can hold open meetings to report data and use focus groups to check researchers' interpretation of the data. Residents and relevant experts can co-interpret survey data and collectively brainstorm actions that might be taken in response to findings. Each of these steps increases research partners' accountability to the larger community and the community's participation in co-constructing the meaning of survey data.

ENVIRONMENTAL MONITORING

Environmental monitoring involves taking samples from one or more locations to measure hazards in any environmental media. CER has documented contaminants in soils and other environmental media near hazardous waste sites (Brown and Mikkelsen 1997; Ramirez-Andreotta et al. 2015). CER has also measured exposure to air pollution (Commodore et al. 2017) from sources such as diesel bus depots (Kinney et al. 2000), ports (e.g., Garcia et al. 2013), and industrial hog farms (Wing et al. 2008). Additional CER has monitored water contamination (Buytaert et al. 2016), including from landfills (Heaney et al. 2013), sewage (Heaney et al. 2011), and multiple threats to Indigenous peoples' water sources

(Cummins et al. 2010; Wilson, Mutter, et al. 2018). Studies using cell phones and other devices as sensors have measured noise pollution in sites such as public housing (Haklay and Francis, 2018).

Environmental monitoring methods have many strengths for CER. As monitoring technology has become cheaper and more sensitive, it has allowed communities to gather their own data, rather than relying on government or industry (English, Richardson, and Garzón-Galvis 2018; Johnston et al. 2020). Community members can collect environmental samples after appropriate training on data collection protocols and labeling (World Health Organization 2014). This increases communities' power to set the research agenda by selecting which contaminants and environmental media are of greatest concern to residents. Many CER studies fill gaps in existing data sets by producing more localized and time-sensitive data about emissions than polluters and government agencies report, forcing them to recognize local knowledge not previously admitted in the regulatory process. For example, in the 1990s, EJ activists adopted simple air monitors using buckets and plastic bags to capture air samples, which could be sent to a laboratory for analysis. Soon, "bucket brigades" were documenting short-term spikes and long-term violations of emissions limits by oil refineries and chemical plants around the world (see chapter 7). Environmental sensors such as these can shift power to communities to pinpoint the sources of pollutants, trace their movements, correlate emissions with health symptoms, and hold polluters and regulators accountable for addressing violations.

However, high thresholds for scientific proof of harm limit the power of environmental monitoring in regulatory forums. Typically, communities must prove that they are adversely affected by environmental hazards by establishing a continuum from contaminant source identification to presence in the ambient environment to exposure and entrance into the human body (Johnston et al. 2020). Regulatory agencies and courts have been slow to accept community environmental monitoring data as valid evidence, sometimes requiring expert testimony to validate the protocols and instruments used in CER (Wyeth et al. 2019). Some contaminants may be unknown or difficult to measure with existing equipment. The most sensitive and accurate sensing technology and the training required to use it are still too costly for many community organizations.

BIOMONITORING

Biomonitoring, sometimes called body burden research, evaluates the presence and concentration of a chemical (or its derivative) in the human body (Paustenbach and Galbraith 2006). As biomonitoring has become more sensitive, affordable, and available, it has become an important tool for documenting the presence and extent of chemicals not normally present in human bodies (Shamasunder and Morello-Frosch 2016). This has expanded researchers' ability to assess the impacts of environmental chemicals and other exposures on human health by supplementing measures of substances external to the body (in food, water, or air, for example) with measures of internal exposures (in breast milk, urine, blood, and tissue) (Morello-Frosch et al. 2015b). The type of biomonitoring used depends on the persistence of the contaminant of concern: for example, a lipophilic chemical with a half-life of two years can be accurately measured in breast milk, while a polar chemical with a half-life of 12 hours will be better characterized in urine. Biomonitoring has provided a more complete picture of the "exposome," an analogue to the human genome that includes all exposures from social and physical environments over an individual's lifetime (Wild 2005). Biophysical monitors, such as skin conductance and heart rate monitors, can provide additional individual-level evidence of the health effects of environmental stressors.

Biomonitoring allows communities to collect their own data about substances related to health conditions that most concern residents, such as risks posed by chemical emissions from industrial sources and consumer products (Adams et al. 2011; Morello-Frosch et al. 2015b). Residents can participate by co-defining research questions with scientists and donating samples of hair, nails, urine, or blood. Studies can respect participants' desire to control their own data by allowing residents to access their personal exposure levels, which prior epidemiological research has generally resisted (see chapter 5). For example, in response to environmental health advocacy, California's biomonitoring program now requires that individual data be communicated to study participants who want this information (Morello-Frosch et al. 2015a). Biomonitoring can also build respect for local knowledge by validating community complaints of environmental health effects that often go undocumented in official public health data. Because biomonitoring can provide objective evidence of substances' presence in the body, it can help communities meet the burden of proof that links exposure with health impacts. By measuring chronic and acute exposures to hazardous substances, and tracing health effects, biomonitoring can be used to question whether acceptable exposure limits in current regulations are in fact safe. It can also assess whether vulnerable communities are exposed to greater risks, stresses, and harms than environmentally privileged communities, building pressure for action.

Several factors limit the use and impact of biomonitoring. Obtaining biosamples depends on building a high degree of trust with research participants, given the sensitive nature of these materials. Analyzing samples requires expert training, and may demand specialized and expensive equipment. Many regulatory and industry scientists have resisted accepting biomonitoring as a legitimate source of data, limiting this methodology's ability to transform environmental health science and public health policy (Shamasunder and Morello-Frosch 2016).

Some studies combine biological and environmental monitoring. For example, in Canada the Aamjiwnaang First Nation community in Ontario, the Occupational

Health Clinics for Ontario Workers, and University of Ottawa biologists collaborated to use bucket brigades and body burden testing among Aamjiwnaang people living near chemical plants, filling gaps in government data collection and building pressure for stronger regulation of emissions (Sabzwari and Scott 2012). Another valuable example is the "Truth Fairy Project," in which East Yard Communities for Environmental Justice collaborated with academic partners to investigate the impact of toxic metal exposures around a closed lead-acid battery smelter in a predominantly Latinx neighborhood of Southeast Los Angeles (Johnston et al. 2020). The study combined analysis of soil in local yards and residents' baby teeth (as biomarkers of lead exposure) to demonstrate an association between soil lead levels and lead ingestion (prenatal and postnatal). The research informed residents about toxic metal exposures and provided evidence to support organizing for legislation that funded removal of lead-contaminated soil from neighborhoods around legacy smelters.

COMMUNITY MAPPING

All environmental exposures entail a spatial component—that is, they exist within, between, around, and across specific social and geographic places. Thus, being able to map out sites and sources of environmental concern, as well as their spatial patterns and distributions, is perhaps the most fundamental component of CER for EJ. How environmental exposures, risks, assets, and opportunities are (mis) represented through map-making—and how maps are then used—plays a critical role in pursuit of EJ. While some EJ research uses mapping and screening tools created by state regulators and environmental scientists (see chapter 7), we focus on mapping that involves primary source data gathering using a CER approach.

Mapping helps communities pursue many goals, such as

- researching and representing cumulative environmental exposures and social vulnerabilities,
- educating the community about historic and current environmental injustices,
- identifying community assets that can help advance EJ,
- targeting health interventions and resources to high-priority places and groups,
- designing local infrastructure,
- mobilizing residents to launch campaigns,
- communicating information to decision makers, and
- supporting advocacy in permitting, development, remediation, and policy processes.

Mapping also presents some dangers for research partners (Corburn et al. 2017). Creating and updating maps can demand significant time and resources, especially if this involves purchasing proprietary mapping tools and learning to use complex software. In some cases, communities may choose not to publicize potentially stigmatizing data (such as levels of pollution or disease) or sensitive cultural information (such as Indigenous sacred sites, which have been subject to vandalism and looting). Monitoring technologies used to generate some data for mapping can undermine participants' privacy rights if researchers do not obtain fully informed consent. In addition, official data used in mapping may be incomplete or inaccurate—a "garbage in, garbage out" problem—so community members may need to ground truth this information by checking it against their own experience and investigations. Other data, including the names and boundaries of the community itself, may reflect dominant outsiders' representations of the community—a "hegemony in, hegemony out" problem—so community members need to be vigilant and reflexive about defining themselves at each step. Because maps, like all data, do not speak for themselves, their ability to contribute to change relies on how well they are used to support organizing and advocacy.

Geographic Information Systems (GIS)

Many CER studies employ geographic information systems (GIS). GIS software acquires, stores, tracks, checks, and displays various forms of data that have geospatial attributes, that is, they can be geographically located and mapped. GIS platforms range from expensive proprietary software (such as ArcGIS), to open-source platforms (like QGIS), to free web-based mapping tools (like MapServer and OpenStreetMap), to platforms for mobile devices (such as Kobo Toolbox). GIS can support a variety of EJ research methods. Studies that employ environmental monitoring can, for example, passively or manually collect samples (such as airborne pesticides, soot, or heavy metals) from geolocated sampling locations via handheld Global Positioning System (GPS) devices (Gibbs et al. 2017) or outfit community residents with sample-collecting devices (such as mobile air monitors) to track participants' exposures across locations (Ma et al. 2020). Researchers can also employ GIS to administer surveys remotely and then manually geolocate the results later via computer-based GIS software; administer surveys in person using mobile devices that automatically record location data; or administer surveys remotely using an ecological momentary assessment approach that prompts respondents via their mobile devices when they are in certain locations (Mennis, Mason, and Ambrus 2018).

GIS using GPS technologies can also take an "activity space" approach to assess an individual's environmental exposures (Cagney et al. 2020). This approach to measuring air pollution, for example, would measure air quality not just at a person's residence, but throughout their entire "activity space," as they travel from home to work, stores, parks, and so on. These methods can also account for the duration and temporality of exposures throughout the day, month, or year, rendering more thorough and accurate assessments of exposures—from pollutants

(Park and Kwan 2020) to greenspace access (Bell et al. 2015), food environments (Widener et al. 2018), and more.

Participatory GIS

GIS methods that allow for deeper community participation in research are commonly called participatory GIS (PGIS), participatory action mapping (PAM), or public participation GIS (PPGIS). PGIS takes a community-driven, user-friendly, and procedurally and epistemically inclusive approach to mapping—one that "ideally places the control of access and use of socially or culturally sensitive spatial data in the hands of the communities who generate it" (Verplanke et al. 2016, 309). PGIS can represent people's local spatial knowledge to inform participatory decision making, communication, and advocacy, and entails "an explicit attempt to use digital mapping technologies to give voice, amplify, and represent local needs—especially of marginalized groups" (Haklay and Francis 2018, 299).

EJ research partnerships have applied PGIS to research topics ranging from conservation and sustainability (Ramirez-Gomez, Brown, and Fat 2013; Nicolosi, French, and Medina 2020), to aspects of urban planning (Boll-Bosse and Hankins 2018). In one promising example, Jelks and colleagues (2018) worked with ten community researchers to examine environmental concerns in an Atlanta watershed, using a customized app with GPS and photo/video capabilities to spatially and visually document concerns in real time. The study filled gaps in official environmental data and generated evidence that residents then used to engage officials to remediate the problems.

Qualitative GIS

Qualitative GIS, or QGIS, also holds promise for EJ research. QGIS integrates various forms of qualitative data—such as photos, audio, and video narratives—within traditional quantitative-based GIS platforms. The goal is to spatialize—and geographically visualize—non-quantitative representations of place-based knowledge and experience that, as described by Jung and Elwood (2010), help address the "inadequacy of absolute Euclidean geometries as a means for representing the abstract, inexact, and socially situated ways that people understand the world" (66).

Expressions of QGIS include geo-narratives (Bell et al. 2015) and geo-ethnographies (Matthews, Detwiler, and Burton 2005) of people's experiences of place, and have included the use of "walk-along interviews" to elicit "spatial transcripts" (Martini 2020). Dennis and colleagues (2009) worked with youth in Madison, Wisconsin, using QGIS to map participants' photos and interview narratives about environmental health and safety issues, producing maps that guided community-based interventions. QGIS can also be combined with augmented-reality platforms, which allow users to position mobile devices to access place-based digital content. For example, Butts and Jones (2021) worked with students and local partners to develop augmented-reality tours to decolonize dominant environmental and social histories of Florida's Paynes Prairie State Park—exposing the history of land dispossession of local Seminole tribes and the slow violence of climate change. Using the project's EcoTour app (www.shannonbutts.com/ecotour), park visitors point their mobile phones at landmarks to encounter information drawn from Seminole oral histories, historical photos and maps, and other archival data, which provide a "deep mapping" of how the park was shaped by a history of environmental injustices.

Counter-mapping

Counter-mapping is mapping that "questions the assumptions or biases of cartographic conventions, that challenges predominant power effects of mapping, or that engages in mapping in ways that upset power relations" (Harris and Hazen 2005, 115). It can involve various forms and practices of spatial representation, whether through PGIS, QGIS, or other approaches, digital or analog. Counter-mapping is generally a community-led mapping process undertaken as a mode of resistance to settler-colonial extractivism, dispossession, and environmental degradation.

> [Counter-mapping] allows a group to combine their own low-tech methods with the state's techniques and manners of representation in order to re-insert themselves and their lived experiences and perspectives, underscore their unique relationship to landscapes, challenge their disadvantaged circumstances, and get their territorial and customary claims to resources recognized by dominant settler societies. (Kidd 2019, 960)

Core to counter-mapping is the understanding that maps, as visual codifications of spatialized power, "are neither neutral nor unproblematic with respect to representation, positionality, and partiality of knowledge" (Harris and Hazen 2005, 101). Importantly, counter-mapping both counters *and creates*—it is productive and generative of new ways of interpreting and representing environmental conditions and experiences. Accordingly, counter-mapping can play an especially critical role within EJ communities enmeshed in the dynamics of exposure (mis)representation and contestation. Often, technocratic and administrative processes and norms for monitoring environmental risk fail to capture the nuanced contexts of daily exposures as experienced by community members. Counter-mapping has contested official processes that omit and obscure—by defect or design—important community knowledges relevant to identifying, contextualizing, and mitigating environmental risks, and to uncovering environmental assets within EJ geographies (Dalton and Stallman 2018). Counter-mapping has been used to document ecological and natural resource conservation and disruption (Harris and Hazen 2005); to visualize Indigenous land rights and dispossession and help communities to resist settler colonialism, extractive industries, and environmental degradation (Hunt and Stevenson 2017; Willow 2013); and to contest and contribute data to policy discourse related to disinvestment and lack of greenspace in Black neighborhoods of Detroit in the late 1960s (Dalton and Stallman 2018) and the spread of gentrification in San Francisco in the 2010s (Maharawal and McElroy 2018).

STORYTELLING

Storytelling and narrative analysis are widely used in CER and organizing for EJ (Houston and Vasudevan 2018). Common expressions of storytelling for EJ include digital storytelling, oral histories, "toxic bios," and counter-storytelling—none of which are mutually exclusive, and all of which can involve other methods discussed here.

Digital storytelling presents data from multiple sources in a narrative format, often using technologies that allow for broad sharing and access. One example of EJ digital storytelling is work completed by Johnston and colleagues (2020). They worked with youth co-researchers, who used personal air-monitoring devices (the AirBeam), PGIS (via smartphone GPS), and photographs to spatially and visually document their daily PM2.5 exposures. In another project, First Nations members in British Columbia used digital storytelling to challenge established policy narratives that divorced health from community interactions with local lands and waters, and demonstrated how residents understood human and natural health as intertwined (Gislason et al. 2018).

Oral histories, when participatory, involve residents as co-researchers in the study design and in gathering, editing, and analyzing individuals' EJ stories—something modeled well by the CER collaboration between DataCenter, Pacific Institute, and the Winnemem Wintu tribe in California. Winnemem researchers gathered and analyzed personal stories and used cell phone GPS to map sacred sites, demonstrating their historical importance for healing and spiritual ceremonies (DataCenter 2015c). In other examples, Adams and colleagues (2018) worked with residents of an Oklahoman "fenceline community" to examine perceptions of long-term petrochemical exposure, and Castleden and colleagues (2017) worked with Indigenous elders in a Mi'kmaw community along the eastern Canadian coast to identify, contextualize, and historicize concerns related to contaminants from a pulp mill. Elsewhere, Armiero and colleagues (2019) engaged EJ storytelling through the curation of stories related to environmental activism and contamination, so-called "toxic bios." They describe their approach as "guerilla narrative," "meaning the sabotage of toxic narratives, which silence injustice, through the coproduction of a counter-hegemonic storytelling" (10).

Counter-storytelling has conceptual roots in notions of counter-narrative or counter-hegemony, and counter-storytelling traditions of critical race theory. As articulated by Delgado (1989), counter-stories "can show that what we believe is ridiculous, self-serving, or cruel... can show us the way out of the trap of unjustified exclusion . . . can help us understand when it is time to reallocate power" (2415). As with counter-mapping, EJ communities often practice counter-storytelling to expose dominant histories and narratives as unjust, oppressive, and self-serving, while offering new stories that point toward justice. The aforementioned digital storytelling project by Gislason and colleagues (2018) with First Nations communities in British Columbia is one such example. In another example, Spiegel and colleagues (2020) worked with adults and youth of the Tsleil-Waututh Nation to examine environmental concerns related to the Trans Mountain oil pipeline in Canada. Tsleil-Waututh researchers developed a counter-story to oil industry narratives of progress, using photography and testimony to narrate the pipeline's threats to local food sovereignty, health, and cultural bonds with the watershed, and to imagine alternatives.

Stories are a grassroots form of making meaning: community members can often contribute to storytelling without extensive training, and EJ stories may be more compelling than academic research for mobilizing people to act (Newman 2012). Storytelling lends itself to communicating complex causality in a form that can be more memorable than scientific data (Griffiths 2007). Part of the power of storytelling lies in its ability to generate collective, relational, and affective narratives of community concerns, priorities, histories, and futures. Ganz (2011) describes how these public narratives can fuel community organizing by connecting a "story of self" (focused on one's calling) and a "story of us" (linking the individual to the community's calling) to a "story of now" (that mobilizes people to take collective action for change). EJ narratives integrate many types of knowledge—personal and collective, local and expert, cultural and scientific, practical and theoretical—into coherent accounts of injustice and justice backed by illustrative evidence. EJ storytelling is therefore a means of providing testimonial evidence—not only for research, but also for organizing, public testimony, and litigation (Evans 2002).

However, in the absence of accompanying scientific data, testimony and other stories may be dismissed as anecdotal evidence drawn from unrepresentative samples. Policy makers and regulators trained in scientific and positivist paradigms may be especially suspicious of stories as overly "emotional" and "irrational." Counter-stories, such as those in the Indigenous examples mentioned above, especially require skillful translation and framing to communicate across cultural and ideological boundaries.

PARTICIPATORY MEDIA, COMMUNITY ARTS, AND PHOTOVOICE

Participatory media and arts-based research methods can be used for data collection or dissemination, or both (Coemans and Hannes 2017; Gubrium and Harper 2013). In data gathering, research participants can communicate their experience through photography, video, and other media. As a vehicle for disseminating data, art can replace or supplement traditional academic publications to express findings through street murals and other public installations, exhibitions of images or artifacts, and dance, theater, music, and other performances. In addition, community arts events can communicate and dramatize information about organizing or public health campaigns. There is a growing literature on using arts-based approaches to CER with marginalized populations (Coemans and Hannes 2017), with Indigenous peoples (Hammond et al. 2018), and for health-related research (Boydell et al. 2016). Additional reviews summarize the use of particular media and approaches in community arts research on EJ issues, such as adaptations of Augusto Boal's Theatre of the Oppressed (Sullivan and Parras 2008), feminist EJ zines (Velasco, Faria, and Walenta 2020), and collaboratively written "policy novels," which weave explanations of environmental policies into fictional storylines (Van der Arend 2018).

In an especially extensive collaboration, informal recyclers in Canada and Brazil represented their work and needs in a long-term participatory video partnership with community organizations, local governments, the University of Victoria, and the University of São Paulo. The project trained participants, who are often stigmatized as "scavengers" and harassed by authorities, to produce brief documentaries for local officials, explaining how informal recyclers perform valuable services by recovering and recycling materials that have been dumped in landfills and streets. Campaigns used these videos to decriminalize informal recyclers' activities in Canada (Gutberlet and Jayme 2010) and integrate this work into the formal recycling sector in Brazil (Tremblay and Jayme 2015).

Photovoice is a particularly well-developed method in CER for EJ, which has informed other uses of media and arts for research. Photovoice is a hands-on, photography-based research method designed to help community residents—as co-researchers—identify and discuss important community issues and take social action (Catalani and Minkler 2010). Residents use cameras/smartphones to visually document aspects of their community that represent—literally and/or symbolically—their concerns and perspectives on a particular topic, then write short narratives that contextualize each photo. While photovoice processes vary, residents typically discuss and analyze their work collectively, curate photography exhibits, and present their research to community and policy leaders (Petteway 2019).

Photovoice has been used broadly for EJ-related research on topics ranging from food and tobacco environments (e.g., Leung et al. 2017; Petteway, Sheikhattari, and Wagner 2019) to built and social environments (Petteway 2019; Sampson et al. 2017). It has also been used to explore more traditional EJ exposures. For example, Madrigal and colleagues (2014) worked with youth co-researchers in a California Latino farmworker community, training them in environmental health and using photovoice to document their environmental concerns and community assets. Similarly, Schwartz and colleagues (2015) used photovoice with Mexican American adults and youth to explore issues related to asthma and pesticide exposure in an agricultural community. In Nevada, Willett and colleagues

(2021) worked with youth scientists to explore the EJ concept of slow violence as manifest in inadequate urban infrastructure, public services, and climate-related disasters (such as wildfires). EJ researchers frequently combine photovoice with other research methods and forms of data. These multimethod studies have paired photovoice with air monitors and PGIS to document particulate exposure (Johnston et al. 2020), with indoor air quality monitors to study risks from woodsmoke (Evans-Agnew and Eberhardt 2019), and with PGIS and X-ray mapping of daily place-based environmental exposures (Petteway et al. 2019).

Reviews of the literature find many potential benefits of using participatory media and arts techniques for EJ research (Coemans and Hannes 2017; Gubrium and Harper 2013; Wilson, Aber, et al. 2018). A core strength is that arts and media offer comparatively accessible and inclusive methods for involving youth and adults across a range of cultural and ethnic communities in conducting and owning their own EJ research. Community media and arts can center and amplify participants' expression of their lived expertise and embodied knowledge of EJ in their communities. These methods excel at communicating the place-based and experiential nature of EJ exposures through research that is simultaneously affective and visceral, and material as well as symbolic. In doing so, media and arts methodology introduces new knowledges that can complement, contextualize, contest, and counter existing EJ data narratives, much like counter-mapping and counter-storytelling. As they discuss their work in progress, many arts and media groups resolve to take collective action to address their conditions. Like storytelling, community arts and media can strengthen community bonds as part of rituals and ceremonies, and imagine alternative futures.

Participatory media and arts also present some challenges similar to those of storytelling methods (Wilson, Aber, et al. 2018). It is difficult to include representative samples of a community in the small groups typical of these projects. Participants often must commit significant time to create, discuss, and present their work. Professional research partners must be careful to avoid imposing their aesthetics and interpretations of residents' work and conditions on community partners (Evans-Agnew and Rosemberg 2016). While research using community arts and media has presented ample evidence that these methods build research capacities and solidarity among participants, this does not always translate easily into transforming policies or practices.

BIG DATA

Big data refers to the growing availability of large data sets produced by a variety of novel sources. This approach is distinguished by its use of complex data analytics to examine an unprecedented volume of records from a variety of sources, often with greater velocity of data gathering and analysis (Grayson, Doerr, and Yu 2020). Given the diversity of these sources, and the fact that they can be combined to yield novel insights, big data is more of a broad methodological approach to research than a focused method. The opening of previously restricted databases, availability of low-cost sensors specifically designed for community scientists, and new open-source data analytical techniques have made big data studies more possible and practical for CER on EJ issues. Examples of big data sources that may be used in EJ research include crowdsourced community science projects, genomic databases, government databases, networks of environmental sensors, satellite remote sensing networks, social media activity, mobile app and web searches and clickstreams, locational data, financial transactions, and records of scanned barcodes.

Many CER projects that involve big data rely on crowdsourcing, "an online, distributed problem-solving and production model that leverages the collective intelligence of online communities for specific purposes" (Brabham et al. 2014). For example, Dodson et al. (2020) used crowdsourcing in a biomonitoring study to track self-reported consumer behaviors related to products containing phenolic compounds (e.g., BPA, parabens). Sun and Mobasheri (2017) crowdsourced volunteered geographic information from a cycling app to examine potential air pollution exposure during active commutes; Picaut et al. (2019) completed similar work using a smartphone app and GPS to crowdsource environmental noise measurements. Crowdsourcing has also been used as a part of multimethod EJ-related work. For example, Barrett et al. (2018) combined crowdsourcing with traditional GIS data to examine asthma hot spots and inhaler use, while Kim, Lieberman, and Dench (2015) used a crowdsourcing approach involving traditional GIS and photos to examine tobacco retail environments.

However, not all crowdsourcing projects aggregate the "collective intelligence" of active crowdsourcing participants, such as the "wisdom of the crowd" model. Instead, many projects revolve around the use of passive surveillance and data collection (e.g., via smartphone GPS) or volunteered reports of environmental behaviors or observations. An example of an EJ monitoring system that has employed crowdsourcing and community involvement is the Identifying Violations Affecting Neighborhoods (IVAN) system in California's Imperial Valley (https://ivan-imperial.org/air). IVAN was created by state regulators to measure particulate matter concentrations and provide real-time air quality reporting to the public. Community members helped to identify air-monitoring sites and learned to maintain the monitors, which are validated and calibrated to official environmental agency reference monitors to ensure reliability. An environmental justice task force made up of regulators and residents reviews the data at monthly meetings to inform their plans to reduce pollution. Over time, the IVAN website began to solicit and map crowdsourced public complaints about illegal dumping, emissions, and other environmental violations, inspiring the launch of additional IVAN networks around the state.

Big data offers many attractions for CER on EJ. Big data can provide CER partners with access to much more specific measurements of household and individual exposures to environmental hazards and benefits, helping communities develop interventions where they are needed most. Large numbers of community members can contribute data, building a critical mass of residents who understand EJ issues and are invested in organizing to address them (Kaufman et al. 2017). Building larger samples of participants who contribute their environmental and health data repeatedly can enable CER to establish the causes of environmental health inequities and harms (Alexeeff et al. 2018), and force regulators and polluters to stop dismissing residents' experiential knowledge of health impacts as anecdotal evidence (Mennis and Heckert 2018). Large samples may also speak to power in another way: officials who know that many of their constituents have participated actively in community science studies may be more likely to pay attention to the results.

Limitations and concerns regarding big data in EJ research have been discussed elsewhere (Mah 2017), with ample cause to be concerned that big data algorithms can function as a discriminatory "weapon of math destruction" without concerted efforts to render them transparent and legible to the public (O'Neil 2016). In this regard, D'Ignazio and Bhargava (2015) introduce the notion of "popular big data" to articulate a vision for how to render big data more inclusive, transparent, and transformative for everyday people—and perhaps counter big data's tendency to (re)produce discrimination and other harms. And no discussion of big data can be had without deep engagement with notions of data justice (Heeks and Shekhar 2019), and concerns of (re)colonization vis-à-vis data extractivism and commodification. Vera and colleagues (2019) draw from feminist, Black feminist, and decolonial theory to outline a reflexive framework for environmental data justice (EDJ) that explicitly calls attention to "extractive data logics" and the "structural whitewashing of environmental data." Mapping the contours of power in database scope and ownership, as well as the bounds of database uses, remains a crucial matter of procedural, epistemic, and distributive justice.

SMALL DATA

Given the challenges of reconciling research involving big data with CER principles, many of the methods we have discussed in this chapter show the value of *small data* in advancing EJ. D'Ignazio and colleagues (2014, 116) describe small data as follows:

> a practice owned and directed by those who are contributing the data. . . . The essence of Small Data is that such communities may not just participate in, but can actually initiate and drive such data investigations towards the better understanding of an important local issue.

Notions of voice, representation, decolonizing, and power are core to small data. A small data approach typically affords communities more control in setting the research agenda, determining data priorities and collection methods, data collection and analysis, and data ownership. This approach presents a counterbalance to what D'Ignazio and Bhargava (2015) refer to as the "empowerment problem" of big data orientations, which exude extractivist and settler-colonial proclivities of epistemic erasure and dispossession that can function to silence and disempower communities.

In regard to investigating environmental factors, D'Ignazio and colleagues (2014) suggest that "a bottom-up, participatory, grassroots approach to . . . data collection addresses the key issues of inclusion, accountability, and credibility, by building public participation into the data lifecycle" (116). The small data approach of *popular epidemiology* was one of the first important methodological innovations of the EJ movement in the U.S. (Brown 1992, 1993). This approach to environmental research is grounded in, animated by, and (co)led by those who are experiencing the exposure(s) in question—with the explicit priority to take local social action based on findings to mitigate and repair harms. Coming to prominence in the early 1990s, popular epidemiology arose from communities' efforts to compile their own evidence of environmental exposures in order to contest—much in the spirit of counter-mapping and counter-storytelling—pervasive governmental and corporate apathy and narratives of harmlessness. Residents became their own scientists, acting as epidemiologists-activists to fight for both epistemic and environmental justice. Privitera and colleagues' (2021) work examining concerns related to petrochemical exposures via use of "toxic autobiographies" is one recent expression of this approach.

Small data orientations deliberately incorporate information and communication technologies (ICTs)—such as mobile phones and web-based mapping platforms—to enhance the democratic and community-led nature of the research process and action based on research findings. Small data studies can take many forms and employ multiple methods discussed in this chapter. The key is that they are community chosen, community led, and focused on (co)producing community knowledges that are excluded from status quo technocratic research practices, and the data are "owned and directed by those who are contributing [it]" (D'Ignazio, Warren, and Blair 2014, 117).

CONCLUSION

Because power and property relations are encoded in research methodologies, choosing methods also involves choices about justice. In CER for EJ, justice is best served by employing methods that shift power and ownership to communities, so they can share fully in the resources, data, and capacities required to do research. Collaborations should maximize community partners' and participants' role in choosing methods, gathering and interpreting data, and determining how information is disseminated. Methodological decisions must recognize the validity and multiplicity of local, experiential, and cultural knowledges, and communities' right to control data about themselves. The ultimate aims are to employ methods that help shift research institutions from extracting and exploiting data about EJ communities to co-creating knowledge with them for environmentally just policies and practices. For most community partners, EJ research is a means to the larger ends of structural and systemic change, especially for health, right relations with nature, cultural and economic flourishing, and racial justice. In this sense, all research is a methodology for transformation.

7

Law, Policy, Regulation, and Public Participation

Carolina Prado, Zsea Bowmani, Chad Raphael, and Martha Matsuoka

Environmental justice (EJ) movements have taken governments to task for failing to regulate environmental risks and harms adequately, denying meaningful public participation in administrative decisions and policy making, and refusing to integrate rights to environmental justice meaningfully into the legal process. Therefore, EJ activists recognize that engaging with state-led or state-controlled processes may not always be the best strategy (Liboiron et al. 2018; Pulido, Kohl, and Cotton 2016). Participating in litigation, policy, and regulatory action requires more resources, expertise, and influence than many EJ communities have. These slow and demanding governmental processes can sap movements' energy and co-opt them into ceding important goals. For example, despite numerous regulatory complaints and lawsuits brought by EJ advocates, the U.S. federal government has consistently refused to apply civil rights law to counter racially discriminatory impacts of siting and permitting of hazardous facilities (Foster 2018).

Critics of pursuing justice through the state argue that EJ movements may be better off challenging the legitimacy of state-led processes, withdrawing from them, and pursuing other strategies, such as direct action against polluters, organizing alternative institutions, and engaging EJ communities in mutual aid (Pellow 2018; Pulido, Kohl, and Cotton 2016). However, for EJ organizers, the question is often when to invest in state-based remedies or to take alternative actions, rather than whether to make a permanent choice between these strategies. Many movements have organized both within and against states to try to transform them over the long run (Purucker 2021). In addition, some Indigenous tribes are sovereign

TABLE 7.1. CER for EJ in Law, Policy, Regulation, and Participation

Dimension of Justice	In CER for EJ in Law, Policy, Regulation, and Participation
Distribution *Who ought to get what?*	Building community capacities to document disproportionate environmental risks and harms to EJ communities, demand remediation, and secure fair access to a healthy environment
Procedure *Who ought to decide?*	Supporting EJ communities' co-production of research to strengthen their influence in environmental regulation, policy, law, and litigation
Recognition *Who ought to be respected and valued?*	Asserting the validity of local knowledge and community-produced research in regulatory, policy-making, and legal processes Recognizing Indigenous sovereignty over environmental decisions on their ancestral lands
Transformation *What ought to change, and how?*	Researching for systemic transformation of administrative, legislative, and judicial processes to acknowledge the cumulative impacts of environmental and social risks, compensate and restore harmed communities, transition to safer substances and practices, and institutionalize community rights to a healthy environment for people and nature

governments, which seek to expand their sovereignty by engaging in intergovernmental relations with colonialist states on equal terms (Nagy 2022). As this chapter shows, there are examples of engagement with state processes that have won significant victories, particularly at the local level, and many EJ struggles approach the state with varying levels of cooperation and confrontation.

When EJ organizers seek justice through the state, they can draw on community-engaged research (CER) to document inequitable harms, legitimize claims, and envision remedies. This chapter discusses how CER has contributed to the development of community-centered paradigms for understanding environmental risks and safer alternatives, efforts to strengthen public participation in the regulatory process, campaigns that build community policy-making expertise, and litigation that complements EJ advocacy and organizing. Table 7.1 relates the chapter's major themes to the dimensions of justice common to CER and EJ.

REGULATION AND PARTICIPATION

EJ organizers and advocates have drawn on CER to inspire foundational changes in frameworks for environmental regulation and public participation. While polluters and officials still resist these changes, they are transforming how some governments assess risks, seek safer alternatives for hazardous substances and industrial processes, and involve the public in regulatory and policy processes.

Cumulative Assessment of Environmental Risks and Social Vulnerabilities

CER for EJ has helped transform approaches to risk assessment used by regulators and policy makers to characterize the nature and magnitude of risks to human health and the environment. EJ advocates and allied researchers have shown how traditional risk assessment abstracts from real-world conditions in ways that understate risks to communities and block remedies (O'Brien 2000), including by

- testing for effects of individual substances and facilities, via individual environmental media, and from individual sources, rather than testing for the synergistic and cumulative impacts of all pollutants to which communities are exposed;
- testing for effects on the "average person" (usually a healthy white male), rather than on more vulnerable groups (such as children, people with compromised immune systems, and people in poverty);
- placing the burden of proof that substances and facilities are harmful on risk bearers (EJ communities), rather than demanding proof of safety from risk generators (such as manufacturers, users, and emitters of hazardous substances);
- requiring high levels of scientific certainty about the causes of harms before acting to prevent them, rather than acting to reduce plausible threats to health and the environment in a timely manner.

Since the 1990s, some jurisdictions have begun to move toward a more just and accurate risk assessment regime that considers cumulative impacts on communities, offers greater protection for vulnerable groups, demands greater evidence of safety from industry, and takes a more precautionary approach to regulating risks even if scientific evidence of cause and effect is not fully established (Corburn 2017).

Creating cumulative risk models that integrate measures of social vulnerabilities (based on socioeconomic factors such as poverty, race, education, and language) with environmental stressors (such as exposure to air and water pollutants, and hazardous chemicals) has been especially important (see box 7.1). These exposure indices quantify a population's risk from aggregated environmental and social burdens over time, and can be employed to create highly localized mapping databases of inequitable risk distributions (Cushing et al. 2015; Morello-Frosch et al. 2011). In the U.S., the data used in these tools are available in many cases because of public right-to-know laws that the EJ movement passed in the 1980s, which required polluters to make annual public reports of hazardous substances in their facilities and of emissions into communities.

Cumulative impacts analysis also engages communities in ground truthing environmental hazards and social vulnerabilities. CER projects organize community residents and researchers to correct and supplement gaps in regulatory

BOX 7.1. NEW TOOLS FOR ASSESSING RISK AND VULNERABILITY

California's EJ advocates, researchers, and state agencies have employed CER to create multiple online mapping tools for assessing cumulative risks and social vulnerabilities to inform policy making (Eng, Vanderwarker, and Nzegwu 2018). Foremost among them is CalEnviroScreen (https://oehha.ca.gov/calenviroscreen), which incorporates data on multiple environmental, public health, and socioeconomic risk factors to create a numerical score of the vulnerability of each census tract in the state. The state's Environmental Protection Agency (CalEPA) consulted with an advisory board of EJ researchers and grassroots leaders to write the definition of cumulative impacts and select relevant indicators, and improve initial drafts of the tool through multiple rounds of public feedback.

Other tools developed by researchers in collaboration with EJ advocates have influenced and supplemented CalEnviroScreen. For example, the Environmental Justice Screening Method includes a broader range of indicators than CalEnviroScreen (including race) and ranks cumulative impacts at a regional level (Morello-Frosch et al. 2015a). The Cumulative Environmental Vulnerabilities Assessment focuses on the state's San Joaquin and Coachella Valley regions (Huang and London 2012, 2016). The California Healthy Places Index (www.healthyplacesindex.org) summarizes social determinants of health at various geographic levels. The Drinking Water Tool (https://drinkingwatertool.communitywatercenter.org) identifies threats to groundwater, such as contaminants and susceptibility to drought, and gives information about how residents can influence groundwater management decisions. One of the most important influences of these projects has been to model how involving community members in ground-truthing data is necessary to ensure accurate mapping and assessment (Sadd et al. 2014).

These tools now integrate cumulative assessment into many policy and regulatory processes, from the local to the state level (Eng et al. 2018). For example, CalEnviroScreen is used to identify communities that receive prioritized funding from California's Greenhouse Gas Reduction Fund, generated by the state's cap-and-trade program, to prioritize areas for targeted enforcement of regulations, and to inform CalEPA's planning of community engagement and outreach (Murphy et al. 2018). Because mapping tools like CalEnviroScreen are publicly available, and their underlying data can be downloaded, researchers and EJ organizations can use these tools to identify inequities, and inform policy proposals and legal actions.

data by checking them on location (Sadd et al. 2014). Ground truthing can also be used to raise EJ challenges to emissions or exposure standards, which are typically set by regulators for a broad geographic area (e.g., using national air quality standards) or a type of pollution source (e.g., coal-fired power plants). When issuing permits for facilities, agencies translate these broad standards into local,

TABLE 7.2. Addressing Challenges of CER on Vulnerabilities Analysis

Task	Challenges	Potential Solutions
Defining relevant pollution sources and their health impacts	Multiple stakeholders have different definitions of sources and impacts	Engage stakeholders in dialogue to reach consensus on sources and impacts
Identifying viable solutions to pollution problems	Possible solutions may have their own secondary impacts	Elicit solutions from community dialogue, rather than determining them during analysis
Addressing tension between pollution parameters and health impacts	Health impacts are experienced below established standards for legal pollution	Foster stakeholder dialogue on pollution limits and impacts
Incorporating ground truthing into cumulative analysis	There is a lack of resources for systematic ground truthing	Identify additional funding sources for ground truthing
Resolving socio-environmental vulnerability	Analyses could reduce but not eradicate impacts of vulnerability	Facilitate improvements, even if incomplete or messy
Incorporating regional uniqueness	Different communities perceive pollution problems and solutions uniquely	Engage communities to adapt best practices to local contexts

source-specific requirements. Ground-truthed data can show how a region might meet standards for ambient air quality, yet contain multiple pollution "hot spots" from sources concentrated in low-income neighborhoods, or how a national standard for mercury in fish designed to protect the average consumer can fail to protect vulnerable groups that rely more heavily on fish in their diets (such as Asian Americans and Native Americans). While cumulative impact analyses have addressed some of the limitations of traditional risk assessment, there are important challenges that need to be addressed, as shown in table 7.2 (adapted from Huang and London 2016).

Alternatives Assessment

Alternatives assessment emerged in the 1990s to protect workers and consumers from chemicals of concern in manufacturing processes and consumer products. Traditional risk assessment was problem focused, aimed at quantifying the risk posed by an individual chemical to cause a specific hazard (such as cancer) at a given exposure level. This process was notoriously poor at informing policy and regulation, instead tending to induce "paralysis by analysis" by demanding years of costly research to establish whether a chemical posed an "acceptable risk." In rare cases in which regulators moved to ban a substance, some manufacturers made regrettable substitutions of one hazardous material for another. In contrast, alternatives assessment is a solutions-based approach that aims "to support the informed transition to safer chemicals by comparing a range of options to substitute a chemical of concern" (Tickner, Weis, and Jacobs 2017, 655). This involves "identifying, comparing, and selecting safer alternatives . . . on the basis of their hazards, performance, and economic viability" (Geiser et al. 2015, 2152). The U.S. Environmental Protection Agency (U.S. EPA), states such as California, and the European Union have begun to adopt this approach.

While alternatives assessment research is mainly conducted by researchers in academia, government, and the largest environmental organizations, CER partnerships with labor unions and frontline workers have translated this research into actionable knowledge used to promote policy and organizing for occupational safety and health. The Chemical Hazard and Alternatives Toolbox (ChemHAT) (www.chemhat.org) offers a good example. Unlike many official and technical databases, ChemHAT draws on global scientific records from many countries and institutions to characterize hazards in plain language and color-coded visuals. ChemHAT reports potential environmental impacts of substances, along with possible acute and chronic effects on human health, including cumulative and synergistic effects, and impacts on children and the immunosuppressed. Importantly, ChemHAT explains where one is likely to be exposed to each chemical, how to protect oneself, safer available alternatives, and links to the underlying, peer-reviewed data sources. ChemHAT is the product of participatory research conducted with workers by labor unions, occupational safety and health organizations, environmental groups, public health scholars, and digital media designers. The tool is designed to empower workers and their organizations to participate in managing risks from chemicals in their workplaces and engage in well-informed advocacy for safer substitutes.

Public Participation

For many EJ communities, procedural justice—the ability to exercise voice and influence over decisions that affect them—is an important goal as well as a method for achieving EJ. Public participation processes can involve the public in agenda setting, creating policy, and making decisions with government agencies (Rowe and Frewer 2004). Community participation can also contribute to better-informed decision making by governments and more effective environmental outcomes by generating policy solutions and increasing community commitment to implementing them (Ford-Thompson et al. 2012; Reed 2008). Public participation is encouraged and even required by many state and federal laws and administrative rules, and by international agreements, such as the European Union's Aarhus Convention, the United Nations Declaration on the Rights of Indigenous Peoples, and the United Nations Sustainable Development Goals. Most of these participatory processes are advisory, but a few are empowered to make decisions directly. For example, participatory budgeting and municipal health councils—which involve community members in setting spending priorities and allocating part of their city's annual budget—have addressed EJ issues of fair distribution of public spending on public health, parks, transportation, waste management, and other services (Baiocchi, Heller, and Silva 2011; Coelho and Waisbich 2016).

However, many governments lack the will or imagination to engage less powerful groups equitably because of industry capture and corruption of administrative agencies and legislatures, inadequate legal frameworks for participation, and reliance on constrained forms of public consultation (such as public hearings) that disempower and alienate community members (Nabatchi and Leighninger 2015). As a result, the extent and quality of participatory processes vary widely—from minimal public notice and comment requirements, to extensive impact reviews and co-production of policies and decisions with residents. For example, hazardous waste siting processes "can be an exercise in democratic deliberation with the proposed host community, an aggregation of pluralistic viewpoints on the proposed siting, or a vehicle for exclusion of citizens most affected by the proposed land use" (Cole and Foster 2001, 106).

EJ organizers aim to increase their communities' power in these formal decision-making processes, moving them up Arnstein's (1969) ladder of public participation. At the bottom of this ladder, officials manipulate participatory processes or merely provide therapeutic opportunities for residents to express frustration, denying them real influence in decisions. The middle rungs describe tokenistic participation, such as expressing priorities or commenting on draft plans, when this does not influence final decisions significantly. At the top rungs, participants share power over decisions with government, either as partners or because decisions are delegated to community committees or given over entirely to the public to decide (through referenda, for example).

Rocha's (1997) ladder of empowerment builds on Arnstein's approach by representing degrees of power for underserved and underrepresented communities. In contrast to Arnstein's understanding of power as the ability to influence others' behavior, Rocha's model especially focuses on power in the relationship between the self and others, highlighting structural and systemic influences on participation in policy making. Table 7.3 shows how CER for EJ can help community members climb this ladder.

To move up these ladders, and to plan and execute CER collaborations well, individuals and groups typically need capacities to deliberate within their organizations and with officials. Deliberative skills include proposing actions or policies, supporting them with reasons and evidence (from systematic data to personal experience and storytelling), listening and responding to others' views, creating inclusive contexts in which all participants can contribute as equals, and arriving at collective agreements using decision rules that all participants can agree are fair and noncoercive (Karpowitz and Raphael 2014). Deliberation is not merely about learning to "talk nicely"; it is about actively countering the power of social status, money, credentials, and intimidation in public discussion so that EJ communities can influence decisions and share power over policy making.

TABLE 7.3. Rocha's Ladder of Empowerment in CER for EJ

Rung	Objective	Contributions of CER for EJ
Political empowerment	Ensuring communities have the resources they need to thrive	Showing the need for more understanding of CER's impact on political empowerment (Salimi et al. 2012)
Sociopolitical empowerment	Building community members' critical consciousness of their relation to power structures, and informed action	Increasing participants' critical understanding of political processes and facilitating collective prioritization of policy priorities (Minkler et al. 2008)
Mediated empowerment	Building the empowerment of individuals or communities to participate in existing decision-making processes	Engaging new residents in community policy making and inspiring some participants to run for office (Minkler et al. 2010)
Embedded individual empowerment	Increasing individual participation through an organizational context	Fostering group-wide identification and empowerment (Stack and McDonald 2018)
Atomistic individual empowerment	Increasing individual efficacy and changing the self-perception of the individual	Fostering participants' skills and self-confidence (Ferrera et al. 2015; Garcia et al. 2013)

The experience of engaging in CER can help EJ communities to develop deliberative capacities as they prioritize issues, and agree on research and policy objectives, deepening the internal democracy of EJ organizations (Minkler et al. 2008). CER has also helped to support deliberation between EJ organizations, government agencies, and the wider public. In some cases, universities have created new public forums for convening environmental deliberation. Some researchers have involved community advisors in designing, facilitating, and evaluating these forums, addressing the EJ aspects of issues such as health and bioethics (Abelson et al. 2013), land use planning (Sampson et al. 2014), climate resilience planning (Schlosberg, Collins, and Niemeyer 2017), and municipal budgeting (Lerner 2014). Sustained deliberative engagement has improved EJ-related policy outcomes, especially at the local level, for climate change adaptation, clean energy, community forest management, sustainable community development, and equitable distribution of public funding (Fischer 2017; Romsdahl, Blue, and Kirilenko 2018).

However, poorly conceived or bad faith deliberation by government officials and public policy makers on EJ issues can exclude disempowered groups, limit discussion to a narrow range of options determined by elites, or fail to affect policy when it challenges dominant political and economic interests (Cole and Foster 2001). In the absence of careful planning and commitment to equity, public discussion can reinforce hierarchies among participants based on their

TABLE 7.4. Core Values and Practices for Public Participation

Values	Practices
Inclusion: Ensuring the rights of those who are affected by a decision to be involved in the decision-making process	Making special efforts to recruit diverse participants and facilitate discussion on equal terms Enabling participation by residents of EJ communities by scheduling meetings in their neighborhoods or offering travel reimbursements, providing translators, and offering child care
Influence: Seeking input from participants about how they participate Ensuring the public's contribution will influence the decision	Welcoming diverse forms of communication and evidence, including personal testimony, stories, cultural beliefs, and emotional expression Adopting process and decision rules that grant participants influence
Recognition: Promoting sustainable decisions by recognizing and communicating the needs and interests of all participants, including decision makers	Avoiding enforced consensus on contested issues Treating oppressed groups' interests as integral to the common good, rather than sectarian or selfish
Information: Providing participants with the information they need to participate in a meaningful way	Full disclosure of accessible information that translates expert thinking into lay terms and languages used in EJ communities
Transparency: Communicating to participants how their input affected the decision	Giving a public explanation of the reasons and evidence for decisions, and how and why they relate to public participants' contributions

socioeconomic status, race and ethnicity, gender, or other characteristics. Table 7.4 lists some central values that CER can use to evaluate participatory processes, which are adapted from the U.S. EPA's National Environmental Justice Advisory Council (2013), along with practical steps to realize these values, as identified in the research cited above.

THE POLICY PROCESS

Policy analysis and advocacy have also provided fertile ground for CER. Unlike policy studies led by professional researchers, CER studies begin with community experience and knowledge, build local capacities to analyze problems and craft solutions, and seek to change the policy process by shifting power to the community level (Cacari-Stone et al. 2014). This section shows how CER can help communities build expertise in each of the major streams in Kingdon's (2011) influential model of the policy process, including defining problems, proposing policies, and practicing politics. Additionally, we discuss how CER for EJ across multiple levels of governance can inform translocal organizing and policy strategies.

The Problem Stream

To define problems, organizers and advocates must understand the issues at stake and legitimize them in the eyes of policy makers and the community. CER can help by enlisting community members in defining and documenting environmental hazards and injustices, and by producing usable knowledge that persuades decision makers to act. CER can help EJ groups overcome the barriers they face at this stage, including scant resources and credibility in policy arenas.

To foster community understanding, CER may assess local awareness of an EJ issue, measuring and elevating a community's environmental consciousness at this initial issue-spotting stage (Rickenbacker, Brown, and Bilec 2019). These data can then be used as a rallying point for community organizing and subsequent goal setting. CER can also play a key role in assessing the feasibility, desirability, and effectiveness of potential organizing and policy strategies. For bold, imaginative strategies that might face pushback, engaging community members in the production of actionable knowledge can help build trust and increase community buy-in.

In the problem stream, community organizations must intervene in a knowledge system that attributes credibility to actors with institutional legitimacy, such as scientists. In these credibility struggles, community members strive to gain recognition as valid knowers and interpreters. CER can help to legitimize community groups' knowledge by generating systematic evidence of the scope, scale, and kind of environmental injustices to command attention and support action. These data can be used as an entry point to gain legitimacy in the policy process by contributing to public comments, securing meetings with elected officials to discuss problems, and identifying policy remedies. However, community-based researchers must decide between using costly, state-of-the-art tools that can produce more valid or reliable data (increasing the data's legitimacy for officials) or using affordable, low-tech tools that may be less precise yet accessible. Often, this research identifies relationships between seemingly isolated instances of environmental harms to reveal a broader pattern of systemic injustice. Box 7.2 describes a groundbreaking

BOX 7.2. THE APPALACHIAN LAND OWNERSHIP TASK FORCE STUDY

Between 1870 and 1930, absentee corporations assisted by local speculators acquired much of Appalachia's natural resources. Many local political leaders collaborated with timber companies to clear-cut forests and with coal companies to dig mines while fighting miners' attempts to unionize. Hundreds of thousands of dispossessed Appalachians became economic migrants to the industrial cities of the North. Changes in land ownership disrupted communal ties and sapped residents' political power, leaving the remaining small landowners as "foreigners on their own land" (Horton 1993, 85).

(*Continued*)

BOX 7.2. *(CONTINUED)*

In 1979–1980, over 60 activists, community members, and academics led by the Appalachian Land Ownership Task Force carried out a massive study of land ownership patterns in six states: Alabama, Tennessee, North Carolina, Kentucky, West Virginia, and Virginia. Researchers revealed the scope of absentee corporate ownership and its effect on the regional economy. They showed how tax giveaways to large landowners deprived local governments of revenue needed to develop and diversify their economies and improve housing, education, and infrastructure. The six-volume report concluded with policy recommendations for creating a fairer tax structure, enacting legal reforms to protect surface land owners and small farmers from mining and logging pollution, empowering local governments to use corporate land for housing and alternative development, and establishing local planning and zoning processes to regulate environmental impacts and land use (Appalachian Land Ownership Task Force 1983).

The Appalachian Land Ownership study changed the public and scholarly conversations about the Appalachian environment and economy (Scott 2012). Researchers publicized their findings in community meetings, the news media, and government forums throughout the region, and in popular pamphlets and academic publications. The study informed years of organizing, policy advocacy, and litigation on tax reform, land reform, and poverty alleviation. Participants in the research went on to form new organizations, like Kentuckians for the Commonwealth, which brought white, low-income residents of the Southeast into the EJ movement. The study also helped to launch the interdisciplinary field of Appalachian studies, introduced many of its researchers to CER, and influenced their understanding of the region as existing in a neocolonial relationship to the corporations that had taken control of the area's natural resources and politics. As one activist put it, the Appalachian Land Ownership study was "a foundational source in understanding the history of Appalachia" (quoted in Scott 2012, 49).

study that developed communities' understanding of EJ and gained participants' entry into state policy-making circles across the U.S. Southeast.

The Policy Stream

Within the policy stream, politically viable solutions to problems are proposed, discussed, and selected. Proposals aim to mobilize public opinion and win public officials' support. Policy proponents must address potential benefits and risks of their proposals, and demonstrate expertise in policy making and policy processes. This stream includes the social relationships in which proposals are embedded, such as the communities of specialists that surround different policy topics. Policy specialists are not easily accessed or persuaded by EJ groups, who are typically seen as inexpert outsiders.

CER can help determine which policy approach to take or whether to engage the state through the policy process or other means. CER can be incorporated in multiple stages of policy strategizing and development, including identifying community priorities (such as pollution hot spots or especially vulnerable residents), identifying policy options, and gaining inclusion in the policy process. This last stage can be especially important as CER builds community members' expertise about the roles and processes of the policy sphere. For example, in a community-engaged mapping project in Tijuana, México, residents were able to learn more about the urban planning process in their city and how to intervene in the community of specialists involved in urban zoning (Prado et al. 2021). CER also helped residents to engage in the interpersonal politics of policy making as they presented street-level environmental data they collected. Table 7.5 summarizes additional examples of how CER contributed to the major tasks in the policy stream.

The Politics Stream

The politics stream focuses on winning passage of policy changes, which may require EJ groups to mobilize public opinion, garner support from other social movement actors, influence policy makers, and engage in electoral politics. To do so, EJ organizations often must overcome limited access to decision makers, the power and resources of polluters and other opponents, and indifference among government agencies and officials. One of the formative urban EJ struggles in the U.S. illustrates how organizers can employ CER in multiple ways to build support for policy changes.

In 1996, West Harlem Environmental Action (WE ACT) confronted air pollution in their largely Black New York City neighborhood, where one in four children was afflicted with asthma, and residents suffered one of the highest asthma mortality rates in the country (Minkler, Vásquez, and Shepard 2006). Children reported that their asthma attacks were often triggered as they walked to school past one of six diesel bus depots in the neighborhood, where a third of the city's buses were garaged. WE ACT suspected that the particles in the diesel exhaust emitted by idling buses was a major contributor to asthma. However, they had no evidence of how much particulate matter the buses emitted, and the city's transportation authority refused to investigate the group's complaints.

WE ACT enlisted epidemiologists from Columbia University's Center for Children's Environmental Health to design an innovative study. Together, the partnership trained youth to measure street-level concentrations of diesel particulates using air monitors clipped to children's backpacks. They also taught the kids to count the number of buses, trucks, cars, and pedestrians that passed through busy intersections. Their research showed that particulate emissions were significantly higher than the recommended limits set by the U.S. EPA, and provided some of the first evidence tracing particulate exposure to bus exhaust (Kinney et al. 2000). Working with their community base, WE ACT developed several policy proposals, eventually convincing the city to convert its bus fleet to cleaner fuels. Next, WE ACT and Columbia expanded their research to examine effects of additional

TABLE 7.5. CER Examples in the EJ Policy Stream

Task and Source	Policy Issue	Partners and Location	CER Methods	Research Application	Outcomes
Identifying Priorities (Prado 2019)	Semitruck traffic and exposure to diesel pollution	Colectivo Salud y Justicia Ambiental and Environmental Health Coalition (Tijuana and Baja California, México)	Particulate matter air quality testing to identify hot spots and their relationship to trucking routes	Results were used in binational air quality forums and in meetings with municipal transport representatives	Research helped pressure officials to install "no truck" public signs along the most impacted streets
Identifying Policies (Minkler et al. 2008)	Food access and needs in low-income communities	Literacy for Environmental Justice and San Francisco Department of Public Health (Bayview-Hunters Point, San Francisco, California)	Community surveys on perceived access to fresh produce and food access needs; semi-structured interviews with residents and merchants about openness to healthy food programming	Survey and interview data informed collaboration between the city supervisor and four city departments on a new program	A good neighbor program was established to persuade corner stores to provide fresh produce
Inclusion in Policy Making (Minkler et al. 2008)	Indigenous children's exposure to lead-contaminated soil	TEAL partnership: Native tribal nations, academic researchers, and government agencies (Tar Creek, Ottawa County, Oklahoma)	Blood lead screenings and caregiver interviews before and after an intervention; organizational network interviews and environmental assessments of homes	Blood lead and environmental data were used to engage with the governor's task force on Tar Creek, individual tribal governments, and the Indian Health Service	Research helped design a program for mandatory blood lead screening and parental notification, and new regulation on containing mine tailings

TABLE 7.6. WE ACT and the Politics Stream

Politics Stream Strategy	WE ACT Application	Policy Impact
Identify key policy representatives	Power mapping to identify air quality policy actors in New York	Results provided key targets for testimony, presentation of research findings, and legal action
Mobilize public opinion	Created health workshops for Harlem residents and an ad campaign on city bus shelters	Increased public awareness and support helped shift public transportation to cleaner fuels
Garner support from other social movement actors	Enlisted the Northeast Environmental Justice Network and the Children's Environmental Health Network to provide expert testimony to policy makers	Testimony influenced policy makers' understanding of diesel exhaust's health impacts
Influence policy makers	Meetings with federal and state air quality regulators, public comments, and litigation against city's transportation authority citing the group's research	EPA initiated its own permanent air monitoring in Harlem and nationwide, expanding the agency's role in gathering data about local air pollution

pollutants on larger samples of Harlem residents (Perera et al. 2002). Table 7.6 illustrates how CER contributed to multiple strategies in the politics stream.

Multiscalar Analysis and Policy

EJ policy making increasingly takes a multiscalar approach to all three streams of the policy process. This approach considers how local environmental injustices arise within larger systems and structures (such as global trade) that shift burdens from environmentally privileged areas to EJ communities (Pellow 2018). Multiscalar analysis also exposes the policies that enable these injustices, from the local to the transnational level, and shows how policy decisions made in one place or level can inflict violence on distant communities (Pulido and De Lara 2018). Struggles against these injustices typically gain strength from translocal information sharing and organizing, in which grassroots EJ organizations collaborate across jurisdictions and borders to address structural causes of harm at multiple points within the system. CER can be instrumental in understanding these complex problems, designing policies to remedy them, and participating in political action to change them.

CER has contributed methods and evidence to inform, enforce, or critique the local impacts of international law, policy, and treaties. For example, organizers and researchers have collaborated to expose how policy failures at the national and local levels have enabled the global trade in e-waste to contaminate workers and fenceline communities (Smith, Sonnenfeld, and Pellow 2006), to evaluate the impacts of the North American Free Trade Agreement on EJ in the U.S.-Mexico border region (Environmental Health Coalition 2004), and to address climate change across the U.S.-Mexico border (Mendez 2020).

CER can also contribute to translocal policy development and advocacy. For example, the Trade, Health and Environment Impact Project (THE Impact Project), a partnership between the University of Southern California, Occidental College, and community-based advocacy groups, emerged from local organizing to address air pollution and other health impacts associated with goods movement through the massive Los Angeles and Long Beach ports complex (Garcia et al. 2013). Residents documented local impacts of increased port activity by gathering data on cargo truck traffic in neighborhoods adjacent to port and freight corridors. Using these data, a coalition of groups pushed local and state agencies to reduce diesel emissions and land use impacts of the ports. The project expanded to include homeowner associations, big green environmental organizations, and a coalition with labor organizations to organize for improved conditions for warehouse workers (De Lara, Reese, and Struna 2016). Recognizing that the global system of trade requires policy interventions at higher levels, the project launched the national Moving Forward Network, which connects coalitions around the U.S. working on port and freight issues to address federal policy affecting their communities.

CER IN THE LEGAL PROCESS

Legal action for EJ is often intertwined with policy and regulation. For example, EJ advocates often bring lawsuits to compel agencies to enforce their regulations, and EJ lawsuits (or the threat of litigation) can also result in new regulations and policies. Communities often pursue legal action when barred from other avenues for influence (such as public participation or policy making), when these avenues fail to achieve a community's goals, or when a regulation is violated. Legal analysis and strategizing can support organizing and political advocacy when used strategically (Kang 2009). For example, a lawsuit can draw media attention to an EJ campaign and prompt opponents to address community complaints, and can force corporations to negotiate with community groups. In some situations, filing a legal complaint is the only route to gain access to regulatory debates with agencies and polluters. The formality of the legal process can result in stronger (i.e., binding) solutions that a court can enforce; this is especially important when government agencies are contributing to environmental harm.

When engaging the state via its legal system, communities face many of the same obstacles that they face in the regulatory and policy arenas. Lawsuits are expensive and take considerable time and effort, drawing resources away from organizing and other forms of advocacy. Litigation also relies heavily on

TABLE 7.7. Traditional versus Community Lawyering Models

	Traditional Lawyering	Community Lawyering
Mode of Representation	Only individual client's needs are considered	Client's needs within the broader context of the community are considered
Source of Expertise	Attorney is the expert who speaks for the individual client	Attorney respects and draws from community expertise and knowledge, and is integrated into the community
Framing of Environmental Problems	Inadequacy of environmental laws or enforcement requires legal responses	Environmental problems may be political ones requiring legal and non-legal advocacy options to build community power
Strategies and Solutions	Attorney retains broad control over strategies and solutions to client's problems	Strategies are co-created by attorney and client in consultation with community to be responsive and accountable to them

professional expertise over community knowledge, limiting who can participate and represent a community. The highly technical nature of the law means an EJ suit may hinge upon the interpretation of a legal term and not necessarily address the root cause of an environmental issue. Even if successful, a lawsuit alone will rarely address the power imbalances that lead to environmental injustices. Given these challenges, legal action must be part of a broader strategy that empowers communities and respects their expertise by complementing EJ litigation with advocacy and organizing.

Community Lawyering and Client Empowerment

EJ lawyers tend to engage in *community lawyering*, a collaborative, community-based model of advocacy that uses the law to benefit marginalized communities with the goal of creating systemic change. It draws from the community-engaged poverty law practices of the 1960s, labor and civil rights movements, and other mass movements for social justice, giving rise to synonymous names like "movement lawyering" and "rebellious lawyering." Community lawyering challenges the traditional top-down, attorney-client approach by situating lawyers and community groups as equal collaborators, respecting community expertise, and advancing community education. Table 7.7 contrasts community lawyering with the traditional model of legal representation.

A key aspect of community lawyering is *client empowerment*. In the context of environmental advocacy, client empowerment "means enabling those who will have to live with the results of environmental decisions to be those who actually make the decisions" (Cole 1992, 661). Attorney Luke Cole (1995) called this the "power model" of legal advocacy because it directly addresses the power (or political) disparity that leads to environmental injustice. When evaluating any legal strategy or tactic, a community lawyer should ask the following (Cole 1992, 668):

1. Will it educate people (including community members, policy makers, the public, and lawyers themselves)?
2. Will it build the EJ movement?
3. Does it address the cause rather than the symptoms of the problem?

In working through these questions, a community lawyer and community group might develop an EJ strategy that includes legal (e.g., litigation) and non-legal (e.g., protest) tactics that tap into the community's strengths, deepen its knowledge, and build its power. Even if a community group chooses not to engage in the legal process, lawyers can help identify and weigh options and give legal advice for particular actions (e.g., participating in public hearings versus direct action). In this sense, community lawyering is the legal profession's equivalent of CER. The following section outlines how community-engaged lawyers can support EJ litigation with CER.

Uses of CER in EJ Law

Environmental legal actions usually fall into one of four types: judicial review of an agency's decision, public nuisance, toxic torts, and citizen suits. EJ lawsuits often challenge the construction of new sources of pollution or the expansion of existing sources. They can also challenge a government agency's decision, rule-making process, or failure to enforce environmental regulations. EJ lawsuits can also be filed against the polluters themselves. Each action requires a plaintiff to prove certain elements, which in turn requires certain kinds of evidence. For example, a community group could file a public nuisance lawsuit against a nearby factory emitting noxious fumes. In such a lawsuit, the plaintiff group must generally prove the defendant's action causes harm to the public, but also causes unique harm to them. CER could generate data that demonstrate elevated rates of respiratory illness from the fumes, but also show that residents who live downwind uniquely suffer from soot deposits in their yards.

CER can also be used in multiple ways in EJ legal advocacy. At the outset, it can generate data to better understand the scope and severity of environmental problems, and identify potential violations. This research can also gather evidence to support a particular legal argument, or inform the overall legal strategy (such as whether to file a new lawsuit or submit a friend-of-the-court brief in an ongoing case). When used strategically, CER can also lessen some of the disempowering aspects of taking an EJ fight from the streets into the courtroom.

Yet, there are obstacles to using CER in EJ litigation. CER-based evidence may not match the elements that must be proven to win an EJ case: evidence of environmental harms alone, such as data collected from typical community monitoring projects, may be insufficient for, or even irrelevant to, a particular legal argument. There are also limits on what evidence is admissible and how favorably a court will view it. For example, according to federal evidentiary rules, scientific data must meet the requirements of expert testimony. This could impact a research project that uses affordable, low-tech tools that are community accessible but may be less precise than costly, state-of-the-art tools that might produce more reliable data.

One way around these challenges is having expert testimony, such as from an academic research partner, affirming that the CER observed known, tested, and approved scientific protocols for data collection; attesting to the quality of the research instruments; or addressing other evidentiary issues (Wyeth et al. 2019). Even without expert testimony, courts may consider lay evidence in cases where the evidence does not require specialized skill or knowledge, such as CER data that establish the presence of contamination that is visible, commonly known, or otherwise readily recognized by the average person.

Timing is also a limiting factor. Designing and executing a CER project and analyzing the data takes time, while statutes of limitations set the deadline for initiating a legal action. Some lawsuits (such as those challenging agency decision making) require a plaintiff to raise all issues beforehand during administrative proceedings. Unaware of such constraints, a community group could easily lose its legal right to sue if it misses a deadline, even if it has the most scientifically robust and legally relevant evidence.

Bucket brigades may be the best-known use of CER in environmental litigation. These are campaigns in which local citizens use inexpensive but technically validated plastic buckets to measure air quality near industrial pollution sites. The first campaign was in 1994 following the release of a chemical from a Unocal refinery in Rodeo, California. An estimated 200 tons of "catacarb," a toxic catalyst used in oil refinery processes, leaked for over two weeks without any public acknowledgement from the company. Although local residents suffered from chronic health issues after the toxic release, they lacked proof that Unocal was responsible. A group of residents hired an environmental attorney, who worked with an engineering firm to design low-cost air-sampling devices for residents to monitor further leaks. These were based on the Summa canister, a standard device used by scientists for taking air samples. By using plastic five-gallon buckets, the engineers reduced the cost of each device from $2,000 to $250. In all, 30 buckets were issued to residents who sampled around the refinery whenever they encountered unusual odors, vapors, or flares. Based on these community-generated data and the public attention they garnered, Unocal eventually entered into a settlement agreement for $80 million with more than 6000 local residents. Other EJ activists and community groups have since adapted the bucket brigade as an organizing model to create more public pressure on firms and regulators, to build community political power, to increase the accountability of polluters to nearby residents, and to improve regulatory compliance. Within a decade, the bucket brigade model spread to over a 100 communities in 13 countries and 16 U.S. states (Overdevest and Mayer 2008).

A landmark 2019 case illustrates several factors that can contribute to successful use of CER. San Antonio Bay Estuarine Waterkeeper sued Formosa Plastics Corporation for repeatedly violating the federal Clean Water Act (CWA) by exceeding the amount of plastic waste it was permitted to discharge into Texas waterways, and for violating state and federal requirements to report such discharges. Because Formosa did not report its unauthorized discharges, regulators lacked evidence of them, so plaintiffs' claims were mostly based on community-collected evidence. After the court found Formosa in violation, the company reached a settlement including $50 million to fund environmental projects in the local area, the largest citizen CWA settlement to date. Suman and Schade (2021) explain the reasons why CER was persuasive. One was the relatively simple type of evidence involved: direct observations and collection of plastic debris by hand, which did not require specialized knowledge or tools to analyze. The evidence also directly responded to the legal elements the plaintiffs needed to prove: Formosa's permit allowed only "trace amounts" of plastic discharge, meaning evidence of a single excess discharge was sufficient to prove Formosa violated the law. The sheer amount of evidence generated by CER—photographs, videos, and 30 containers containing 2428 samples of plastic waste collected during the three-year period—demonstrated the magnitude of the violations. Yet, as the attorneys explained, citizen science alone was not enough; key experts and testimony admissions were fundamental to the court's acceptance of CER. While the Formosa case is unique, it offers lessons in how to use CER to support EJ litigation effectively within broader advocacy and organizing.

Law and Legal Aid Clinics

Academic legal clinics and community law offices are two other important legal providers that frontline communities often turn to when facing environmental struggles. Both have unique roles to play in CER.

Environmental Law School Clinics. Environmental law clinics (ELCs) are law school programs that provide legal services to clients and often hands-on legal experience to law school students. Some ELCs practice client empowerment and community lawyering approaches. Most ELCs train law students in representing real-world clients under the supervision of experienced attorneys, expanding access to justice for individuals and organizations that otherwise could not afford legal assistance (Babich 2013).

ELCs are ideal places for law schools to develop programs for community-based research, as these clinics often have strong connections to community groups. Linda F. Smith (2004) identified three methodologies that clinicians can use to incorporate CER into their law school clinical programs. *Action research* is a three-step process of developing a plan, implementing the action, and assessing the results of the action. This is often a useful approach for scholars to use in working with community members who seek to address real problems with focused interventions. In *problem-based service learning*, students work in teams to solve real problems in community settings by researching the issue and applying their theoretical understanding to the community concern. While this approach may not lead to "new knowledge" that is suitable for faculty publication, it does result in new knowledge for the community partner. Finally, *academically based community scholarship* is applied research guided by faculty and often carried out with the assistance of a class of students. This form of scholarship should provide the community partner with answers or solutions to an identified problem, and the faculty researcher should be able to convert the project to new knowledge that is appropriate for publication as legal scholarship.

Several ELCs stand out in their achievement of civic engagement and CER. The Environmental Law and Justice Clinic at Golden Gate University School of Law is one such example, which, in addition to providing legal representation and research for low-income community groups, has made important contributions to community-based environmental law scholarship. Others, like the Emmett Environmental Law and Policy Clinic at Harvard Law School, produce self-help guides and other advocacy tools developed from community partnerships. As law schools continue to grapple with fulfilling educational and public service goals, ELCs will remain important infrastructures to contribute to CER for EJ.

Legal Aid Clinics. Legal aid clinics, or community law offices (CLOs), are well positioned to serve low-income communities. CLOs develop long-term working relationships with community groups and an understanding of local power relations to identify potential allies. Most CLOs are also trusted by the communities in which they work and are sensitive to those communities' needs (Cole 1992). This unique position makes CLOs important sites for CER, as they can connect researchers directly with community members. CLOs may themselves be subjects of research that seeks to better understand client needs and improve services.

The Escambia Project in Florida provides one such example. Led by local community services organizations and design experts, the year-long experiment launched in 2017 with the goal of increasing access to legal assistance. The Escambia Project is one of the first instances of using participatory design methods to reform the civil justice arena, ultimately engaging more than 100 community members, with support from dozens of local volunteers and organizations. Community members were equal partners and decision makers throughout the design process: they identified which ideas would be piloted and took part in their prototyping, testing, and evaluation. The project generated tools to help intake workers identify whether a prospective client has a legal issue and, if so, what kind, making it easier to provide pro bono legal assistance to low-income neighborhoods, and to coordinate the delivery of legal help with other services offered in a single location (Moss 2020). Increased access to legal assistance can improve community members' ability to respond to environmental injustices and intersecting problems caused by poverty and oppression.

FUTURE RESEARCH

In addition to conducting CER for particular legal actions, policy campaigns, and regulatory struggles, future collaborations could promote transformative justice by strengthening the infrastructure of tools, processes, and institutions for conducting CER for EJ. It would be valuable to develop more screening tools that represent cumulative impacts and social vulnerabilities, like the tools developed in California. Research partners can enlist communities in additional ground truthing, to improve the accuracy and comprehensiveness of public data sets and the usefulness of these mapping tools. CER can support campaigns to require regulators to use these data to consider cumulative risks in permitting decisions, and employ these tools to develop additional policies and laws to address issues such as climate resilience in EJ communities (Roos, Pope, and Stephenson 2018). Collaborative research on how to expand the role of community lawyering, and academic law clinics and community-based legal aid clinics, for EJ is also needed.

CER can also look beyond particular campaigns and lawsuits to help develop broader frameworks for EJ law and policy work by enlarging the scope of impacts and vulnerabilities that shape people's environments. Jason Corburn has suggested that EJ research should examine the interactive effects of multiple "environments" that shape well-being, including

> (1) the material and physical environment (e.g., housing, streets, parks, air pollution, wealth, etc.), (2) the social and political environment (e.g., social cohesion, networks, political power, etc.), (3) the institutional and policy environments (e.g., the administrative decisions that shape places such as zoning rules, environmental impact thresholds, public participation procedures, etc.), and (4) the cultural environment (e.g., the meanings, interpretations, narratives, perceptions, feelings, and imaginations that get attached to places). (Corburn 2017, 63)

The goal of this kind of CER would be to involve residents of EJ communities in creating policy directions based on a common vision of "the kind of society we want to live in, whose lives are valued, and how restorative justice can address the damage already done to communities" (67).

Finally, for transformative justice, we need a better understanding of how residents in grassroots EJ communities can use CER to climb Rocha's ladder and share power in policy making and regulatory decisions. How can public participation processes be designed to increase grassroots EJ organizations' ability to participate meaningfully and influence decisions? How can participation in CER most effectively build individuals' and groups' capacities to advance EJ through policy and legal action, especially to address complex, multiscalar impacts such as global trade in goods, services, and waste? What resources do EJ organizations need to engage more effectively in these struggles and how can CER help to provide them?

8

Community Economic Development

Miriam Solis, Martha Matsuoka, and Chad Raphael

Because the economic structures of colonialism, capitalism, and racism have powerfully shaped environmental injustices, advancing environmental justice also requires transforming economic structures and relationships (Faber 2018; Pulido 1996). Dominant approaches to community and economic development impose top-down planning that extracts wealth, excludes local interests and cultures, and extinguishes nature (Agyeman, Bullard, and Evans 2003). Similarly, much development research is conducted in a top-down manner by experts aligned with the perspectives of governments, intergovernmental agencies, philanthropies, and nongovernmental organizations (Munck 2014). In response, activists and researchers aligned with the environmental justice (EJ) movement have promoted different conceptions of development and community-engaged research (CER) that emerge from and prioritize local knowledge, priorities, and power over decisions.

This chapter begins by contrasting dominant approaches to development with alternative visions of the economy and nature advanced by EJ activists and researchers. We then highlight CER's contributions to four strategies for promoting just and sustainable development, including re-localizing economies, community-led worker education and training, just transitions for labor and communities to a decarbonized economy, and community ownership of production. Brief case studies illustrate each strategy, some involving professional researchers and some conducted largely or wholly by lay experts, from whom researchers can learn as well. We conclude by sketching some recommendations for how future CER can support environmentally just community economic development. Table 8.1 summarizes how the main themes of the chapter relate to the dimensions of justice common to CER and EJ that are employed throughout this book.

TABLE 8.1. CER for EJ in Community Economic Development

Dimension of Justice	In CER for EJ in Development
Distribution *Who ought to get what?*	Strengthening investment in community capacities to conduct research for equitable and sustainable communities and economies, including local ownership of production
Procedure *Who ought to decide?*	Asserting EJ communities' participation in and influence over the design and conduct of research to support economies driven by local priorities, control, and cultures
Recognition *Who ought to be respected and valued?*	Applying local knowledge and values to educate and train workers, and re-localize production in response to neoliberalism and globalization
Transformation *What ought to change, and how?*	Transforming economic structures and relations to enact just transitions for workers and communities to a sustainable, regenerative economy

DOMINANT AND ALTERNATIVE ECONOMIC VISIONS

Neoliberalism, Sustainable Development, and Participatory Development

CER and EJ activism challenge the dominant development paradigms of recent decades, including neoliberalism, sustainable development, and participatory development. Many governments around the world pursued neoliberal prescriptions for growth from the 1980s onward, based on liberalizing trade, deregulating markets, reducing protections for labor, and privatizing management and ownership of public services and natural resources (Harvey 2005). The environment and EJ communities bore the brunt of these policies as energy and infrastructure megaprojects destroyed traditional landscapes and livelihoods, speculative real estate investment displaced residents from urban neighborhoods, trade policies weakened environmental and labor protections, and public disinvestment in services worsened inequalities of wealth and living conditions (Apostolopoulou and Cortes-Vazquez 2018).

At the local level, neoliberalism undermined community economic development agencies' ability to meet residents' needs as governments adopted market-based development approaches and partnerships with private sector corporations and finance institutions. In the U.S., for example, these community- and place-based agencies had emerged from local political organizing, such as by the Black Power and neighborhood democracy movements of the 1960s, to strengthen grassroots control over development and social services, fight displacement of working-class communities by urban renewal projects, and demand an end to financial redlining of Black and brown neighborhoods (DeFilippis 2012). As the federal government cut funding for movement-run development organizations, community development agencies increasingly answered to local governments, banks, and private real estate developers, who reasserted their dominance over urban investment and planning. In many cities, nonprofit community development corporations struggled to fill the vacuum left by federal and municipal disinvestment in public services and affordable housing.

Activism and research have also sought to elevate EJ in the sustainable development paradigm, defined initially as development "that meets the needs of the present without compromising the ability of future generations to meet their own needs" (World Commission on Environment and Development 1987, 16). Proponents of sustainable development call for a green economy, which promises to raise environmental protection to a top priority, coequal with economic vitality and social inclusion. A green economy is defined as one "that results in improved human wellbeing and social equity, while significantly reducing environmental risks and ecological scarcities" (United Nations Environment Programme 2011, 2). After the 2008 financial crisis, green economy ideas influenced national stimulus plans in countries such as the U.S., South Korea, and Ethiopia, and policy discussions in intergovernmental bodies such as the World Bank, United Nations Environment Programme, and World Business Council for Sustainable Development. Proposals included private and public investment in ecosystems management, waste management, clean technologies, renewable energy, green cities, and sustainable agriculture (Affolderbach and Krueger 2017). Other policy levers included incorporating the value of ecosystem services into economic policy decisions; green subsidies, ecotaxes, and pricing strategies that encourage environmentally and socially beneficial shifts in consumption and investment; and regulations that foster technology innovation and diffusion (Fiorino 2017).

Some local development agencies have incorporated sustainable development themes in their planning (Wheeler and Beatley 2014). They have pursued changes to the built environment—such as denser housing, more and rehabilitated green spaces, and transit-oriented development—to reduce greenhouse gas emissions and improve residents' health and quality of life. In addition, these measures are often promoted on the basis of the "green jobs" they will create for local residents. Local governments and universities have also fostered ecopreneurialism, or initiatives to incubate start-up companies that focus on improving environmental performance (Levenda and Tretter 2020).

However, the more that sustainable development policies bend toward the logic of the market and the priorities of the state, the less likely they are to fulfill EJ goals of distributing environmental benefits more equitably, recognizing alternative worldviews of humans and nature, and democratizing control over decision making. Sustainable development programs led by states and intergovernmental agencies have often promoted extractive and exploitive growth strategies at the expense of local ecosystems and equity—for example, by imposing expensive and culturally damaging megaprojects on local communities, supporting conservation schemes that deny Indigenous people access to their ancestral lands, and sacrificing the interests of future generations and nonhuman nature for short-term growth (Agyeman, Bullard, and Evans 2003; Atapattu, Gonzalez, and Seck 2021). In cities, many development plans touted as advancing equity have ended up spurring speculative investment that reinforces patterns of economic disenfranchisement (Campbell 1996). EJ researchers and activists increasingly question whether efforts to build more sustainable urban communities are displacing residents rather than ameliorating urban planning's consequences (Agyeman et al. 2016; Anguelovski 2015). Community gardens, farmers markets, bike lanes, and other public and private investments can also fuel gentrification and ecotourism that displace local people, overtax local ecologies, and increase transportation-related greenhouse gas emissions.

CER's values and methods inherently question who has knowledge, whose knowledge is valued, and who gets to decide on the scope, scale, purpose, and process of research. Participatory development emerged as an approach among those who acknowledged that development projects conceived and imposed from above had often failed to reduce poverty and inequality, and to distribute environmental benefits fairly and democratically. At its best, participatory development engages grassroots communities to develop projects, technologies, and organizations that respect local self-determination and cultural specificity (see Levidow and Papaioannou 2018; Pansera and Sarkar 2016; and the example of Barefoot College discussed below). At its worst, this approach is coopted by state and intergovernmental organizations to make false promises of influence to communities, design manipulative and time-consuming consultation processes, and use them to legitimate decisions foisted on communities from above (Cooke and Kothari 2001).

Alternatives to Development

Movements and researchers concerned with EJ have proposed a variety of alternative approaches to the economy and relations with nature that are best understood as alternatives *to* development. These approaches share some common themes, including local control, culturally appropriate economic relations and technologies, greater equity, reduced consumption, and liberation of people and nature from exploitation. For example, the founding *Principles of Environmental Justice*, adopted at the First National People of Color Environmental Leadership Summit (1991), called for "securing our political, economic and cultural liberation that has been denied for over 500 years of colonization and oppression" by promoting "economic alternatives which would contribute to the development of environmentally safe livelihoods"; economic self-determination; "the right of all workers to a safe and healthy work environment"; "oppos[ition] to the destructive operations of multi-national corporations"; and "consum[ing] as little of Mother Earth's resources and [producing] as little waste as possible." Around the world, a host of alternative visions to development address EJ (see table 8.2).

TABLE 8.2. Alternatives to Development Relevant to Environmental Justice

Approach	Definition
Buen Vivir (Latin America)	An umbrella term for multiple Indigenous life philosophies, *buen vivir* (living well) encompasses "harmony with nature (as a part of it), cultural diversity and pluriculturalism, co-existence within and between communities, inseparability of all life's elements (material, social, spiritual), opposition to the concept of perpetual accumulation, return to use values and movement even beyond the concept of value" (Kothari, Demaria, and Acosta 2014, 367–368)
Ecological Swaraj (India)	Also known as radical ecological democracy, ecological swaraj ("self-rule" or "self-reliance") is a framework that emerged from local civil society organizations, which "respects the limits of the Earth and the rights of other species, while pursuing the core values of social justice and equity"; embraces direct, grassroots economic democracy; and has a "holistic vision of human well-being encompass[ing] physical, material, socio-cultural, intellectual, and spiritual dimensions" (Kothari, Demaria, and Acosta 2014, 368)
Ubuntu (Southern Africa)	Ubuntu philosophy recognizes the communal constitution of identity ("I am because we are") and prescribes an ethics of caring for other humans and natural beings, and restrained use and sharing of natural resources (Etieyibo 2017)
Degrowth (Europe, North America)	Degrowth rejects the goal of economic growth as destructive, calling instead for "a democratically led redistributive downscaling of production and consumption in industrialized countries as a means to achieve environmental sustainability, social justice, and well-being" (Kothari, Demaria, and Acosta 2014, 369)
Social and Solidarity Economies (Europe, North America)	These grassroots economic initiatives foster cooperation, mutual aid, relationship building, local self-reliance, and environmental sustainability, such as community development credit unions; land trusts; urban gardens, community-supported agriculture, and worker, consumer, and producer cooperatives (Miller 2009)
Regenerative Economy (United States)	A vision of the economy embraced by the U.S. EJ movement and labor allies, the regenerative economy includes restoring nature, local economic control and democratic workplaces, respect for local cultures and traditions, and racial and economic equity (Movement Generation Justice and Ecology Project, n.d.)

Rooted in community knowledge, experience, and questions, a community-engaged approach to research aligns well with these alternative visions. Like these approaches, CER helps endow community organizations with resources to conduct research (distributive justice), promotes communities' participation in directing and conducting research for their own ends (procedural justice), draws on their local knowledge and values (recognition justice), and can help create long-term research and economic relationships that are locally controlled, democratically governed, and in harmony with nature (transformational justice).

Having provided this broad overview of how EJ relates to mainstream development, we turn now to highlight four emergent strategies for using CER to advance economic alternatives, including efforts to re-localize economies, rethink worker education and training, create just transitions to renewable energy economies, and promote community ownership of production. These strategies are not mutually exclusive nor exhaustive of the alternatives to dominant economic approaches. In addition, the examples of CER we present vary in the degree to which they challenge neoliberal and sustainable development models. Our aim is to illuminate the strengths and limitations of how these efforts employ CER to enact EJ, rather than provide simplistic "success stories" or "replicable models."

RE-LOCALIZING THE ECONOMY

Some approaches to CER for EJ advance economic alternatives by drawing on local knowledge and values to promote local production of goods, services, food, and energy. These re-localization initiatives are not new—they have taken various forms since the emergence of industrial capitalism's corporate takeovers (North 2010a). Many recent re-localization efforts have also emerged in response to the precarious working and economic conditions created by neoliberalism and globalization. While a few re-localization models pursue complete self-sufficiency, most efforts promote diverse and connected localized economies where residents can equitably benefit from and democratically participate in deciding trading terms. To achieve these goals, communities have organized their own currencies, alternative exchange models, and community banks (North 2010b), as well as local supply chains, as discussed below.

The re-localization of the economy stands in opposition to neoliberalism's intensification of globalized processes of production based on profit maximization and economic efficiency principles. Neoliberalism has involved the relocation of manufacturing jobs to places where the state fails to impose regulations that protect communities from industrial environmental harms and exploitative labor practices. Globally, free trade policy supports this approach, by doing away with tariffs, quotas, and other restrictions. In re-localization efforts, regulation is a powerful way to support multiple economies, as opposed to one global economy (Cato 2006).

CER efforts for re-localizing the economy have highlighted and created a stronger case for local production and consumption. In Oregon, for example, researchers at Portland State University partnered with the local makers movement and the Portland Made collective. The partnership generated information on the economic impacts of artisans and makers through survey and interview methods. Findings also highlighted the diversity and complexity of Portland's artisan and maker community (Heying and Marotta 2016). Seeking to push back against valuing local economies strictly in monetary terms, the New Economics Foundation based in London collaborated with trade associations and university researchers to conduct research on the social and cultural value of traditional retail markets. The research project built on the trade associations' long-standing role of advocating for traditional retail markets that provide relatively healthy and affordable products to historically marginalized communities (Bua, Taylor, and Gonzalez 2018).

Case Study: Cleveland's Anchor Institution Framework

A CER partnership between the Democracy Collaborative, a nonprofit think tank and research center, and the Cleveland Foundation provides a valuable case study of strategies to re-localize the economy. In 2007, the foundation invited the Democracy Collaborative to discuss community-based approaches to building wealth in Cleveland's Greater University Circle, an area that includes large educational and medical institutions, as well as predominantly Black communities that have experienced persistent poverty and disinvestment (Dubb and Howard 2012). Based on interviews with 200 community members, the research highlighted the city's University Circle area wealth and $3 billion in procurement spending (Wright, Hexter, and Downer 2016). The identification of these resources and influence prompted the Democracy Collaborative, community-based organizations, civic leaders, and residents to brainstorm ways to establish sustainable cooperative business models that created and kept wealth in the local community. This led to the creation of Evergreen Cooperatives, a group of worker-owned green businesses, to "[focus] on economic inclusion and building a local economy from the ground up" (Dubb and Howard 2012, 10).

This "Cleveland model" of linking worker-owned cooperatives to achieve market scale and viability took hold. Evergreen's first three cooperatives were the Evergreen Cooperative Laundry, Ohio Cooperative Solar, and Green City Growers (at the time the nation's largest hydroponic greenhouse). Each has been operating for more than ten years, and together they employ approximately 100 people, 40 percent of whom are from the Greater University Circle neighborhood (Howard and Camou 2018). University Circle area universities and medical institutions have committed to long-term contracts with the cooperatives, strengthening their long-term viability, increasing their access to credit, and enabling economic re-localization. Employing workers and making them company owners allows capital to flow and stay in the local community, and large institutions are able to source locally.

Another key outcome of these efforts is evident in the Democracy Collaborative's role in conceptualizing and disseminating an anchor institution framework. Anchor institutions are large organizations rooted in place, like universities, hospitals, cultural institutions, and municipal governments that have the potential to shape local markets (Dubb and Howard 2012). The Democracy Collaborative has disseminated the framework, including insights on implementation and potential roadblocks, through toolkits, online resources, and capacity-building support services that cover workforce and inclusive hiring; purchasing and inclusive local sourcing; and investment and place-based investment (Koh et al. 2020). They emphasize that community-wealth building is necessarily participatory; it requires "stakeholder mobilization" and local interpretations of community wealth (Howard and Camou 2018, 280). Long-term collaborative structures with historically marginalized communities can preempt power imbalances that elite institutions—including anchors themselves—can generate. Many anchor institutions have since advocated for and adopted aspects of the framework, contributing to a growing understanding of challenges and possibilities involved in re-localization efforts.

EDUCATION AND TRAINING

Increasing the knowledge, skills, and capacities of workers or organizational members is central to alternative economic models. Building the capabilities of people based on community members' existing knowledge and local priorities challenges corporate models in which technical experts with little connection to a place drive decision making. Ongoing education and training of workers has the goal of ending communities' reliance on harmful work conditions in extractive industries, which contribute to the morbidity and premature death of members from historically marginalized communities (Pollin and Callaci 2019). In industrial capitalism, workers often face unhealthy and unsafe environments, including exposure to toxic chemicals, dangerous equipment, or overwork—all at little pay. Workers often have limited access to work opportunities beyond their current positions, or they may experience employer harassment and intimidation.

CER for EJ has brought attention to the environmental risks workers face, as well as the critical role of education and training in improving work environments and communities. Academically based labor centers in the U.S., for example, have collaborated with unions to conduct research on worker protections, skills, and education. For example, Cornell University's Worker Institute leads research collaborations that emphasize workers' rights and collective representation. The Institute's Labor Leading on Climate Initiative advances the leadership role of labor in responding to the environmental and climate crisis through a wide range of activities, including research, leadership development, and technical assistance. Among their initiatives are climate jobs studies pursued in partnership with labor unions in different states, as well as workshops to train workers to advocate for expanding the clean energy sector while providing unionized, well-paying jobs. In 2021, this work helped a labor coalition to persuade the New York state legislature to pass wind power subsidies and a first-in-the-nation set of labor standards for construction of clean energy projects, including prevailing wage requirements for workers (ILR Worker Institute 2021).

At the University of Texas, Austin, a CER project by Miriam Solis in collaboration with EcoRise, an environmental education organization, aims to spur educational and career pathways in the green building field with and for youth of color (Solis, Davies, and Randall 2022). The project employs critical race theory and the concept of community cultural wealth (Yosso 2005) to engage youth participants' understanding of and connection to place. By identifying their concerns about, priorities for, and ideas regarding environmental injustices in their neighborhoods, the education programming is being used by participating youth and city officials to assess whether local climate action plans are responding to these insights.

Case Study: Barefoot College

Barefoot College, founded by Bunker Roy in 1971 in Rajasthan, India, is one of the best-known institutions in the Global South that takes a participatory approach to workforce development, emphasizing local knowledge and self-reliance.[1] Barefoot provides an alternative to top-down worker training programs that rely on external experts and resources, which typically impose statist and corporate development paradigms on people in poverty. We highlight this example *not* because its success stems from partnering with research institutions, but because it has become a closely studied model for bottom-up participatory development, which could inform how university-community collaborations might recognize local knowledge and capacities more fully.

Barefoot College's inclusionary model is founded on the idea of bricolage—making the most of what is at hand (Westley 2013). Barefoot's use of bricolage places an emphasis on both appropriate technology and human capital. It values using resources already available to the community, rather than seeking external resources, and respects local knowledge and empowers local people in order to make progress. Barefoot prioritizes recruiting women and low-caste people as employees to elevate their financial status and subsequently their well-being. Common barriers to success, such as the inability to read and write, are not barriers to success at Barefoot, where very poor and illiterate women have become successful water engineers, solar engineers, designers, architects, and so on (Roy 2011).

Barefoot's model has four key components: alternative education, valuing traditional knowledge and skills, learning for self-reliance, and dissemination (Roy and Hartigan 2008). Its approach to education aligns with Mahatma Gandhi's philosophy, championing practical skills and traditional knowledge while emphasizing a humble way of life—everyone works, sits, and eats on the floor. The college teaches students to unlearn the significance of degrees and qualifications by placing no importance on them and instead underlining the importance of traditional knowledge and skills, which teaches the villagers to value the skills they already possess and thus serves to empower them. Through providing learning opportunities that enhance villagers' ability to serve their communities, Barefoot bolsters their confidence and self-reliance. When illiterate or semi-literate villagers are trained to be accountants, educators, and engineers by Barefoot, they learn that certifications and degrees are not a requirement to do these jobs successfully, which improves the self-sufficiency of the community as a whole. Barefoot has disseminated its

1. Thanks to Skyler Kriese for research assistance and for writing an initial draft of this case study.

model across India and throughout the world with a few nonnegotiable core principles: the absence of hierarchy, a living wage, and collective decision making (Roy and Hartigan 2008).

Other factors contributing to Barefoot's success include its ability to build social capital, mobilize local resources, and achieve financial sustainability (Kummitha 2017). Providing a platform to the excluded creates social capital through consistent interaction among community members, which has opened up doors to new development. Barefoot's evolution from a voluntary organization to a social enterprise has facilitated its mobilization of local resources to become self-sustaining. It increases its internal resources by securing funds from the communities in which it works, and it acquires resources from external sources and agencies that adhere to Barefoot's principles. Employing this approach helped Barefoot to have a transformative impact in a short amount of time in its home district of Rajasthan and beyond. CER projects that want to take a participatory approach to developing education and job training can learn a great deal from Barefoot about how to institutionalize local participation and respect for local knowledge, and spread a model of education and training to involve large numbers of workers around the world.

JUST TRANSITION

The concept of a just transition has been central to shaping community, worker, environment, and climate-centered economic development models. In current research and practice, just transition strategies typically refer to the pursuit of decarbonization in ways that mitigate and redress the inequities experienced by people whose lives are dependent on a fossil fuel economy and/or who lack reliable access to energy supplies (Newell and Mulvaney 2013; Mascarenhas-Swan 2017). The concept of just transition emerged from the labor movement and broadened as labor-community alliances formed in the EJ movement. For example, the Oil, Chemical and Atomic Workers union leader Tony Mazzocchi called for a "just transition," to provide "a new start in life" for workers threatened by environmental policies (Córdova, Bravo, and Acosta-Córdova 2022; Labor Network for Sustainability and Strategic Practice Grassroots Policy Project 2017). With a focus on protecting needs of workers in the transition away from the fossil fuel economy, the just transition framework addresses the potential harm caused by decarbonization, while remaking the economy in ways that prioritize the well-being of the environment and the people who live there (Cha et al. 2019). These efforts focus especially on places that industries have exploited through extractive practices, including the degradation of the natural environment and poor working conditions (Newell and Mulvaney 2013). Environmental justice requires not just prohibiting these harmful practices, but making these places front and center in both building a post-carbon future and enabling self-determination. Just transition approaches challenge the tenets of neoliberal economic development by bringing attention to how a green economy can reinforce the fossil fuel–based economy's patterns of social, political, and economic disenfranchisement.

CER efforts for a just transition involve identifying and prioritizing the groups most affected by climate injustice and a changing economy. In the U.S., for example, the Alliance for Appalachia is building on a long local history of using participatory action research methods to document the region's extractive and exploitative coal industry, as well as to envision a post-carbon future. It is "building power through knowledge" (Tarus, Hufford, and Taylor 2017, 156) by asking, "Who owns Appalachia?" through multiple projects, including one that gathers data on self-bonding practices that enable coal corporations to eschew their commitment to reclaim lands they have damaged. Alliance for Appalachia members used the information to advocate for a rule change on the corporate bonding practices. (For urban planning research projects guided by just transition goals, see chapter 11.)

Case Study: Black Mesa Water Coalition

The case of organizing and economic development in Black Mesa, Arizona, presents an example of just transition strategies in the face of powerful institutional entrenchment, informed by research conducted largely by lay experts. The Black Mesa Water Coalition (BMWC) was founded by Indigenous (Navajo and Hopi) and Chicano students at Northern Arizona University to address exploitation and extraction by coal mining and the impact on water supplies in Navajo land (Liu 2010). In the early 2000s, 12 extracting industries were operating in the Navajo territories, including Peabody Energy Corporation, the largest coal mining company in the U.S. (Smith and Black Mesa Water Coalition 2007). The coalition grew from a student-run organization to a broader coalition of organizations rooted in communities affected by mining, collaborating to build power to transform the fossil fuel economy in ways that benefit the Black Mesa and Indigenous communities. Recognizing that many local people relied on hazardous mining jobs in a region where the unemployment rate was 48 percent, the coalition's strategy centered alternative employment opportunities for community residents and workers. Through local and regional campaigns, BMWC developed solar and wind projects and created the Navajo Green Economy Coalition, which organized a green jobs campaign that sought to change the Navajo economy and in the process democratize tribal government (Liu 2010).

Recognizing the impacts of coal mining were not just environmental and economic but spiritual and cultural, the campaign began by "translating green into Navajo" (Curley 2018, 61) in an intentional effort to generate a community-informed conceptualization of green priorities. The BMWC collected and synthesized this information by leading dialogue circles with Navajo Nation chapter presidents, reservation residents, college students, and allies from other organizations to discuss what a green Navajo Nation could look like and how the Navajo Nation could transition its economy. Through these efforts, participants and coalition members decided that good green jobs for the Navajo Nation would mean adopting traditional ways of subsistence, such as green manufacturing through wool mills (Curley 2018; Liu 2010). The campaign also helped identify and realize several policy and governance changes needed to activate a participatory and green economy, resulting in the passage of green jobs legislation through the Navajo Tribal Council, the first green economy legislation passed by any tribal government (Chorus Foundation 2014). By 2020, the Navajo Nation produced enough solar energy to power its territories and much of the Southwest (LaDuke and Cowen 2020).

The BMWC and its members identified several important lessons from their research. Among them is the importance of "[tailoring] the definition of green to your community" (Liu 2010, 14). What constitutes good green jobs are thus contextual and ought to reflect local history, concerns, and priorities. Nonetheless, Curley (2018) points out that despite the BMWC's efforts, its forward-thinking conceptualization was appropriated by tribal governments and extractive industries to justify a simultaneous reinvestment in coal production as part of the Navajo Nation's energy transition plan. This aftermath provides an important lesson on how transitions are nonlinear and require a complete detachment from oil and coal companies (Curley 2018).

COMMUNITY-BASED OWNERSHIP

CER for economic development has also informed the pursuit of community governance over land and the means of production. Local control and governance are alternatives to corporate ownership models that commodify natural resources for private gain. Communities have pursued alternative organizational models that reflect these principles. For example, in worker-owned cooperatives, members participate in decision making and equitably share its benefits. In community land trusts (CLTs), a nonprofit organization holds land "in trust" to support the community; the land can be used for many purposes, including housing and agriculture, and land is kept affordable in perpetuity by removing it from the speculative market (Axel-Lute 2021). CLTs emerged through civil rights activism in the 1960s to promote asset ownership (Meehan 2014). Another strategy includes community benefits agreements whereby communities negotiate and secure social uses for private development. To varying degrees, these strategies enhance local decision-making power in community economic development matters. Mascarenhas-Swan (2017) points out that "[w]hile solutions will be applied locally, communities' ability to wrest control of the economy from the current governing forces requires these local communities to band together in ways that build movement muscle." Community-based ownership thus requires movement building across places and issues.

Local organizations have used CER to set their own priorities for advancing community-based ownership and to influence public investment. For example, a nonprofit collaborative of hospitals formed the One Brooklyn Health System to create an integrated healthcare system for the borough's predominantly low-income central and northeast neighborhoods. The system employed participatory action research (PAR) to design its community health needs assessment and community services plan processes every three years. Three rounds of PAR conducted in 2016–2018 included surveys, focus groups, and interviews, some of them led by local youth, in collaboration with community organizations and academic partners (the MIT-affiliated urban planning consultancy NextShift Collaborative, Pratt Institute, and University of California, Berkeley). These studies identified local priorities and developed a holistic analysis of the neighborhoods' assets and needs. The resulting recommendations informed a $1.4 billion state investment in community-based healthcare infrastructure, but also in affordable housing and other local capital improvements the community now recognized as necessary to address social determinants of health (One Brooklyn Health System 2019).

Case Study: Jackson Cooperative's Community Production Initiative

The Cooperation Jackson network in Mississippi is challenging capitalism and white supremacy by advancing a model of eco-socialism. The Black-led network owns 7.4 acres of land where members operate a community land trust on reclaimed and repurposed areas. The CLT is the site of worker-owned cooperatives focused on urban farming, cooperative housing, and sustainable energy (Akuno 2017). The network of cooperatives supports new cooperative conceptualizations, formalization, and growth, and it runs worker education and training programs.

Among the cooperatives is the Center for Community Production (CCP), an initiative to democratize the ownership, control, and use of technology. The worker-controlled small-scale manufacturing center opened in 2019 (Cooperation Jackson 2019). Its worker-owners are identifying production needs via community engagement and local market and industrial production trends. Among the CCP's projects is the Ewing Street Eco-Village Pilot Project to create sustainable urban housing. The CCP collaborated with City College of New York's Advanced Design Studio to generate prefabricated modular home concept designs to inform the final project (Bagchee 2019). The research-practice partnership's final report was also an educational tool on collaborative design. The CCP will grow to have a commercial manufacturing division, a training center, and an innovation hub that prototypes products, such as toys, tools, and medical equipment (Akuno 2017).

The CCP's engagement and analysis of new information from within the cooperative network and outside of it reflects the relevancy and influence of research activities for radical community development. Large corporations generally drive and own technological innovation; even "open source" approaches systemically exclude the priorities and concerns of community members. The CCP provides an alternative to exclusionary expert-driven approaches to sustainable development by offering a "radical vision of technoscientific practice" (Ludwig 2021, 12). In addition, these efforts aim to head off anticipated job losses due to automation that threaten communities of color. The project adapts technological innovation to establish new social freedoms for Cooperation Jackson members and collaborators.

SCALING UP, OUT, AND DEEP

CER is playing a critical role in advancing the economic alternatives EJ leaders have identified as necessary for healthy, regenerative communities. The strategies discussed here—re-localization, education and training, just transition, and community-based ownership—point to several important lessons. Among them is the potential role of large educational or independent research institutions. In the case of the Cleveland model, the Democracy Collaborative was a critical convener. It brought resources to the local effort, including funding, staff with formal training in research design and methods, and the capacity to disseminate information in wide-ranging formats and to multiple audiences. It also leveraged the power of anchor institutions' purchasing policies to support local producers and EJ. Research institutes also often adhere to ethics protocols for conducting research with historically marginalized communities. In order for these institutes to be effective, however, they must have the institutional flexibility required to follow the lead of community-identified priorities. Researchers are often limited in their ability to grant control and oversight to communities, due to administrative constraints and their own underlying logics of efficiency and productivity. CER on the social economy also needs to pay heed to how change can be achieved, including an understanding of existing governance systems and strategies based on the issues that most matter to communities (Downing 2009).

However, large institutional research partners are not always necessary. Communities can design, implement, and synthesize research on their own accord. For example, the BMWC met with community members to translate *green* into Navajo, and Cooperation Jackson members worked with residents to conceptualize the CCP's focus. These efforts certainly reflect participatory approaches to problem identification and deliberation (Forester 1999), but are they "research" in a traditional sense? On the one hand, we must question the imposition of a "research" classification; in these examples, members may not have used this designation. Communities might reject such a descriptor for their information-gathering efforts, due to research's exploitative and extractive role in communities (Tuck and McKenzie 2014). On the other hand, the research questions and methods communities are using increasingly resemble those used by professional researchers, often in combination with normative theoretical frameworks that challenge structures of power. In addition, community-based organizations often build their own capacities by hiring personnel dedicated to research and writing. We seek to elevate these critical community-led efforts as contributions to larger bodies of knowledge and as examples of the potential that comes from justice-oriented approaches to research.

CER that advances economic alternatives also disrupts conventional notions of how we define and measure economic well-being. Credentialed experts from the fields of economics, urban planning, and public policy often conduct evaluation and cost-benefit analyses to assess economic performance (Hufschmidt et al. 1983). These approaches can be useful when they reveal disparities among groups; communities often use this information to create a basis for stronger regulatory action. However, the efforts for economic alternatives and liberation discussed in this chapter are presenting new forms of understanding what an economy is. As LaDuke and Cowen (2020) point out, if the economy is "how we live," we cannot separate economic from social and environmental well-being. Similarly, Movement Generation Justice and Ecology Project (n.d.) uses economy's root words to extrapolate that its meaning is "management of home," requiring that we tend to the web of relationships, or ecosystem, within which our home is nested.

At the same time, this definition of economic well-being points to a central challenge for future CER for EJ: bringing isolated examples to scale. Recent research and practice emphasize the need for community economic alternatives to expand their impact across three different scales (Moore, Riddell, and Vocisano 2015). Often, local initiatives seek to *scale out* by extending the size of their organizations and the reach of their solutions to more places and people, as Barefoot College has done. However, many local practices cannot be replicated or diffused mechanically across diverse communities, and many efforts to scale out produce incremental change that is too slow and piecemeal to reach large populations. Thus, we also need strategies for *scaling up* by influencing laws, policies, and institutional practices that can help spread innovations faster and farther, for example by partnering with social movements to influence governments and intergovernmental agencies. Additionally, community economic innovators can aim for *scaling deep* by shifting underlying cultural norms, beliefs, and narratives, forming new relationships within civil society that can create more profound transformations for equity and sustainability over time. Even less is known about how to scale up and deep than about how to scale out (Moore, Riddell, and Vocisano 2015). It would be especially useful for teams of engaged researchers and community-based partners to illuminate these three processes of scaling through cooperative research with the many players involved—innovators, movements, legislators, and so on—across multiple sites. Otherwise, promising local economic alternatives may remain vulnerable to resistance from outside, above, and within.

9

Public Health

Ryan Petteway, R. David Rebanal, Chad Raphael, and Martha Matsuoka

The field of public health has made major contributions to community-engaged research (CER) for environmental justice (EJ). This is especially true of research that takes a population health perspective, as opposed to clinical, behavioral, or biomedical approaches. Public health has a deep and rich history of engaging matters of social and health equity at the community and population levels, especially as related to racial, class, and place-based environmental inequities. As this chapter shows, public health research is well positioned to address EJ issues because of the field's practical commitments to applying and translating research for social action and policy change. In addition, public health researchers' leadership in developing community-based participatory research methods has influenced CER in many disciplines.

This chapter summarizes some important ways in which CER for EJ has emerged from public health. We begin with an overview of the recently updated 10 Essential Public Health Services, a framework that puts health equity at the center of the field. We present an overview of the core areas of public health research and practice that have especially advanced CER for EJ: community-based participatory research, social epidemiology, place-health research, and health impact assessments. For each area, we summarize core conceptual and procedural groundings, citing some of the key literature and exemplary studies. Next, we identify three broad directions public health research can take to strengthen CER to advance EJ. These directions include engaging more explicitly and purposefully with antiracism and decolonizing praxis and principles; redefining what counts and gets counted as "environmental"; and centering notions of placemaking and power in the (re)production of spatialized and racialized environmental injustices. Table 9.1 shows how the chapter's major themes relate to the dimensions of justice common to CER and EJ.

TABLE 9.1. CER for EJ in Public Health

Dimension of Justice	In CER for EJ in Public Health
Distribution *Who ought to get what?*	Centering health equity by addressing the structural determinants of health, and their roots in historic and ongoing environmental injustices
Procedure *Who ought to decide?*	Strengthening EJ communities' participation in and influence over community-based participatory research and health impact assessments, and in policy making and practices that affect health
Recognition *Who ought to be respected and valued?*	Expanding antiracist and decolonizing approaches to knowledge and research, and recognizing sociospatial exposures to policing, spatial stigma, White spaces, and Indigenous health as environmental justice issues
Transformation *What ought to change, and how?*	Transforming public health through community-based participatory research focused on health equity, and employing antiracist and decolonizing praxis, to promote environmental justice

CER FOR EJ IN PUBLIC HEALTH

The 10 Essential Public Health Services

As a foundation for public health, the 10 Essential Public Health Services (EPHS) is a particularly relevant framework for CER to advance EJ—especially in the U.S. context, where the EPHS is widely used in public health education and accreditation, is cited in some state statutes, and helps define the field to the public. Federal agencies and public health experts developed the EPHS doctrine in 1994 to help distinguish the work of public health agencies and organizations from health care.

The EPHS framework was updated in 2020 to describe the field of practice more fully, center essential activities around equity, and identify the structural injustices that cause health inequities (see figure 9.1) The influence of community-based participatory research (CBPR) can be seen in the shift from the original framework's focus on the field of "public health solving community problems" (U.S. Centers for Disease Control, n.d.) to the current version's call to "strengthen, support, and mobilize communities and partnerships to improve health" (10EPHSFITF 2020). Equity is now a goal of each of the 10 essential services, from creating community partnerships, to engaging in policy and legal advocacy, ensuring access to health and health care services, and developing a diverse and competent workforce. An accompanying statement highlights the need to "remove systemic and structural barriers that have resulted in health inequities . . . includ[ing] poverty, racism, gender discrimination, ableism, and other forms of oppression" (para. 3). These updates to the EPHS provide a stronger rationale for engaging in CER for EJ, although, as we argue below, the field has more work to do to fulfill this promise.

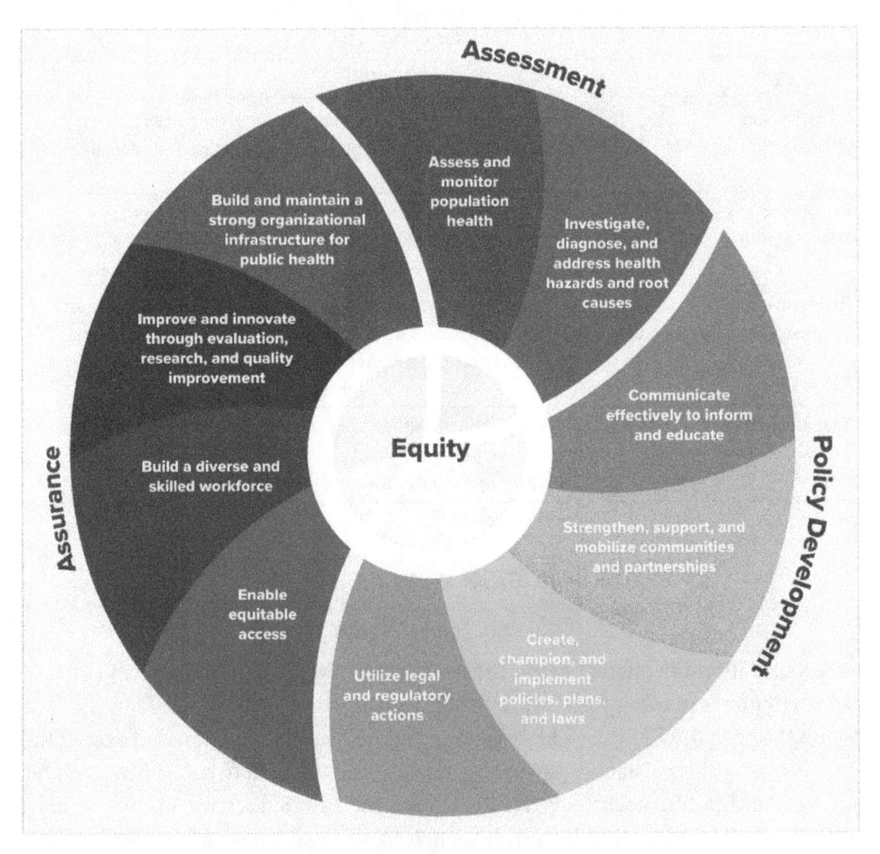

FIGURE 9.1. The 10 Essential Public Health Services framework.
SOURCE: 10EPHSFITF 2020.

Community-Based Participatory Research

Public health scholars developed CBPR to engage community partners in the research process and share power with them, strengthen research with local knowledge, ensure that communities benefit, and produce research that results in meaningful actions through interventions or policy change (see chapter 2). Since the 1990s, the CBPR tradition has been a major contributor to the theory, methodology, practice, and institutionalization of CER for EJ across many disciplines and research topics.

Public health scholars authored textbooks and handbooks that taught community-engaged theory and methods to EJ researchers in many fields (see, e.g., Blumenthal et al. 2013; Israel et al. 2013a; Minkler and Wakimoto 2022; Wallerstein et al. 2017). Researchers trained in public health helped forge an interdisciplinary approach to CER for EJ—individually and in research teams spanning multiple research institutions and community organizations. This research provided evidence used in early EJ struggles in the U.S. by conducting epidemiological studies in fenceline communities and industrial hygiene studies in workplaces (see chapter 1). This EJ research has expanded in scope to address law and policy (see chapter 7), food justice (see chapter 10), and urban planning (see chapter 11). Public health researchers and their community partners have also led reflexive research on the CBPR process itself, advancing understanding of power and justice within knowledge production (e.g., Chávez et al. 2008; Muhammed et al. 2015; Shepard et al. 2002; Wallerstein et al. 2019) and demonstrating the value of CBPR methods for increasing the rigor, relevance, and reach of research (Balazs and Morello-Frosch 2013).

Public health has also played a major role in developing the institutional infrastructure for CER, especially in the U.S. From the 1990s onward, schools and programs of public health launched new curricula, centers, and initiatives devoted to CBPR, built long-term relationships with community partners, and recruited a critical mass of graduate students of color committed to environmental and social justice. Professional associations—from the renowned American Public Health Association to newcomers such as Campus-Community Partnerships for Health—promoted CBPR and promulgated standards for conducting and evaluating this kind of research to increase its acceptance in the field. Health researchers secured foundation and government funding streams for CBPR on EJ from the mid-1990s onward, including a 13-year federal interagency program that supported over 50 CER projects for environmental and occupational health, led by the National Institute of Environmental Health Sciences, Environmental Protection Agency, and National Institute for Occupational Safety and Health (Baron et al. 2009). In the 2010s, the National Institutes of Health, and some state environmental and public health agencies, prioritized funding for CBPR to combat health inequities (Blumenthal et al. 2013).

Social Epidemiology

Social epidemiology scholars and practitioners tend to be less concerned with any one specific disease or illness, or any one specific cause. Rather, they are most interested in explicating how broader societal power relations (re)produce the inequitable sociopolitical, economic, legal, and environmental contexts that structure population distributions and patterns of health and illness (Krieger 2020). Central to much of this scholarship are the health effects of various forms of social exclusion, oppression, and inequality, including, for example, structural racism (Agénor et al. 2021; Bailey et al. 2017), gender inequality and sexism (Borrell et al. 2014), aspects of class inequality (Bor, Cohen, and Galea 2017; Fujishiro et al. 2021; Muntaner et al. 2015), and considerations of intersectionality therein (Agénor 2020; Bowleg 2012).

Critical contributions of social epidemiology relevant to EJ-related research also include work that explicates how these outside social and political exposures "get under our skin" to affect physiological functioning across our lifespans. This research has contributed several key concepts that help to illuminate EJ and health. For example, *allostatic load* is a measure of the cumulative burden of chronic stress and life events, as identified by biomarkers and clinical criteria (Seeman et al. 2010). *Weathering* provides a metric of premature decline in health from the cumulative impacts of experiencing social and political marginalization and economic adversity (Geronimus et al. 2006). The concept of *embodiment* describes the process through which social and physical environmental exposures work their way inside of our bodies, revealing patterns of structural inequality that are built into societal arrangements of power and risk (Krieger 2005; Vineis et al. 2020). *Life course* approaches account for the origins of health inequities by tracing how social, economic, and physical environmental exposures at each stage of human development affect health within and across generations (Gee, Walsemann, and Brondolo 2012; Jones et al. 2019).

Informing much of the work in these areas are broader theories and frameworks that situate health within its wider social, political, and economic contexts and power relations, which fundamentally shape who is exposed to what, and when. Core theories and frameworks for EJ include social production of health and political economy orientations (Harvey 2021; McCartney et al. 2019), ecosocial theory (Krieger 2001), fundamental causes (Phelan and Link 2015), and models of social, macro, and commercial determinants of health (de Lacy-Vawdon and Livingstone 2020; Naik et al. 2019). Non-CER studies informed by these frameworks have explored EJ exposures, often in relation to the broader structural foci of social epidemiology (e.g., structural racism, gender inequality, class inequality). This has included, for example, work demonstrating links between ambient air pollution and racial residential segregation (Jones et al. 2014; Morello-Frosch and Jesdale 2006); air and noise pollution and neighborhood deprivation (Saez and López-Casasnovas 2019); noise pollution and racial and economic segregation (Casey, Morello-Frosch, et al. 2017); neighborhood racial composition and annual exposures to toxic waste emissions (Hipp and Lakon 2010); intersectionality and cancer risks related to air toxics (Alvarez and Evans 2021); neighborhood racial composition, income, and urban greenness (Casey, James, et al. 2017); and neighborhood racial composition, tree canopy, and cardiovascular and respiratory health (Jennings et al. 2019). These currents in social epidemiology have influenced and inspired CER studies of EJ, which can add a valuable complementary approach to the statistical analyses of large data sets mentioned here. Integrating CBPR and social epidemiology offers an especially promising avenue for applying CER to advance EJ, especially when employing a place-health approach to research (Petteway et al. 2019a; Wallerstein, Yen, and Syme 2011).

Place-Health Research

As a subdiscipline of social epidemiology, place-health research focuses on place-based exposures as encountered within specific geographies and sociopolitical spatial contexts, and represents a well-developed area for advancing EJ through CER. This research draws on complementary disciplines—such as human geography, health geography, urban planning—to understand the natural, built, economic, and social environmental contexts of specifically defined places (Arcaya et al. 2016). Often, outside researchers work collaboratively with residents to uncover and address potential EJ-related concerns. The place-focused and environmental-oriented nature of this particular public health work lends itself well to adopting core CER principles for advancing EJ knowledge production and social action. As Petteway, Mujahid, and Allen (2019) discuss, such work can leverage the "practical and procedural translational advantages of much place-based research (e.g., space-bound, locality- and/or jurisdiction-specific), while simultaneously capitalizing on the scientific and political translational advantages of harnessing place-based knowledge, insight, and expertise of the people whose lives unfold within the 'place' being studied" (6).

CBPR in this area has examined issues related to neighborhood food environments (Breckwich Vásquez et al. 2007), parks and greenspaces (Peréa et al. 2019), tobacco environments (Petteway, Sheikhattari, and Wagner 2019), and aspects of neighborhood built and social environments (Petteway, Mujahid, and Allen 2019). Other work has focused on more traditional EJ exposures. For example, Madrigal et al. (2014) worked with Latinx youth in a farmworker community to examine environmental concerns using photovoice. Johnston et al. (2020) worked with youth co-researchers who used multiple participatory methods, including participatory GIS and personal air-monitoring devices to document exposure to airborne particulate matter, while Nolan et al. (2021) completed similar work with youth researchers to study nitrogen dioxide and sulfur dioxide exposures. Other scholars have conducted participatory survey-based environmental research within fenceline communities (Cohen et al. 2012), and survey and water sampling work with residents of a heavily polluted Latinx community (Sansom et al. 2016). This body of work not only offers valuable empirical evidence, but also enhances community participants' agency, strengthens the transparency and accountability of the research to the community, and disseminates the results to residents and leaders in ways that facilitate their efforts to remedy EJ concerns.

Even so, significant conceptual, methodological, and procedural challenges remain for place-health research (Arcaya et al. 2016; Petteway et al. 2019a). Documenting environmental threats may contribute to stigmatizing places and the people who inhabit them (discussed below). This research can also be limited by choosing short-term temporal measures, and narrow and static spatial designations (such as census tracts), that do not adequately measure long-term and cumulative exposures across the spaces people actually traverse. An important response to these problems is to measure environmental exposures across a person's activity space, which includes all of the places they go to, pass through, and encounter on a routine basis. Unlike most exposure-related research that focuses on one spatial location (e.g., air pollution in one's residential neighborhood), an activity space approach can provide a more comprehensive picture of exposures based on people's mobility patterns—between home, work, school, places of recreation, shopping locations, transportation routes, and so on. Park and Kwan (2020) have applied this approach to studying air pollution, while others have applied it to research on noise pollution (Tao et al. 2021), greenspace (Bell 2015), and aspects of local food, alcohol, and tobacco environments (Lipperman-Kreda et al. 2015; Widener et al. 2018). While promising, this activity space work would be greatly enriched in rigor, relevance, and reach by taking a more participatory approach that more thoroughly centers community knowledges, experiences, and spatial perceptions of exposures, and enlists community partners in disseminating the findings and implementing responses.

Health Impact Assessment

Another area of public health that plays a promising role within EJ-related research and practice is health impact assessment, or HIA. HIA is an analytic process and tool developed to generate evidence regarding the potential health harms and benefits of proposed policies, programs, projects, or plans (Harris-Roxas and Harris 2011). Originating in and extending the use of environmental impact statements (EIS) in construction and development projects, HIA is

> a systematic process that uses an array of data sources and analytic methods and considers input from stakeholders to determine the potential effects of a proposed policy, plan, program, or project on the health of a population and the distribution of those effects within the population. (National Research Council 2011, 5)

HIA generally consist of six stages: (1) screening whether the decision-making process can benefit from an HIA, (2) scoping potential health effects of the proposal and parameters of the study, (3) assessment of the health impacts, (4) recommending mitigations and alternatives to protect health, (5) reporting and communication to stakeholders and decision makers, and (6) monitoring decisions and health outcomes (Bhatia 2011). A core feature of HIA is that it can be used to assess any type of policy, program, project, or plan—including zoning, land use, community development, transportation, and housing—and all elements that shape distributions and patterns of place-based environmental exposures, experiences, and opportunities. Ideally, HIAs are completed prior to any final decision making regarding a potentially harmful environmental change, policy, or practice so that potential health impacts are assessed by health officials and policy makers. Thus, by its very nature, HIA is a tool designed to promote EJ by providing evidence to preempt environmentally detrimental actions before they can produce health-harming exposures.

While HIA has been practiced for decades, explicit connections to notions of health equity, racial equity, and environmental and social justice have only become core aspects of HIA work more recently (Buse et al. 2019; Heller et al. 2014), prompting increased community engagement and centering community knowledge(s) within all assessment activities. While much HIA work has focused on topics like transportation and housing (Cole, MacLeod, and Spriggs 2019; National Center for Healthy Housing 2016), applications have evolved to examine a more expansive range of EJ-related topics, including racism, community policing, and mental health (Human Impact Partners et al. 2015), and tobacco licensing (Upstream Public Health 2015).

While HIA has done well to advance EJ in public health, HIA remains relatively limited outside of academic and university-led contexts. For example, in a review of all documented HIAs conducted in the U.S. between 1999 and 2020, Petteway and Cosgrove (2020) found just 71 of 2532 (3 percent) in which local health departments served as a lead or authoring partner—suggesting that public health has far to go in making HIA part of routine practice to advance EJ. HIAs can also expand community participation by welcoming local organizations and residents more fully into the research process.

RE-(EN)VISIONING CER FOR PUBLIC HEALTH

Public health—especially through the prism of place-health research—can further embrace and refine CER principles and praxis to advance EJ in three ways. First, building upon the complementary conceptual groundings and goals of CBPR and social epidemiology, we call for deeper engagements with *antiracist and decolonizing praxis and principles*. Second, we encourage deeper, more deliberate and explicit engagement with *placemaking and power* in historic and present processes and practices that make, unmake, and remake our daily place-health contexts. Third, we invite reflection and dialogue regarding *what counts as "environmental"* within EJ-related work in public health, briefly highlighting some promising areas that deserve closer attention.

These directions amplify strengths of place-health research by deepening engagements with notions of power, inclusion, and representation within knowledge production processes—re-(en)visioning place-health research as a site of resistance, contestation, and transformation to change embodied contexts and consequences of environmental injustices. Moreover, public health research needs to engage more fully with the *theories* mentioned here, which may be widely known but are not yet deeply practiced. Faced with pressures to conduct ever more empirical research, while appearing to address pressing issues of justice and community participation, empirical researchers can be tempted to poach theoretical concepts and apply them shallowly. In the mid-1990s, Green et al. (1996) issued a similar critique of the co-optation of participatory research by many studies that failed to develop substantive community partnerships and co-conduct research on equal and mutually beneficial terms. Today, we see the need for a comparable reckoning with antiracist, decolonizing, and EJ theories, to achieve a more deeply transformed focus and practice of CBPR in public health, rather than a hurried and transactional relationship to these theories. The mid-1990s critique led funding agencies and others to adopt stronger and more specific requirements for community participation in health research, and we hope that the kind of thorough reflection that we can only sketch out here will prompt a similar response.

Engaging Antiracism and Decolonizing Praxis

While CBPR researchers have considered racism and power dynamics within research collaborations (e.g., Chávez et al. 2008; Muhammad et al. 2015; Wallerstein et al. 2019), public health can move further towards a CBPR that centers antiracist and decolonizing praxis and principles. We noted earlier that revised EPHS implores the field to address structural inequities and their causes. As Alang et al. (2021) write, dismantling the upstream barriers to delivering essential public health services "requires building alliances across systems to address the range of social determinants of health caused by White supremacy" (818). This much-needed reckoning can be oriented by frameworks such as Ford and Airhihenbuwa's (2010) articulation of a public health critical race praxis (PHCRP) and Alang and colleagues' (2021) explication of strategies for how the EPHS can contribute to dismantling White supremacy. Each draws from critical race theory and merges it with theories and concepts from social epidemiology. While the entirety of these frameworks demands concentrated attention from the field, several elements are particularly relevant to CER for EJ.

Most broadly, these frameworks call for opening avenues of "disciplinary self-critique"—understood as "the systematic examination by members of a discipline of its conventions and impacts on the broader society" (Ford and Airhihenbuwa 2010, 1394). Alang et al. (2021) recommend many strategies to this end, including the need for the field to incorporate critical race theory and antiracist methodologies across the public health curriculum, and set measurable goals for faculty and student racial equity competency. These are certainly prerequisites for faculty and students who plan to do CBPR, along with learning to assess their own individual, institutional, and disciplinary positionality in relation to the community (see chapter 3). Public health can also prioritize research and policy development that explicitly targets indicators of White supremacy and structural racism (Adkins-Jackson et al. 2021; Hardeman et al. 2022, Agénor et al. 2021).

Another core principle is honoring "voice"—that is, "prioritizing the perspectives of marginalized persons"—to enable the (co)production and inclusion of new knowledges (Ford and Airhihenbuwa 2010, 1394). This must extend beyond the traditional practice of including community "voice" on advisory boards, to more intentionally and thoroughly "center the margins" within all aspects of EJ research and knowledge production. Public health can also "ensure equitable allocation of resources and redistribution of power in community partnerships" (Alang et al. 2021, 816) by moving from models in which community organizations are junior partners toward fully collaborative and even community-owned and community-led approaches (see Wilson, Aber, et al. 2018).

Taken together, principles of "voice" and "disciplinary self-critique" can help bring techniques of *counter-storytelling* and *counter-mapping* into the fold of CBPR, policy, and public communication for EJ (see chapter 6). As Delgado (1989) explains, counter-stories "can show that what we believe is ridiculous, self-serving, or cruel . . . can show us the way out of the trap of unjustified exclusion . . . [and] can help us understand when it is time to reallocate power" (2415). Counter-mapping "challenge[s] dominant ways of conceiving the landscape and the socio-political interests they represent" (Willow 2013, 872). These approaches are both destructive *and* productive: they help us to interrogate and dismantle narratives that curate and incubate exclusion and oppression, and (re)imagine and act to pursue just and anti-oppressive alternatives. For example, these approaches can reframe the structural determinants of environmental health as the product of ongoing colonization, racism, and exploitation, rather than individual genes, lifestyles, and bad fortune.

This capacity for counternarratives could enable deeper engagement with the PHCRP principle of "social construction of knowledge"—referring to "the claim that established knowledge within a discipline can be re-evaluated using antiracism modes of analysis" (Ford and Airhihenbuwa 2010, 1394). And in this regard, public health researchers working on EJ projects would do well to reflect more on Smith's (2021) work on decolonizing knowledge production and curation. Particularly, Smith's reflections on notions of (mis)representation and commodification of knowledge(s), which resonate with PHCRP, offer guidance on how to "unsettle" research power dynamics that often function to silence, erase, or co-opt community knowledges for outsider benefit. Core areas for decolonizing considerations include decisions about which EJ research topics get studied (i.e., who sets EJ research agendas), which methods are chosen and who choses them, which forms of data are prioritized, whose knowledges and perspectives are centered/valued, who owns and/or has access to EJ research data, and who materially benefits most from the research, for example, financially, professionally, socially. In short, decolonizing demands consideration of far-reaching changes in control over research agendas, methodologies, and research ethics, as well as reconciling dominant and traditional ecological knowledges and reconceiving just relations among people and other nature.

Simultaneous with these considerations is the imperative of more expressly and thoroughly orienting CBPR for EJ around *intersectionality*. The ten PHCRP principles emphasize intersectionality within EJ, which requires that researchers not only "center the margins," but *center the intersections*. This means recognizing that varying configurations of overlapping environmental and social oppressions—for example, along race, class, and gender lines—necessitates varying configurations of "voice," methods, and knowledges to be centered within any one specific EJ concern. Engaging the antiracist and decolonizing principles discussed here can help public health researchers become more responsive to EJ scholars who have called for greater attention to matters of intersectionality (Alvarez and Evans 2021; Ducre 2018; Malin and Ryder 2018). Deeper consideration of these concepts should prompt CBPR to pursue new research designs, methods, and forms for communicating results and recommendations.

Centering Placemaking and Power

As discussed above, CBPR-oriented place-health research represents perhaps the best expression of public health research for EJ. However, much place-health research tends to *de*-place EJ relationships, failing to examine how they are rooted in economic, political, and social processes that shape the spatial distributions of environmental risks and opportunities. For example, de-placing research might measure cross-sectional exposure to air pollution but not track historic and present policies and practices related to environmental deregulation, land use, transportation policy, greenspace, and housing. Cross-sectional research that ignores the mechanisms and manners through which place is actively made, unmade, and remade over time presents as ahistoric, apolitical, and power blind—ignoring critical aspects of how environmental exposures are (re)structured over time and space.

In response, recent theorizing emphasizes how the process of *placemaking* is shaped by physical, material, symbolic, and discursive policies and practices, with "place" understood as an inherently political site of continual contestation (Allen, Lawhon, and Pierce 2019; Petteway 2022). Thus, placemaking must be understood as social, political, material, and symbolic/representational, with processes that structure fundamental relations of space, property, and capital that undergird place-health contexts across communities and geographies. In settler-colonial states such as the U.S., the (un/re)making and taking of place are highly racialized, which shapes the spatial sorting and organization of environmental privilege and risks in residential, occupational, and recreational places (Kent-Stoll 2020; Neely and Samura 2011; Powell 2007). These interrelated notions can help guide CER in naming power and explicating the many factors that shape the place-based contexts of health inequities and EJ over time.

Reimagining What Counts as "Environmental"

Public health can further advance CER for EJ by expanding its focus on deficits-oriented physical and chemical exposures to include more sociospatial exposures, including positive "exposures" to places and spaces of joy, inclusion, love, healing, and resistance. Sociospatial exposures are inclusive of a broad range of social interactions and relations that can act as environmental stressors or destressors, from experiences of discrimination based on gender, race, disability, and sexuality, to aspects of gentrification, displacement, dispossession, and place-attachment and memory. We limit ourselves to discussing just a few potentially important EJ-related examples here.

Policing. As Simckes et al. (2021) outline, the population health impacts of exposure to various aspects of policing can be quite substantial—especially given historic and present contexts of racialized police violence. The near omnipresence—or potential/threat of presence—of police within neighborhood, work, retail, recreation, and education environments makes policing a rampant, even continuous, environmental exposure. The physical and psychological harms of racialized policing—both direct and indirect—are well-documented in public health scholarship (Bor et al. 2018; Lett et al. 2021; Turney and Jackson 2021), as are harms from policing of racialized immigration status (Asad and Clair 2018; Patler and Laster Pirtle 2018). If people of color can be surveilled, harassed, pursued, apprehended, and killed in any place for any reason, then policing must be recognized as a toxic environmental exposure—one that harms health, for example, via stress pathways related to anticipatory anxiety and allostatic load.

Alang et al. (2021) urge public health to integrate measures of exposure to police brutality and other indicators of structural racism and White supremacy into routine health surveillance research. We can imagine the development of a policing-related version of the well-known Toxic Release Inventory (TRI)—a toxic police inventory, which maps, tracks, and monitors spatialized practices of (racialized) police surveillance and aggression as duly acknowledged environmental exposures. There would be an important role for CER in creating this inventory, which could include crowdsourced maps of street-based police harassment, GIS data that show routes and locations of experiences of "driving while Black," and crowdsourced location data for mapping police encounters in residences, workplaces, and recreational and educational spaces.

Spatial Stigma. Public health researchers would also do well to closely examine spatial stigma (Halliday et al. 2020; Keene and Padilla 2014). Notions of stigma are well-known and researched within public health in relation to issues such as HIV, obesity, smoking, sexuality, and disability. Spatial stigma, however, presents a particularly important form of stigma for EJ because stigma associated with a place or space can act as an environmental stressor (Keene and Padilla 2014; Tran et al. 2020). Moreover, the ways a place or space is (mis)represented in research can function to amplify or counter such a stigma (Cairns 2018; Graham et al. 2016). This last point is especially important within public health research, which has a proclivity to focus on deficits and problems of places. In research on Black communities, for example, the representation of place can be "swallowed up by the very death and decay that is bolstered by the hard empirical evidence of Black geographic peril" (McKittrick 2011, 951).

Some CER partnerships have grappled with the dangers of stigmatization by prioritizing community partners' control over how potentially damaging information is disseminated (Minkler, Pies, and Hyde 2012), or by choosing projects that actively destigmatize communities (Gutberlet and Jayme 2010; Tremblay and Jayme 2015). At a minimum, public health research needs to begin each CER project by exploring potentially stigmatizing impacts on relevant communities with community partners, and incorporating their considerations to shape the research agenda, questions, and dissemination plan from the start.

Related to, yet distinct from, spatial stigma is the notion of "the white space," which Anderson (2015) describes as "settings in which Black people are typically absent, not expected, or marginalized when present" (10). The racialization of spaces in countries such as the U.S. means that Black, Brown, Indigenous, and other people of color will often be seen *as* the potential environmental threat when moving through White-dominated or White-associated spaces. The White gaze of fear and stigma attaches to and travels with people of color, who are often well aware of this surveillance when moving through space. This of course has direct implications for considerations of policing as an environmental exposure, but also for considering White space itself as a discrete exposure. Here, we can imagine community-engaged place-health research at the intersections, for example, of structural racism, intersectionality, allostatic load, and life course—making use of activity space approaches to assess White spaces as an EJ exposure, building on the work of Kwan (2013), Wong and Shaw (2011), and Candipan and colleagues (2021), and using community-led methods like participatory GIS and photovoice.

Indigenous Lands and Spatial Healing. Ancestral and Indigenous knowledges reveal that connections to land and nature are healing (Redvers 2020). However, due to colonization, Indigenous peoples now endure some of the gravest health disparities in the U.S., which include cancer, cardiovascular disease, infant and maternal mortality, substance abuse, and depression (Echo-Hawk 2019; Paradies 2016). Public health CER can recognize historical and ongoing injustices for Indigenous people, and work to reclaim and reimagine their relationship to land, food, medicinal plants, and sacred sites. According to the Urban Indian Health Institute

(UIHI), EJ and health equity efforts have overemphasized Western cultural norms, focusing on the role of institutional and structural barriers to health care with little attention to cultural and traditional knowledge systems (Echo-Hawk 2019). Instead, UIHI is working toward health equity for American Indian / Alaska Native populations by "breaking barriers, building beauty, and restoring culture," by supporting tribal communities in "exercising self-determination and reclaiming their unique cultural knowledge systems for the health of the future generation." In their work, "data, research, and evaluation are cultural values and ancestral practices, and we are reclaiming them to be used for Indigenous people, by Indigenous people" (para. 9).

As one of 12 Tribal Epidemiology Centers providing research services to tribal governments and U.S. governmental agencies, UIHI is one example of the growing Native American health infrastructure. Within this infrastructure, tribes and intertribal organizations have developed their own extensive research capacities, including tribal institutional review boards with their own research ethics protocols. Native and other researchers in academia and government can collaborate with these organizations, and should expect to do so as junior partners or co-principal investigators.

CONCLUSION

This chapter has sketched out several ways in which public health can evolve into a more courageous, politically attuned partner to communities struggling for EJ. The field has established a solid base for this work in the newly centered goal of equity in the 10 Essential Public Health Services, and traditions of CBPR, social epidemiology, place-health research, and health impact assessments. Now, public health CER must engage in deeper and more creative thinking about how to enact antiracist and decolonizing principles; enrich social epidemiology with the study of placemaking and activity spaces; expand conceptions of environmental health to include EJ issues provoked by sociospatial exposures to policing, spatial stigma, and White spaces; and take inspiration from Indigenous efforts to reclaim their lands, cultures, and health infrastructures.

This requires imagining new futures for both the science and practice of public health for EJ—including research translation and political engagement (e.g., Galea and Vaughan 2019; Morgan-Trimmer 2014; Schwartz et al. 2016). This involves remembering that public health research is ultimately about healing bodies, lives, and communities, not merely analyzing samples and specimens. This will be facilitated by recruiting and training a new generation of researchers whose lives are rooted in embodied experiences of environmental injustice. This also demands that all researchers develop capacities to question their own positionality in relation to the EJ communities with whom public health should collaborate reciprocally and respectfully, and to the field. Who is producing EJ knowledge, taking up the discourse space, and driving (or stifling) policy and research priorities? Who has the power to use, (mis)represent, and discuss whose bodies and lives in research? Do researchers possess the care and courage—not just the scientific curiosity and capital—to fight for environmental justice?

10

Food Justice and Food Sovereignty

Vera L. Chang, Teresa Mares, Martha Matsuoka, and Chad Raphael

Community-engaged research (CER) connecting environmental justice (EJ) with issues in the food system spans multiple disciplines to address a diverse array of topics, such as Indigenous communities' demands for food security and sovereignty, farm and food workers' struggles against contamination and exploitation, urban neighborhoods' efforts to challenge food apartheid and revitalize urban agriculture, and rural communities' battles to protect themselves against toxic farm runoffs and concentrated animal feeding operations.

Food justice, as both a social movement and an area of academic research, is firmly rooted in concerns raised within the EJ movement (Alkon and Agyeman 2011; Gottlieb and Joshi 2010; Sbicca 2018). Kristin Reynolds (2020) defines food justice as "a concept and related movement that considers the social and political roots of inequities in the food system and holds that these structural issues must be addressed to solve problems such as disparate access to healthy food and exploitative or unfair labour practices" (180). While food justice has been a primary framing for activism confronting structural racism in the U.S. food system, the food sovereignty movement spread from peasant struggles in the Global South to the Global North. Food sovereignty is "the right of peoples to healthy and culturally appropriate food produced through ecologically sound and sustainable methods, and their right to define their own food and agriculture systems. It puts the aspirations and needs of those who produce, distribute, and consume food at the heart of food systems and policies rather than the demands of markets and corporations" (Forum for Food Sovereignty 2007). Both food justice and food sovereignty push for more equitable and dignified relationships between people,

TABLE 10.1. CER for Food Justice and Food Sovereignty

Dimension of Justice	In CER for Food Justice and Food Sovereignty
Distribution *Who ought to get what?*	Uncovering the social and political roots of injustices in the food system—especially exploitation of immigrants and other workers, racism, and colonialism—to promote equitable access to land and farming, safe working conditions and fair labor practices for farm and food workers, regeneration of land, and healthy food
Procedure *Who ought to decide?*	CER partnerships for building local capacities and power for organizing, movement building, and worker participation in decision making across the food system
Recognition *Who ought to be respected and valued?*	Foregrounding worker, producer, and community experiences and knowledge of the food system Respecting traditional ecological knowledge about agriculture and rights to culturally appropriate food
Transformation *What ought to change, and how?*	Decolonizing, antiracist CER and community-led movements to support structural transformation of the food system that advances food justice and food sovereignty for Black, Indigenous, and people of color (BIPOC) farmers, farm and food workers, consumers, and the land

food, and land—framing the environment as the spaces where we live, work, play, *grow, and eat.*

The projects described in this chapter illustrate the political, strategic, and imaginative role CER can play in identifying and resisting food injustices, recognizing and respecting multiple forms of knowledge and expertise, and building more equitable and sovereign food systems. CER on food justice exists on a continuum from projects in which communities provide input or other contributions (e.g., BAMCO and UFW 2011) to projects in which communities themselves define, design, and direct the research on issues that directly affect their lives (e.g., Fox et al. 2017). Because food-related disparities intersect with other forms of oppression and injustice—based on race/ethnicity, indigeneity, class, ability, sexual orientation, gender identity, and citizenship status—all CER must integrate decolonizing and antiracist approaches (Bang and Vossoughi 2016; Bradley and Herrera 2016). Table 10.1 summarizes the dimensions of justice relevant to this research.

CER is particularly important for challenging corporate and political efforts to resist regulation, minimize the importance of pollution and human rights violations, and silence scientific evidence (Nixon 2011). While industrial agriculture producers and food processors have strong ties to government and academic institutions, impoverished communities of color seldom have access to researchers, are underrepresented in the research profession (Wing 2002), and lack political clout to defend themselves (Nicole 2013).

However, it is important to interrogate the institutional hierarchies that may be embedded in CER. For example, the U.S. Department of Agriculture's Cooperative Extension (CE), operated through the nation's land-grant universities, has a mandate to address local and national environmental and agricultural issues through collective development of research and educational programming with local communities. Extension specialists ideally serve as problem solvers, educators, and collaborators working with communities to translate research into action and share knowledge with people who depend on it for their livelihoods. Yet CE and land-grant universities have been critiqued for persistent exclusion of social and political factors, inequitable policy making and resource allocations, and privileging production models and the interests of industrial producers (Henke 2008); imposing top-down technology transfer from "experts" to farmers (Warner 2008); and ignoring the needs of low-income, BIPOC, and female farmers (Ammons et al. 2018).

The food system raises more EJ issues than we can address in a single chapter. The Real Food Challenge, a national campaign to promote environmental and social justice across the U.S. food system, provides a helpful overview of these issues (see figure 10.1). In this chapter, we focus on three aspects of the food system that have provoked especially robust programs of CER on EJ issues: agricultural pollution of fenceline communities, demands for food sovereignty and security, and farmworkers' and food workers' rights.

AGRICULTURAL POLLUTION OF FENCELINE COMMUNITIES

Much of the initial research on EJ and the food system focused on disproportionate impacts of agricultural waste and chemicals on fenceline communities (Rhodes et al. 2020). CER contributed to the EJ movement's opposition to contamination of low-income and BIPOC communities' air and water from pesticides, herbicides, fertilizers, and noxious odors. Air- and water-monitoring studies have demonstrated community exposure to pesticide drift from nearby fields (Harrison 2011; Freese and Lukens 2015; Marquez et al. 2016). CER studies have monitored the impact of agricultural irrigation and runoff on nitrate and arsenic levels in fenceline communities' water systems (Balazs and Morello-Frosch 2013) and coastal acidification from agricultural emissions (Gassett et al. 2021). CER has also explored agricultural pollution's damage to local culture as well as environments, such as Mitchell's (2018b) photovoice project on an American Indian community's experience of river contamination on their homelands. CER has addressed the toll of confined animal feeding operations (CAFOs), or "factory farms," on communities surrounding industrial-scale dairy, poultry, cattle, and hog farms (Carrel, Young, and Tate 2016; Johnston and Cushing 2020).

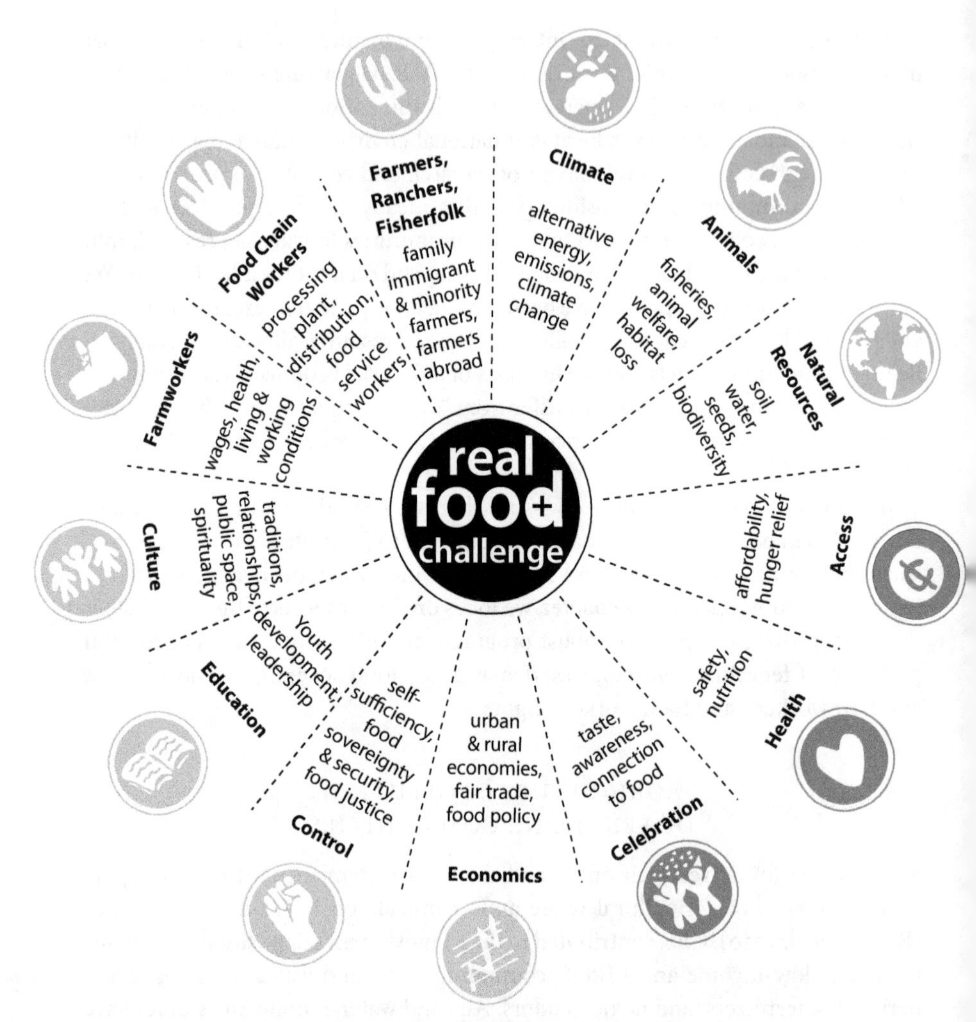

FIGURE 10.1. The Real Food Challenge movement's summary of justice issues in the food system. SOURCE: www.scu.edu/sustainability/operations/food/rfc/.

Health Impacts of Hog Farms in North Carolina

An influential and sustained body of CER and organizing on community impacts of hog farming in North Carolina informed research in other regions and on additional CAFOs. In the 1990s, a shift to large-scale hog farming released unprecedented levels of air and water pollutants and malodors from manure lagoons the size of football fields, spray fields used to disperse additional hog waste, and decomposing hog carcasses. Community leaders understood this was an EJ issue.

"The pork industry came to eastern North Carolina because we are Black, poor, rural and have no political clout," explained Gary Grant, executive director of Concerned Citizens of Tillery (CCT) (quoted in Vanderwarker 2012, 72). CCT collaborated with a research team from the University of North Carolina led by epidemiologist Steve Wing to conduct some of the first studies of the health effects of hog operations on surrounding communities, which supported landmark regulatory action and legislation (Rhodes et al. 2020).

The research collaboration began by showing that hog CAFOs were located disproportionately in communities with high levels of poverty, Black and Latinx residents, and households dependent on well water, which was vulnerable to groundwater contamination (Wing, Cole, and Grant 2000). Next, the research partners co-designed an innovative study of the hog farms' effects on neighbors' health and well-being. CCT recruited residents who lived near CAFOs from 16 communities to participate in a two-week sampling study. The study deployed trailers to conduct real-time monitoring of weather patterns and multiple air pollutants, combining these data with residents' reports of the strength of malodors, respiratory problems, blood pressure, and lung function measurements. Residents also reported their perception of their quality of life, a psychological measure rarely included in EJ studies. CCT members led the study design and recruiting efforts and provided background knowledge on regional politics and history and on the industry's tactics (Rhodes et al. 2020). To build trust with participants and protect their anonymity, researchers held training sessions at participants' homes, churches, and other local meeting places.

This unique data set yielded multiple studies of physical and psychological effects of living near hog CAFOs and documented frequent malodor and elevated levels of multiple pollutants—including ammonia, volatile organic compounds, and particulates (Guidry et al. 2018). Researchers found an association between increased exposure to hydrogen sulfide gas (a hog waste biomarker) and elevated blood pressure among participants (Wing, Horton, and Rose 2013). Residents also reported increased levels of stress and changes to their daily activities in response to malodor (Horton et al. 2009). The study provoked additional research showing hydrogen sulfide exposure among children in schools near hog operations (Guidry et al. 2018), downstream contamination by swine waste (Heaney et al. 2015), and health threats to employees on hog farms, including potential transfer of antibiotic-resistant bacteria from hogs to workers (Davis et al. 2018).

These studies supported successful organizing campaigns led by CCT, the North Carolina Environmental Justice Network (formed by Wing, Grant, and other activists), and allied organizations representing communities and workers. In 2007, organizers celebrated a victory over the multinational pork corporations when the state adopted a permanent statewide moratorium on industrial hog operations. In 2018, three EJ organizations won a settlement from the North Carolina Department of Environmental Quality for failing to regulate hog facilities to protect Black, Latinx, and Native American communities from pollution, one of the few successful complaints against environmental racism under federal civil rights law in the U.S. (Rhodes et al. 2020).

DEMANDS FOR FOOD SOVEREIGNTY AND SECURITY

Defined by the United States Department of Agriculture (USDA) as "access by all people at all times to enough food for an active, healthy life," food security entails much more than possessing the financial resources to purchase food (Coleman-Jensen et al. 2021). Examining these deeper and more complex meanings of food security, or what is understood as the four pillars of food security—availability, access, utilization, and stability—opens up myriad possibilities for researchers who want to collaborate with communities plagued by food injustices. As a social movement, food sovereignty builds upon demands for food security to engage more fundamental questions of agency and control over land and other agricultural resources. Borrowing from Hannah Arendt, Raj Patel frames this as the "right to have rights" (Patel 2009).

The following case studies demonstrate the necessity of recognizing the particular and shared concerns that marginalized communities face in accessing food that is meaningful and conducive to well-being. These cases also demonstrate the importance of considering the historical and structural contexts that have shaped access to land and other food-related resources. These examples honor the deep, local, embodied knowledge or traditional ecological knowledge (TEK) connected to food and agriculture that resides in marginalized communities (Nelson and Shilling 2018), allowing researchers to connect food insecurity to systemic patterns of racialized injustice and exploitation. Although these cases focus on the U.S., they are linked to other parts of the world through diasporic connections and networks of migration.

Food Security and Gardening with Farmworkers in Vermont

For U.S. farmworkers, food insecurity stems from poverty, language barriers, fear of detention and deportation, and long work hours that leave little time to access and prepare healthy food. While there is a wealth of research on farmworker food insecurity, few studies have utilized CER to simultaneously document and ameliorate food disparities in farmworker communities (see, e.g., Brown and Getz 2011; Kresge and Eastman 2010; Villarejo et al. 2000). These studies have primarily focused on seasonal workers in traditional destinations of migration.

Teresa Mares, in collaboration with the Huertas Project (connected to the University of Vermont Extension's Bridges to Health program), addressed food insecurity within a community of year-round farmworkers in New England's dairy industry. Most of Vermont's estimated 1000–1200 Latinx dairy workers live and work in isolated dairy farms in rural areas. Most workers are young men who moved to Vermont from central and southern Mexico and are living on their own, separated from families in their countries of origin.

Beginning in 2009, the Huertas Project addressed migrant farmworker food security and sovereignty concerns, identified through years of the university's CE outreach to Vermont's dairy farms, by addressing disparities in access to fresh food. Huertas began collaboratively designing and planting kitchen gardens at farmworker homes, prioritizing the cultivation of culturally familiar foods that are often inaccessible in northern Vermont. For farmworkers from agrarian backgrounds, the gardens are a place to employ forms of agroecological knowledge learned in their home communities. For those from urban areas, the gardens are a place to become more deeply connected to foods they have enjoyed, but perhaps have not grown on their own. Since 2011, Mares has served as the co-director of the Huertas Project, integrating research findings and farmworker perspectives into a continual redesign and evaluation of the project, and sharing findings to better inform social service providers and other stakeholders in the local food system.

Over nine years of fieldwork, Mares and colleagues found that 18 percent of the 100 farmworker households surveyed were food insecure, with 4 percent experiencing very low food security (Mares 2019). These data were collected by administering the Household Food Security Survey Module (HFSSM), a tool designed by the USDA. However, Mares soon realized that the instrument is not well suited to farmworker households because of its heavy dependence on financial measures as a proxy for food security and the restrictive manner in which the survey module defines a household. For transnational farmworkers who are contributing economically to households on both sides of the border, the HFSSM fails to capture the complexity of their daily food access struggles.

CER can help to supplement inadequate measures of food insecurity such as the HFSSM, which have often been developed by government entities with little input from affected communities. Incorporating grounded theory and mixed methods, Mares supplemented these surveys with in-depth interviews that included questions and themes that were more relevant and rooted in the everyday experiences of farmworkers. These interviews revealed that for a majority of farmworkers, a lack of money was not the primary obstacle to obtaining food. Rather, a combination of limited time for grocery shopping given the timing of work shifts, language barriers, fear of Border Patrol and ICE personnel, and transportation challenges resulted in farmworkers having little agency over the sources of their food, or the means to access it. Additionally, the need to support families in their countries of origin often limited the amount of money farmworkers felt they could spend on their own food needs. Many of these interviews were conducted with Huertas participants, revealing that the gardens they planted increased their access to foods conducive to health. The strong relationships Mares developed through Huertas were key to understanding the limitations of the HFSSM and the more relevant and pressing issues confronting farmworkers.

Food Insecurity and Food Sovereignty for Indigenous Communities in the Klamath River Basin

Sowerwine, Mucioki, et al. (2019) note that "[u]nder settler colonialism, dramatic changes in the management of the lands and waterways related to mining, hydroelectric dams, agriculture, logging, and fire suppression have resulted in the near loss of Native fisheries, and drastic reduction in the abundance and availability of Native foods" (587). Alongside this ecological devastation came the structural violence linked to genocide and forced assimilation policies that disrupted traditional relationships of reciprocity and kinship and the knowledge systems connected to the natural world. Limited access to healthy food and high rates of diet-related disease are of serious concern in Native communities across the U.S. (see, e.g., Bauer et al. 2012), yet few studies employ a CER approach guided by the principles of environmental and food justice (Jernigan et al. 2012, 2017; Sowerwine, Mucioki, et al. 2019; Sowerwine, Sarna-Wojcicki, et al. 2019).

CER is especially valuable for revealing the connections between food security and food sovereignty within Indigenous communities, as illustrated by the collaborative work of Lisa Hillman, a member of the Karuk Tribe and the manager of its Píkyav Field Institute (PFI), and colleagues from the Department of Environmental Science, Policy, and Management at the University of California, Berkeley (UCB). This team has investigated barriers to food access among tribal members in the Klamath River basin. The Karuk word *píkyav* translates as "to repair" or "to fix," and at the center of these reparative efforts are the Karuk's intertwined social, cultural, and ecological systems on their homelands in northwestern California and southern Oregon.

The research partnership resulted from a long and deliberate process of co-creating principles to ensure "protection of intellectual and cultural property and recogniz[e] tribal sovereignty" (Karuk-UCB Collaborative, n.d.). These guidelines stem from the Indigenous Research Protection Act and were adapted to local needs and priorities. To better understand barriers to food access, the team employed a community-based participatory research (CBPR) approach, conducting more than 711 surveys, 115 follow-up interviews, and 20 focus groups with members of the Yurok, Hoopa, Klamath, and Karuk tribes. Tribal members and communities were engaged as "active and equal participants throughout the research process" (Sowerwine, Mucioki, et al. 2019, 588).

Data revealed that 92 percent of respondents were food insecure to some degree (one of the highest rates of food insecurity among Indigenous communities in the U.S.), compared with roughly 12 percent of the overall U.S. population (Sowerwine, Mucioki, et al. 2019). However, like Mares, the research team found significant limitations with the USDA's HFSSM, including a narrow framing of food security that does not include attention to deeper cultural and spiritual meanings of food or the ecological relationships between people and the food that sustains them.

To address these limitations with input from the tribes, the research team developed an indicator for "Native foods security" to examine the relationship between access to Native foods and household food security. Using this indicator, researchers found that only 7 percent of households were Native-foods secure and 70 percent of households never or rarely had access to Native foods on a consistent basis. The study demonstrated that improving access to Native foods is key, and that this requires revising laws and policies that limit access to ancestral lands and resources. Among the many applied outcomes of this research is the development of 89 lesson plans for K–12 students that "center content relevant to tribal identity and the traditional food system" (Sowerwine, Sarna-Wojcicki, et al. 2019, 177). Additionally, the project resulted in the development of a Karuk and Yurok Tribal Herbaria housed at the Karuk Office of Historic Preservation and the Karuk People's Center, wherein tribal members "collected, pressed and mounted, and preserved hundreds of plant species of cultural and regional significance" (Sowerwine, Sarna-Wojcicki, et al. 2019, 178).

Black Farming, Resilience, and Agency

Like Latinx and Indigenous communities, Black communities in the U.S. disproportionately experience food insecurity and food injustice. Barriers to Black Americans' access to food and farmland cannot be separated from the violent histories of slavery, disenfranchisement, and the systematic denial of land and agricultural lending. A number of studies have pointed to elevated rates of diet-related disease and food insecurity in Black households and communities that are connected to these forms of violence (e.g., Burke et al. 2018; O'Reilly et al. 2020). Some studies use a CER approach to examine these inequities (Carlson, Neal, and Magwood 2006; Paschal et al. 2020; Rollins et al. 2021). The loss of Black-owned farms has been dramatic, declining from a high point of 14.3 percent of all farmers identifying as Black in 1920 to 1.5 percent in 2012 (Taylor 2018).

In response, movements for Black food justice and food sovereignty have gained traction in recent years. While their priorities vary, a primary goal has been to cultivate Black resilience and freedom through re-establishing connections to both rural farmland and urban food systems (McCutcheon 2021; Penniman 2018; White 2018). The research on Black farming and resilience, most of it done by Black women, has often used decolonial forms of ethnographic research that leverages both deep emic knowledge of structural racism in the food system and close community connections (Garth and Reese 2020; McCutcheon 2013; Reese 2019). These studies challenge narrow definitions of CER wherein the lines between insider and outsider are often seen as static, rather than fluid and tied to intersectional identities.

Monica White is one example of a community-engaged researcher who has helped to connect and advance the intertwined movements for Black food sovereignty, land and environmental justice, and civil rights. Researching urban farmers in Detroit (White 2011), and their connections to cooperative practices of Black farmers in the U.S. South (White 2018), White has illuminated the collective agency and resilience that is embodied by Black farmers and their role in ensuring food security for their communities. White's approach to CER incorporates a historical perspective, showing how Black struggles for land and food sovereignty are not new, even if they are responding to new challenges.

White is the founding director of University of Wisconsin's Office of Environmental Justice and Engagement, which supports faculty and students working on CER connected to environmental issues. In this role, she draws upon her community engagement as past president of the board of the Detroit Black Community Food Security Network, on advisory boards of Southeastern African American Farmers' Organic Network and the Institute for Agriculture and Trade Policy's Food Justice Task Force, and as a fellow with Food First. Her CER approach also inspires her commitment to publishing open-access scholarship that reaches beyond academic readers, such as her columns in the *Journal of Agriculture, Food Systems, and Community Development*, underscoring that openly sharing research findings can be as valuable as co-producing those findings.

FARM AND FOOD WORKERS' RIGHTS

The long-term, underlying causes of exploitation and environmental injustices faced by food system workers globally are varied and complex, but a key driver is uneven value distribution along industrial supply chains, with power consolidated at the top that squeezes suppliers and workers as they compete. Inequities of race/ethnicity, gender, sexuality, citizenship, class, and ability have enabled unequal power relations to flourish in the production of goods consumed worldwide. This is particularly the case where early world markets for industries were entwined with colonialism, such as in tea production in India and chocolate production in Ghana, where forced labor persists today (LeBaron 2018).

Few consumers or food industry professionals understand farm and food workers' conditions and characteristics, because there has been little data gathered about them (BAMCO and UFW 2011; LeBaron 2018). Farm and food workers can be difficult to "count" in standard employment statistics because seasonal, contract, and undocumented workers are less likely to be reported to government agencies; small farms are often excluded from official statistics; and regulatory bodies can withhold data from the public for confidentiality reasons. Thus, farmworker rights and needs are frequently overlooked in policy and academia.

CER on farm and food workers fulfills the dual goals of EJ to deepen democratic processes and support workers' rights. Through CER, workers, academics, advocates, and even industry have teamed up to make empirical data on the conditions of workers more visible to the public and food industry, help advance human rights, and prove what is otherwise invisible: workers' marginal earnings, economic uncertainty, and harsh and often exploitative working conditions (BAMCO and UFW 2011; Fox et al. 2017; Gray 2013; Kline and Newcomb 2013; Mares 2019). CER outreach projects also combat exposure, injury, illness, and poverty of workers due to abusive and hazardous workplace environments. CER can provide factual bases for the need for greater attention, resources, and legal protections for workers and regulation of working conditions. The findings of CER projects on food workers point to an urgent need for enforcement systems that can uphold labor rights at the bottom of the supply chain.

To illustrate these ideas, we point to projects that exemplify how CER can address questions of labor and human rights in food systems. These examples demonstrate the need to recognize particular and shared barriers to marginalized worker communities' basic health and safety. The cases also demonstrate the importance of considering historical and structural contexts that contribute to workers' impoverishment, vulnerability, and exploitation. This literature recognizes the natural and built environment—including where people work and live—as intricately connected with people's well-being and with EJ.

Farmworker Issues and Protections in the United States

As corporations seek to increase their profits and power in the food system, food production becomes a source of economic, political, and cultural contention (Howard 2016). Through corporate consolidation, the most powerful and dominant corporations can generate downward pressure on wages and labor standards, and produce environmental inequalities that result in institutional violence. This leads to suffering and even lethal consequences for suppliers, workers at the bottom of the supply chain, and other marginalized communities. Simultaneously, some corporations may ameliorate some of their negative effects on communities and use their resources to raise awareness of and support for EJ goals.

An example of the latter is a for-profit and nonprofit CER partnership between Bon Appétit Management Company (BAMCO), a subsidiary of the largest food service company in the U.S., and the United Farm Workers of America (UFW), the country's largest farmworkers' union. In 2011, BAMCO and UFW (2011) collaborated to publish a fact-finding document, *The Inventory of Farmworker Issues and Protections in the United States*, which provided the most comprehensive and bleak picture of the few legal protections farmworkers had at the time.

The *Inventory* authors gathered, synthesized, and translated data on farmworker conditions into easily accessible formats for the public and food industry. BAMCO was responsible for the majority of research, data collection, and drafting of the *Inventory*. UFW provided project direction and legal expertise, Oxfam America provided insight into the status of farmworkers, and an independent sociologist analyzed data.

Focusing on health, safety, and enforcement from federal, state, and private sources, the *Inventory* cataloged key laws and regulations for the United States and the six states with the largest farmworker populations. *Inventory* researchers compiled data on farmworker well-being from the U.S. Department of Labor's National Agricultural Workers Survey, USDA's Census of Agriculture, USDA's National Agricultural Statistics Service, state regulatory bodies, farmworker organizations, and academic research. Fifty-two farmworker advocacy groups, nonprofit legal organizations, and governmental agencies made contributions to the *Inventory* by providing background information, data, and other input.

The report illustrated rampant disregard for workers' well-being. It cataloged the many forms of occupational hazards and toxic exposures that farmworkers face resulting from loopholes in health and safety protections, lack of regulatory oversight, and widespread unreported labor violations. A major finding was the significant missing data on farmworker conditions and issues—due to poor, untraceable, and nontransparent labor law monitoring and record keeping by state and federal regulators. The absence of adequate data makes it difficult to publicize and remedy the health and safety problems rampant in farm labor.

The *Inventory* advocates for farmworkers to have the same legal protections in the workplace that apply to other occupations in the U.S. By establishing a baseline of conditions, the *Inventory* has been useful in calling for improvements for farmworkers, such as more legal protections against child labor, reproductive justice for farmworkers, better protection of women and girls against sexual violence, expanded regulations against pesticides and heat stress, greater accountability for pesticide reporting, comprehensive healthcare of farmworkers, and increased awareness of structural racism in the food system. The *Inventory* has also contributed to governments' understanding of farmworkers' legal needs (Legal Services Corporation 2015).

CHAMACOS

Another area of CER supports protecting farmworker women, children, and their communities against pesticide exposure. In 1998, the Center for the Health Assessment of Mothers and Children of Salinas (CHAMACOS) launched the world's largest and longest birth cohort study of pesticides and environmental chemicals in pregnant women and children living in an agricultural community. The longitudinal study incorporated CER to disseminate findings creatively with community partners and inspired youth-led research on pesticide-related health, safety, and EJ issues.

The study is part of the Center for Environmental Research and Children's Health (CERCH), which investigates environmental exposures to families and helps translate research findings into strategies to reduce environmental disease.

Supported by environmental and health agencies and nonprofit organizations, the study is run by University of California, Berkeley (UCB) professor of public health Brenda Eskenazi; community partner Clinica de Salud del Valle de Salinas; and an advisory council of farmworkers, growers, youth, and scientists.

CHAMACOS measures environmental exposures and assesses children's growth, health, and development in California's Salinas Valley, one of the country's most productive farming regions. Methods include biological samples, environmental samples, neurodevelopmental tests, lung function tests, anthropometric data, neurodevelopmental and physical assessments, questionnaire data, and factors such as diet and school performance. Over 800 children were enrolled in the study, with over half tracked prior to birth.

Among many findings, the study has linked pesticides sprayed on fruit and vegetable crops with respiratory complications, developmental disorders, and lower IQs among children of farmworkers (Eskenazi et al. 1999, 2004, 2007, 2013). CHAMACOS research contributes to knowledge about the impacts of pesticides on children's brain development and respiratory health, the interaction of stress and early life adversity on health in chemically exposed populations, and methods to reduce pesticide exposures. This research underscores the urgent need for public policy to target economic, social, and gender disparities (e.g., improved wage and hour laws, access to healthcare, and occupational safety protections), and to address the material needs and protect the health of marginalized communities.

Over time, the project has focused on community engagement and bidirectional learning. CERCH developed outreach programs to address pesticide exposure prevention for farmworker families unlikely to receive formal training otherwise. The center educated more than 30,000 farmworkers and community members and distributed thousands of materials accessible to farmworkers for redistribution within their communities, such as graphic novellas, educational puppet shows, and hotline cards. CERCH also created a train-the-trainers model that teaches migrant farmworkers how to educate others about pesticide safety practices in their community. CERCH and its CHAMACOS Youth Council—Latinx youth learning about and addressing environmental health concerns—have collaborated with worker organizations, such as the California Department of Education's Office of Migrant Education, and arts organizations, such as Hijos del Sol, to communicate the study's health findings and conduct trainings on pesticide safety with the wider community.

Between 2016 and 2018, CHAMACOS Youth Council implemented a follow-up study called Chamacos of Salinas Evaluating Chemicals in Homes and Agriculture (COSECHA) to empower the next generation of environmental health leaders, researchers, and activists. COSECHA studied pesticide exposures associated with hormone-disrupting and carcinogenic effects among 100 teen girls (Harley et al. 2018, 2019). Led by a UCB reproductive epidemiologist, Kim Harley, Clinica de Salud del Valle de Salinas and 11 paid local youth research assistants collaborated in each phase of the study. Methods included using GPS devices, environmental sampling bracelets capable of detecting over 1500 chemicals, indoor dust samples, in-person questionnaires, urine samples, and a catalog of crops grown on nearby fields. Drawing on experiential knowledge, the youth researchers designed strategies to communicate their public health findings through television segments, tabling at local events, community presentations, a Radio Novella "edu-tainment" series, and a community mural. They also distributed 800 doormats printed with tips for reducing pesticides in homes, which COSECHA research indicated had a protective effect. COSECHA also strengthened youth researchers' professional skills. All but one member of the youth research cohort went on to college, in an area where only 59 percent of people (aged 25 years or older) have graduated from high school, and 13 percent have graduated from college (Town Charts 2021).

Immigrant Dairy Farmworkers in New York State

The report *Milked: Immigrant Dairy Farmworkers in New York State* revealed New York dairy farmworkers' working and living conditions by highlighting these workers' rarely heard voices (Fox et al. 2017). *Milked* was co-authored by a team of community leaders at two grassroots organizations advocating for institutional justice and change for low-wage workers—Worker Justice Center of New York (WJCNY) and Workers' Center of Central New York (WCCNY)—and researchers from Syracuse University and Cornell University.

Farmworkers participated actively in the study, helping develop interview questions, lead focus groups, transcribe and analyze data, and contribute photographs. Additional researchers analyzed the dairy industry structure, and health and safety challenges on farms. The research team conducted 88 semistructured interviews with immigrant farmworkers on 53 dairy farms across the state. No source has compiled the full population of dairy farmworkers in New York from which to draw a sample, so the study demanded time- and labor-intensive direct outreach to workers. The interview's 225 questions covered participants' demographic information, work histories, wages, working and housing conditions, social integration, interactions with immigration enforcement agents, and interests in organizing for change.

WJCNY and WCCNY used *Milked* to support immigrant dairy farmworkers' organizing to resist workplace violence and harassment, recover stolen wages, and lobby for improved farm housing and working conditions. *Milked* provides an empirical basis for advocating for federal and state agency intervention, as well as dairy processing company policy changes. For example, the report argues that New York State should no longer exempt farmworkers from basic labor rights, such as the rights to organize, to a day off, and to overtime pay. The report also presents evidence for state policy changes to enable undocumented immigrants to get driver's licenses, provide state oversight of workplace health and safety for dairies, and ensure that all farmworkers have safe and dignified housing with a right to receive visitors. The report calls upon dairy companies to adopt and enforce worker-led codes of conduct for their fresh milk suppliers to ensure they follow ethical labor practices, and to buy only from farms that participate in rigorous and independently conducted labor rights monitoring. The report also urges milk consumers to hold dairy companies accountable for working conditions.

Conducting this survey strengthened WJCNY and WCCNY outreach as researchers made contact with workers on farms and involved them in organizing efforts. The organizations had weekly conference calls with workers to strategize about how to respond to issues documented in the research, such as wage theft, workplace violence, and health and safety conditions. These workers' networks were key for developing leadership and solidarity for action, including farm protests and a campaign to implement occupational safety and health measures across the New York dairy industry.

NEW DIRECTIONS FOR RESEARCH

Within the food justice and sovereignty movements, building collective power, diversifying strategies, and forging solidarities across social boundaries are priorities (Sbicca 2018). CER can contribute to these goals in three broad ways. One involves CER practitioners forging broader collaborations with each other and with communities to build and sustain grassroots power. For example, the Agroecology Research-Action Collective has developed *community of practice* principles and protocols for researchers that describe horizontal nonexploitative learning with food movements and mechanisms for multidirectional accountability among research partners (Montenegro de Wit et al. 2021). Collective efforts such as this can help develop long-term collaborative projects across communities that build greater strength and relevance than isolated CER projects can do on their own. In addition, CER partners can move beyond documenting food injustices to *develop and disseminate policy solutions in public forums.* For example, Vera Chang (2020) has published findings from her CER with the Coalition of Immokalee Workers, a farmworker-led human rights organization, as examples of solutions journalism, which evaluates responses to social problems rather than simply describing them. These publications focus on effective worker-designed responses to problems in the agricultural workplace, educating the public and policy makers about the potential for constructive change. Third, academic researchers can *involve more of their students in CER on food issues* as a contribution to transforming public consciousness. Goldberg and Minkoff-Zern's (2021) research on a CER collaboration between an undergraduate class focused on labor and the food system and Restaurant Opportunities Centers United, a network of worker organizations fighting to raise restaurant wages and labor standards, found that participating in CER can shift students' viewpoints and values from a purely consumer-based perspective to include workers' perspective on the food system.

CER in this area can also *deepen decolonial and antiracist research.* First, CER can help protect immigrant rights advocates, including farmworkers and their advocates, from retaliation and arrests aimed at silencing dissent. A mapping database that shows incidents of harassment and detention of immigrants who speak up for their rights across the U.S. created by the New Sanctuary Coalition and New York University School of Law's Immigrant Rights Clinic offers one promising response (www.immigrantrightsvoices.org). Second, there is a need for additional work on BIPOC food sovereignty that addresses policy barriers, such as a recent CER study of how USDA's Farm Bill Conservation Title programs hinder Black farmers' ability to mitigate invasive species on their farmland (Fagundes et al. 2020). Third, we need to learn from new partnerships between academic institutions and Indigenous natural resource managers that center tribal food sovereignty and prioritize trust-building processes rather than maximizing research publications and products (e.g., Matson et al. 2021). Fourth, CER for food justice and sovereignty needs to expand to neglected constituencies and places. In part because urban communities are accessible to many researchers, we need more CER with rural communities (e.g., Cannon 2020; Engle 2019) and collaborations that bridge the urban-rural divide (e.g., Soergel 2021), which can help to build stronger ties and movements for food justice. In addition, prison food systems are significant sites of food insecurity, malnourishment, contamination, and exploitation of incarcerated labor by corporations for farming and manufacturing (Pellow et al. 2019).

CER is desperately needed to *strengthen community responses and resilience to disasters.* CER is starting to show that many effects of climate change on food and farming communities are disproportionately borne by women (van Daalen et al. 2020), as well as communities of color and low income, linguistically isolated people, and outdoor laborers (Aneesh et al. 2020; Castillo et al. 2021). For example, a research partnership conducted over a decade by researchers at Santa Clara University with smallholder coffee and corn growers in Nicaragua has documented and developed solutions to climate-induced drought and farming communities' seasonal hunger—an example of the "hungry farmer paradox" found in rural areas throughout the global food system (Bacon et al. 2014, 2021).

CER can also strengthen EJ communities' resilience to disasters by drawing lessons from rapid research on the COVID-19 pandemic. This research showed how the pandemic exacerbated intersectional forms of environmental injustice, such as poverty, discrimination, disease exposure, and other hazards (e.g., Ammons et al. 2021). Studies such as the multipart COVID-19 Farmworker Study, produced by a coalition of academics and community-based organizations based on data gathered with and by farmworkers, provided timely data to support immediate policy recommendations for strengthening safety net resources and ensuring safer working conditions (CBDIO et al. 2021). Additional research offers lessons for integrating CER and EJ principles into disaster and resilience responses, such as a

COVID-era study of how USDA-funded emergency food relief programs, which typically distribute processed foods supplied by agribusiness companies, can instead purchase local fresh produce from small farmers of color (Environmental Justice and the Common Good Initiative 2021).

Finally, CER can advance *restorative justice to transform academic institutions' relationships to BIPOC communities and the food system.* There is a need for more CER to serve the needs of underfunded tribal, historically Black land-grant, and Hispanic-serving agriculture colleges and universities (Valley et al. 2020). New initiatives can learn from promising examples—such as Michigan State University's Racial Equity in the Food System Workgroup, the First Americans Land Grant Consortium, and some Cooperative Extension programs—of how to resource BIPOC-led and BIPOC-serving institutions, and build bridges between them and predominantly white institutions, to advance food justice and sovereignty through research, teaching, and community outreach.

11

Urban and Regional Planning

Ana Isabel Baptista, Martha Matsuoka, and Chad Raphael

Many of the most significant environmental justice struggles in the United States concern land use conflicts that implicate urban and regional planning efforts. These place-based struggles reflect how our built environments have been shaped over time by histories of racism, inequality, colonization, and ecological exploitation. Thus, environmental injustice and environmental racism are in part products of urban and regional planning systems that have resulted in not only the maldistribution of harm, but also a lack of access to vital resources and decision-making processes that form our cities and towns.

This chapter explores the history and role of urban and regional planning in relation to community-engaged research (CER) and environmental justice (EJ). Highlighted are key examples of how CER has influenced EJ struggles in a variety of planning applications. These examples reflect the contributions of researchers, activists, and community-based, frontline and grassroots groups to urban and regional planning efforts across a diversity of issues, such as air pollution, climate resilience, energy, and water infrastructures. Planners and communities have integrated CER into planning practices through a variety of approaches—from participatory to radical planning—to address a host of challenges and opportunities for advancing dimensions of EJ. Table 11.1 summarizes how planning can address the four dimensions of justice common to CER and EJ discussed in this chapter.

TABLE 11.1. CER, Urban Planning, and EJ

Dimension of Justice	Community Engaged Research in Urban Planning for EJ
Distribution *Who ought to get what?*	Combining community knowledge with public data sources to document cumulative impacts of environmental harms, such as the siting of polluting industries, and inequitable distribution of environmental burdens and benefits Prioritizing equity and social justice outcomes of planning processes
Procedure *Who ought to decide?*	CER to support community-based planning, especially led by non-state EJ organizations using radical planning approaches
Recognition *Who ought to be respected and valued?*	Centering local grassroots knowledge and intersectional analysis in the planning process, rather than professional and official expertise applied from outside the community
Transformation *What ought to change, and how?*	Developing new systems and structures of planning and development that practice restorative justice for EJ communities and ecologies (abolitionist ecology, reparation ecologies, just transitions, rights to the city, etc.)

HISTORY AND TRADITIONS OF PLANNING IN RELATIONSHIP TO ENVIRONMENTAL JUSTICE

The history of planning is as old as human settlements. Populations around the world have continuously evolved methods of settling or organizing land for human survival within their physical and cultural contexts. Contemporary planning in the U.S. has antecedents in western European traditions of planning that arose out of particular conditions of 19th-century industrialization and urbanization. This Eurocentric tradition of planning was spread across the globe through the processes of colonization and imperialism. The justification for planning is contested. Some tout it as a means to check the "free market" and exert state interventions in the public interest, such as public health and the separation of incompatible land uses, while others argue that planning's primary purpose is to serve as an instrument of capitalism, controlled by experts and elites to make cities conducive to capital flows and profit (Fainstein and DeFilippis 2015). Ambe Njoh describes the spread of European planning models as a rapacious vehicle for the acculturation of racial others (Njoh 2010). Most of the traditional planning models (described in table 11.2) represented a Western, rational approach grounded in utopian visions of cities laid out according to principles of efficiency, order, and beauty, and imbued with the racist, imperial, and colonial assumptions of the day. These utopian city planners entrenched patterns of inequality, erasure, and market logics that were reevaluated and reckoned with later by progressive planning models (see table 11.3).

TABLE 11.2. Traditional Planning Models and Critiques

Planning Model	Characteristics	EJ and CER Critiques
Western, Rational Planning Radiant City (LeCorbusier) Garden City (Howard) City Beautiful (Burnham)	Utopian and rational goals Based on liberal notions of "free market" economics Driven by professional planners, typically situated within the state, or working with elite actors Favors top-down processes	Reflects entrenched white supremacy, settler-colonial values Drove patterns of racial segregation Equity not a priority Public participation not a focus
Euclidean Zoning	Regulates physical form and location of land uses (residential, commercial, industrial) Sets guidelines for physical layouts (building heights, permitted uses, etc.)	Means for racial segregation Exclusionary forms of zoning expelling affordable housing and industrial uses from whiter, wealthier areas Captured by property-owning elites for profit maximization
Rural Planning Provincial planning Town and country planning (Dandekar)	Addresses rural economic development and resource management Linked to key sectors of rural development, including agricultural and natural resource-based economies (fisheries, forestry, etc.)	Failure to acknowledge low-wealth, marginalized, and Black, Indigenous, and people of color (BIPOC) populations occupying rural areas, and to meet their basic needs (sanitation, clean water, farmworker protections, etc.) Based in economic patterns shaped by slavery, settler colonialism, and nativism
Globalized, Neoliberal Planning	Global market interests dominant in development and planning practices Serves to manage competition for urban land with an emphasis on technology and efficiency International financial institutions and private real estate sector as key actors in planning processes	Contests the state's role in planning practices, which are seen as captured by market interests Disfavors formal planning practices, which serve the interests of private capital, in contrast to more informal bottom-up processes Tamps down the role of insurgent, rights-based social justice movements in planning

Many traditional planning models stand in stark contrast to EJ approaches to planning and community-engaged practices, both in form and function. For example, CER approaches to planning emphasize (1) direct democratic ideals of participation from the ground up or by people directly impacted; (2) centering equity and social justice concerns in both the process and outcomes of planning;

TABLE 11.3. Alternative Planning Models

Models	Emphases	Examples	Characteristics
Progressive Planning (Fainstein and DeFilippis 2015)	The role of learning, social justice advocacy and equity are central to planning's purpose Emphasis is on a bottom-up approach to planning with greater attention to multiple forms of public participation in the planning process Planning is largely within the purview of the state and professional planners	**Advocacy Planning** (Davidoff 1965)	Privileges the interests of the most disadvantaged and places the public planner in the role of advocate
		Equity Planning (Krumholz 1982)	Focuses on the goal of redistribution, with public planners promoting progressive policies from within state-centered planning
		Communicative Planning (a.k.a. discursive or deliberative planning) (Healey 2012)	Focuses on social learning, and more inclusive and democratic processes of understanding social conflict and planning
Radical Planning	The goal of planning focuses on liberation and realization of a just society The model problematizes formal or state-led participation processes and focuses more on direct, participatory democracy or self-determination Social movement actors, grassroots groups, and marginalized, dispossessed peoples are central to the planning process	**Insurgent Planning** (Tactical urbanism, Right to the City, favelados, Slum Dwellers International, etc.) (Gonsalves et al. 2020; Miraftab 2012)	Has citizens acting directly through self-determined oppositional practices that claim urban spaces Aims to address specific forms of oppression Focuses on counter-hegemonic, transgressive, and imaginative planning practices
		Black Radical Tradition (Jacobs 2019; Pulido and De Lara 2018)	Challenges racial capitalism and state-centered planning Centers Black experience, solidarity across identity Focuses on community knowledge, intersectional oppressions, and activism in the formation of plans Emphasizes emancipatory and abolitionist goals, outside the state
		Indigenous Planning (Jojola 2008; Porter et al. 2017)	Centers Indigenous knowledges, identity aspirations, worldviews, and cultural practices Focuses on decolonized, transformative, and epistemic justice

(3) de-centering professionalized planners and elite actors as the main drivers of planning, and instead putting communities and activists in the role of experts; and (4) encouraging collaboration and interdisciplinarity in planning methods. Most importantly, the goals of planning using a CER approach in an EJ context also differ dramatically by prioritizing transformative forms of justice and well-being over the goals of efficiency, order, or profit seeking. Table 11.2 summarizes some of the traditional planning models and how they contrast with or are critiqued by CER- and EJ-informed planning practices. These models and their respective critiques are represented in simplified terms to highlight the distinctions between them. But there are also overlapping characteristics and diverse expressions of traditional approaches that can be found in a variety of contemporary planning practices, including some that involve CER and attend to EJ concerns.

Approaches to CER in Planning

In the decades after World War II, the era of traditional, top-down planning driven by private sector interests and state planners was forcefully contested. During this period, the rise of the Civil Rights movement coincided with the resurgence of social reform-minded planning practices that included greater consideration of issues of social equity, democratic ideals, and diverse public interests. This pivot introduced various models of planning that served as a foundation for many CER practices in use today in EJ communities. Table 11.3 summarizes the dynamic continuum of alternative planning practices, from progressive planning models that attempted to reform traditional planning to more critical and radical planning practices drawn from the Global South, Indigenous struggles, and Black radical traditions.

Along this continuum of alternative planning models, there are diverse perspectives on planning's goals, approaches to public participation, and the situatedness of planners. In the progressive planning model, "progressive social change results only from the exercise of power by those who previously had been excluded from power" (Fainstein 2000, 466). In this view, planning is not just a process mediated by public and private interests and controlled by the state, but a process of active engagement with and by social movements to produce a more just city. The more radical strains of planning put these social movement actors in the driver's seat to envision alternative futures that take back cities and land from the exclusive control of propertied elites (Harvey 2008, 24). The visions that these different forms of planning produce can often overlap, such as promoting equitable access to resources or community well-being. But they can also diverge, as many radical planning traditions seek to go beyond distributive or procedural forms of justice and state-centered planning to enact abolitionist or transformative forms of justice in the form of liberation or reparations. For example, a radical approach to planning in a community facing food insecurity like Detroit would plan around the development of autonomous, community-owned food production and distribution based on cooperatively owned land and markets. This is precisely what D-Town Farm and the Detroit Black Community Food Security Network (n.d.) set out to do to meet the community's food needs—rather than pursuing state-subsidized or privately controlled food markets. This is just one example of how the EJ movement shares many of the same goals that radical planning proposes (Griffin, Cohen, and Maddox 2015).

Participation is important in both the progressive and radical traditions. The norms of participation are embedded in the professional planning code of ethics (American Institute of Certified Planners 2021), which calls for the "meaningful involvement" of communities. However, some challenges and critiques emerge around the role of participation in progressive planning models. Participation without power is meaningless and frustrating (Arnstein 1969). Communicative forms of planning attempt to grapple with the uneven power dynamics often at play in state-led planning models. For example, John Forester (1989) offers pragmatic ways in which professional planners can influence the conditions that shape a community's ability to participate in formal planning processes, such as (1) notifying less organized groups early in planning processes, (2) supplying critical technical and political information to communities, (3) anticipating the political and economic pressures that will shape plans, and (4) sharing those issues with groups early and through open as well as informal processes, etc.

While participation in communicative planning is still driven by professional planners, in radical planning traditions participation often falls well outside of state forums, such as public hearings, planning charrettes, or public meetings. Instead, participation in traditions such as insurgent planning can be unorganized and spontaneous, and sometimes includes illicit acts by residents attempting to reclaim, control, or shape spaces at the center of planning contestation. One example that embodies the transgressive and imaginative forms of action that insurgent planning can produce is the case of community communicators working in the favelas of Rio de Janeiro, known as the Frente de Mobilização da Maré (Friendly 2022). During the COVID-19 pandemic, and in the face of repressive federal government actions, these communicators planned creative uses of local media and outreach to promote prevention actions, distribute mutual aid for the provision of basic services, and disrupt the presence of the police state inside favelas (Friendly 2022). This exemplifies how radical planning transgresses the norms of formal participation, as communities resist oppressive state actions and reclaim control to shape the conditions of their lives.

Progressive and radical planning models can also involve differing views of the roles of the state and professional planners. Radical planning de-emphasizes the roles of professional planners and the state in favor of activist-led forms of planning that engage deeply with social movement actors, situating planners as within and aligned with movements (Huq 2020). These approaches increasingly call for planners' training to include active engagement with resistance movements, and learning more critical and liberatory practices that de-center whiteness and employ decolonial, antiracist, and revolutionary practices and tools (Urban Planners for Liberation 2021). Deshonay Dozier (2018) suggests introducing students to BIPOC planning voices, histories, and readings that draw on diverse disciplines and tools, such as adrienne maree brown's (2017) *Emergent Strategy* and the *Abolitionist Planning for Resistance* guide (UCLA Abolitionist Planning Group 2018). Planning education can also learn from activist and practitioner sources, including the Center for Urban Pedagogy (n.d.), BlackSpace (n.d.), and the *Urban Green Policy Toolkit* (Oscilowicz et al. 2021).

In many examples of CER-based approaches to planning in EJ communities, EJ groups partner with professional planners and bring multiple actors, including state actors, into the planning process. In this sense, CER-based planning can be more collaborative, intersectional, and open to multiple forms of expertise and knowledge than traditional or even progressive forms of planning. One example can be seen in the case of the Ironbound Community Corporation (ICC) in Newark, New Jersey (see table 11.4). As part of local efforts to reclaim the waterfront from industrial and real estate speculation for the development of public parks, the community spearheaded the Ironbound Open Space and Recreation Plan. ICC initiated this plan with a committee of residents who identified their vision for a public waterfront and mapped out assets, park needs, and potential threats. The group also partnered with public planners to draw up renderings, and together they implemented a campaign to stop the privatization of the waterfront. Ultimately, this plan was the foundation of the city's Riverfront Park design that was implemented in 2013.

CER approaches are particularly relevant for planning related to emergent, intersecting, and multiplying threats in EJ communities. These threats pose both acute and chronic impacts in the form of legacy pollution, health disparities, climate risks, disasters, and displacement that formal planning processes ignore or are ill equipped to address. Thus, many locally based, grassroots EJ organizations have found themselves applying radical planning tools to ensure their survival and resurgence.

CASES OF CER AND EJ IN PLANNING

There is a rich array of examples in which EJ communities have used CER in the context of planning. These examples include community efforts to draw attention to and collect data on harmful conditions; prepare for or respond to disasters; advocate for greenspace; push back against displacement; ensure healthy, safe, affordable places to live; reimagine economic prosperity; respond to climate impacts; and most importantly to lay out visions of an environmentally just future. The examples discussed below also reflect a variety of progressive and radical planning traditions that have involved community groups, EJ activists, residents, and professional planners both within and outside the state. So many of the EJ struggles that plague communities in the U.S. have their origins in the legacy of racist planning and zoning imposed on BIPOC communities from above. In contrast, the cases described here exemplify planning practices that emerge from the lived experiences and leadership of EJ communities. The plans also depict the richness of community-led visions for alternative, reimagined future possibilities of a more just and free world.

Some of the most prominent examples of CER in planning practices happen at the local level, where residents and grassroots EJ organizations have led efforts to carry out community-led planning and land use zoning reforms. While substantive regulatory reform at the state or federal level is often slow, EJ groups have been more capable of exerting their organizing power to impact regional, municipal, and county planning. Recent research on local land use policies and zoning regulations in the U.S. identified a total of 40 policies from across the country that had an explicit focus on EJ (Baptista 2021). The measures were adopted by more than 20 municipalities, two counties, and two local utilities, from Los Angeles to New York, largely as a result of local EJ advocacy. These policies spanned a diverse range of approaches, including (1) outright bans on unwanted, noxious land uses; (2) EJ policies embedded in general plans or explicit EJ policies or programs adopted by municipalities; (3) environmental justice reviews, often tied to the development process; (4) proactive planning measures or comprehensive approaches; (5) phase-outs, fees, or enforcement activities aimed at mitigating existing noxious land uses; and (6) use of local public health codes to prevent noxious or nuisance activities in EJ areas. Box 11.1 details how several California EJ communities employed CER to

BOX 11.1. CALIFORNIA GREEN ZONES

Green Zones emerged from EJ activists who sought relief from repeated struggles over siting of facilities that concentrated pollution in communities of color and low wealth. Despite decades of attention to EJ concerns in California, little progress was made to mitigate existing toxic hot spots. Many EJ activists reacted to the opposition and the complexity of regulating cumulative impacts of multiple pollutants at the state level by turning to local planning venues, which might address the concerns of EJ communities more proactively.

Green Zones are specific areas within a locality designated by the local government and identified by residents for improvements in economic development and public health through the reduction and prevention of existing burdens, and direction of investments to greener development projects (California Environmental Justice Alliance 2011). Typically, this process includes (1) greater regulation of polluting land uses through the creation of special use or overlay zones by local planning offices, (2) community decision making to identify the zones and targeted interventions, and (3) collaboration with the public and private sector to direct investments to local green businesses with local employment opportunities.

(*Continued*)

BOX 11.1. (*CONTINUED*)

By 2015, 13 organizations in 11 EJ communities were using the Green Zones approach (California Environmental Justice Alliance 2011). Municipalities including San Francisco, Los Angeles, Richmond, and Commerce, as well as the County of Los Angeles, have also adapted this approach to their local zoning and development processes.

CER has made significant contributions to the development of Green Zones. The community organizations involved in creating Green Zones conducted extensive ground-truthing exercises with local residents, using their knowledge of the area to identify previously undocumented hazards, confirm or highlight particular hot spots for pollution, identify vulnerable or sensitive areas of the neighborhoods, and then fact-check the existing state and local databases. This form of local data collection not only helped to identify the areas for Green Zones, but also shaped the types of planning controls and incentives residents in each area needed to address local concerns. In addition, EJ organizations and residents collaborated with volunteer or professional planners to help develop proposed planning ordinances and overlay zones. These collaboratives also worked with local city planners to engage them early in the process of developing the scope of zoning changes and target neighborhoods for Green Zones. Communities not only engaged in research, but also led the visioning and implementation of Green Zones campaigns that persuaded municipal and county governments to adopt model ordinances.

implement a proactive planning approach called Green Zones—a model that has since been adopted by EJ communities in other parts of the country.

Cases such as those summarized in table 11.4 show how community-based, grassroots organizations take planning into their own hands to guide the future development of their communities. In many cases, social movement activists operating outside state processes initiate planning, articulate transformative visions for the future of their communities, and counter neoliberal values of efficiency and profit seeking by emphasizing community well-being, health, and equity. Some plans are developed collaboratively with multiple stakeholders—including planners, residents, and state and private actors—yet these stakeholders often use data from CER grounded in local knowledge and experiences of residents to map out existing conditions and identify opportunities.

CER also features prominently in planning to address air pollution through local monitoring or ground-truthing efforts. Some of the earliest and ongoing EJ struggles centered on addressing the cumulative impacts of multiple sources of air pollution in overburdened fenceline and frontline communities. These communities searched for ways to raise the alarm about local conditions to skeptical government officials, who put the burden of proof of harm on residents. Without empirical evidence of emissions and exposure data, and lacking regulations that required polluters or regulators to gather these data, residents were left to

TABLE 11.4. Community Planning Initiatives by EJ Organizations

Organization	Characteristics of CER	Plans and Resources
WE ACT for Environmental Justice New York, NY	Planning processes led by WE ACT organizers include multiple community meetings, charrettes, and development of public education materials for advocacy campaigns to implement community visions and goals WE ACT also has planners on staff to lead community planning efforts	WE ACT for Environmental Justice (n.d.) plans: Northern Manhattan Climate Action Plan Harlem on the River: Making a Community Vision Real Green Renaissance: A Guide to Healthy, Sustainable, Urban Development in Harlem
Ironbound Community Corporation Newark, NJ	ICC staff lead and initiate community planning processes They hire professional planners to assist in plan development and lead community charrettes and meetings to identify future visions and goals for plans	Ironbound Community Corporation (n.d.) plans: ICC Community Master Plan Ironbound Open Space and Recreation Plan East Ironbound Revitalization Ironbound Riverfront Park Plan, 2004–2011 East Ironbound Neighborhood Revitalization Plan, 2018
Environmental Health Coalition San Diego, CA	EHC's community planning tools: Community action teams with residents trained to serve as spokespersons for campaigns and plans Leadership training programs, which provide residents with skills in planning and land use rules Community surveys to collect and document local needs Community visioning with residents to develop neighborhood plans Support from land use planning firms to work with residents in the development of plans	Community land use planning initiative, EHC planning (Environmental Health Coalition, n.d.)

fend for themselves to protect against exposures. In some cases, data about the source of hazards were incomplete, lacking granular information about conditions on the ground, such as smaller polluting facilities or unregulated, illegal activities present in EJ areas. EJ communities took responsibility for monitoring, data collection, hazard identification, and enforcement—functions commonly left to

TABLE 11.5. Community Science on Air Pollution for Planning

Organization	Characteristics of CER	Plans and Resources
Los Angeles Collaborative for Environmental Health and Justice Los Angeles, CA	A coalition of EJ organizations worked with residents to identify local air quality hazards The coalition developed a list of land uses and facilities considered sensitive or hazardous Residents were trained to locate and map facilities by walking in the community, using maps and air photos, to verify accuracy of regulatory databases	*Hidden Hazards* report (Los Angeles Collaborative for Environmental Health and Justice 2010) Clean Up, Green Up ordinance (City of Los Angeles 2016)
El Puente for Peace and Justice Brooklyn, NY	Residents conducted a door-to-door asthma prevalence survey in Williamsburg, Brooklyn (Ledogar, Acosta, and Penchaszadeh 1999) Local residents sampled and interviewed people fishing in the East River to estimate the number of fish caught and consumed; the data improved the U.S. EPA's risk estimates related to consumption of contaminated fish (Corburn 2002) Local youth and residents used mobile phone apps to record levels of air pollutants, conduct field observations of park usage and vehicle counts, and develop GIS maps of data sources (Ramírez et al. 2019)	Our Air! / ¡Nuestro aire! plan (El Puente, n.d.)
Community Air Mapping Project for Environmental Justice (CAMP-EJ) **NYC Environmental Justice Alliance** New York, NY	Residents from two EJ communities in the South Bronx and Brooklyn used low-cost, portable air quality monitors to measure local air quality and characterize air pollution exposures locally (Gilmore et al. 2021) Community groups and residents developed recommendations in response to the data	*CAMP-EJ: Findings and Recommendations Report* (Gilmore et al. 2021) HabitatMap, Aircasting (HabitatMap 2021)

government entities. This led to a diverse set of efforts, from community science and do-it-yourself sampling techniques to community mapping and ground-truthing activities, some examples of which are summarized in table 11.5.

Climate and disaster planning are also critical areas of concern for EJ communities, which often face disproportionate disaster-related burdens and have underlying conditions that can make them more susceptible to disaster impacts. This has become especially evident over the last two decades, as natural and man-made disasters have laid bare environmental racism and injustice. Cases of climate resilience planning have increasingly been taken up by EJ communities to respond not only to the climate crisis but to the threat of gentrification and displacement that can result from climate adaptation investments. The addition of greenspaces or investments in green infrastructure can lead to speculative real estate developments that have been referred to as "disaster gentrification" or "climate gentrification." For example, Greenberg (2014) examined the examples of New Orleans after Hurricane Katrina, Lower Manhattan after the attacks of September 11, 2001, and the New York region after Hurricane Sandy to demonstrate how disaster recovery can initiate cycles of displacement and disinvestment for EJ communities.

Planning scholars have increasingly turned their attention to this wicked problem: residents of EJ communities who struggle to improve conditions in their communities then find themselves priced out of their own communities as they become more attractive (Anguelovski et al. 2019). Efforts to respond to green gentrification have produced some interesting proposals, such as the "just green enough" approach, which favors smaller-scale greening projects tied to local social and ecological needs (Wolch, Byrne, and Newell 2014). Pearsall and Anguelovski (2016) give examples from Brooklyn, Boston, and Seoul to demonstrate how EJ and anti-displacement activism can use complementary tactics, such as initiating collaborative projects to integrate affordable housing measures with small-scale greening projects in line with local community needs and desires. There are also powerful community-led planning efforts to characterize and respond to neighborhood-level impacts of gentrification (Matsuoka 2017; Matsuoka and Urquiza 2021). Table 11.6 presents a variety of examples of EJ communities using CER in the process of responding to climate risks and disasters through their preparedness planning and recovery efforts, often in direct opposition to more traditional, top-down or state-led climate initiatives.

The EJ movement has long taken up the contestation over both wanted and unwanted land uses that invoke a collective voice to shape communities as more inclusive and healthy places for all people to thrive. Similarly, the Right to the City is both a demand and a movement that calls for low-income, marginalized people to have a say in all aspects of shaping the city, turning away from capitalism's rapacious cycles of investment and profit that benefit the real estate developers and speculators (Harvey 2008, 24). The EJ movement's efforts to shape community control of land redevelopment and housing apply radical and insurgent forms of planning to reimagine our relationship to economic prosperity, housing, and community development.

For example, the use of community land trusts to achieve permanent affordability and protect land for collective uses (such as farming) is increasing in many EJ communities, to regain local control over development processes overtaking community spaces (Blumgart 2015). One of the most exciting and revolutionary examples of this type of CER planning is the Jackson-Kush Plan developed in Jackson, Mississippi. The plan is the grounding document for the organization

TABLE 11.6. Climate and Disaster Planning

Organization	Characteristics of CER	Plans and Resources
The Green Resilient Industrial District Plan (The GRID), UPROSE Brooklyn, NY	Community-proposed alternative to private, real estate–driven development of "Industry City" for luxury retail use Reflected community vision to transform the neighborhood and industrial waterfront to integrate climate adaptation, mitigation, and resilience Focus on alternatives based on Just Transition values, including analyses of existing conditions, plans, and policies related to neighborhood opportunities for climate adaptation, and green industry and clean energy sectors	Collective for Community, Culture, and Environment (2019) *Sunset Park Green Resilient Industrial District* report (Collective for Community, Culture, and Environment 2019) *The Grid* (UPROSE n.d.)
Sandy Regional Assembly and Recovery Agenda, NYC EJ Alliance New York/New Jersey Metropolitan Region	Initiated regional convenings with labor, environmental, EJ, social justice, and service organizations to identify short- and long-term recovery and disaster response needs (Sandy Regional Assembly 2013) Focused on grassroots-led recovery prioritizing low-income people, communities of color, immigrants, and workers Centered bottom-up approaches to resilience planning and investments	Climate Justice and Community Resiliency plan (New York City Environmental Justice Alliance, n.d.)
Community-Driven Climate Resilience Planning: A Framework, National Association of Climate Resilience Planners U.S.	Bottom-up processes driven by residents of vulnerable and impacted communities to define challenges and solutions (Gonzalez 2017) Climate solutions that consider relevant, unique assets and threats in communities (Kresge Foundation 2019) 3 key capacities for climate resilience: (1) assert a community vision and priorities (2) assess community assets and vulnerabilities (3) build community voice and power	National Association of Climate Resilience Planners (n.d.) *Climate Resilience and Urban Opportunity Initiative* (Kresge Foundation 2019)

Cooperation Jackson, which was formulated by the New Afrikan People's Organization and the Malcolm X Grassroots Movement. The plan drew on government data to map conditions facing the Black Belt South and reflects the rich legacy of the Black Liberation Movement in its goals to "advance the development of the New Afrikan Independence Movement and hasten the socialist transformation of the territories currently claimed by the United States settler-colonial state" (Akuno 2017, 3). This plan was based on three fundamental pillars: (1) building people's assemblies, (2) building an independent Black political party, and (3) building a broad-based solidarity economy. This type of people-led plan demonstrates the possibilities for radical forms of CER planning to articulate emancipatory ideals of a free and just future.

Urban and regional planning also plays a key role in the development and access to a variety of public and private infrastructures. Typically, these infrastructures serve populations across a wide geographic area and are sometimes considered locally unwanted land uses (LULUs) due to the related pollution, risk, and nuisances (odor, traffic, noise, etc.). LULUs are concentrated in areas where industrial development corridors were developed along racially and class-segregated residential patterns. In this way, regional infrastructures, including highways, wastewater treatment plants, energy production facilities, goods movement centers (i.e., seaports, warehouse hubs, airports, railyards, etc.), and waste facilities, are often sited in EJ communities. Additionally, many EJ communities throughout the U.S. lack basic infrastructure, including sanitation, clean drinking water, public transportation, broadband, and energy services. CER plays an important role in EJ struggles to mitigate effects of these infrastructures and transform them over time. Table 11.7 highlights examples in EJ communities in Baltimore, Puerto Rico, and Los Angeles.

CONCLUSION

While urban planning's origins in the United States gave rise to problematic planning models, contemporary practices have evolved with the advancement of more progressive and radical approaches led by planners, community activists, and social movements. Throughout the country, EJ communities have redefined planning's purpose and created new tools to meet their needs and reimagine their collective futures. Community plans that are informed by CER share noticeable similarities, such as privileging local knowledge, prioritizing more equitable benefits and well-being, and a collaborative and democratic approach to planning. The cases highlighted in this chapter demonstrate the depth of expertise and experience in community-led CER for planning oriented to EJ goals. Many EJ organizations today have planners on their staff and build their planning around resident-led efforts. These groups often integrate popular education and organizing into work with residents in ground truthing, visioning, and implementing community-based planning efforts.

There are also exciting new opportunities for pushing CER planning practices to new areas of focus. One of these emergent areas can be found in abolitionist

TABLE 11.7. Infrastructure and Greenspace Planning

Organization	Characteristics of CER	Plans and Resources
Community Solar Energy Initiative, Resilient Power Puerto Rico Puerto Rico	Engages community groups most impacted by Hurricane Maria to deliver direct donations for the installation of solar energy systems (Funk 2021; Resilient Power Puerto Rico, n.d.-a, n.d.-b) Matches funding with community centers that agree to become community energy hubs and provides technical installation support Once installed, communities identify post-disaster needs and priorities, develop a collective operations and maintenance plan, and define community energy resilience agenda	Resilient Power Puerto Rico Lookbook (Resilient Power Puerto Rico, n.d.-b) *Energy Independence in Puerto Rico,* Community Solar Projects, StoryMaps (Funk 2021)
Baltimore's Fair Development Plan for Zero Waste, Fair Development Roundtable Baltimore, MD	Participatory approach with the leadership of grassroots organizations, youth-led groups including Free Your Voice, the United Workers, Institute for Local Self-Reliance, and other partners in Baltimore Focus on replacing waste incineration with local economic opportunities in zero-waste industries, such as food waste composting, repair work, and recycling	Institute for Local Self-Reliance, *Baltimore's Fair Development Plan for Zero Waste* (Liss et al. 2020)
Community Alternative 7, Coalition for Environmental Health and Justice Los Angeles, CA	Presents alternatives for goods movement projects, including the I-710 freeway expansion (Karner et al. 2018) Project alternatives developed by coalition of local residents along the freeway, legal organizations, EJ and community groups Alternatives included a list of key elements, such as public transit, community benefits, pedestrian and bike investments	I-710 campaign (East Yard Communities for Environmental Justice, n.d.) *I-710 Corridor Project HIA* (Human Impact Partners 2011)

or reparative forms of radical planning, which are gaining more attention among EJ communities and allies (Sze 2020, 29). This approach centers on the struggles for freedom from violence and the abolition of prisons, border walls, the police state, and other expressions of the carceral state that perpetuate violence against BIPOC and low-wealth people (Dozier 2018). EJ communities seeking freedom from both the extractive economy and the prison-industrial complex can use abolitionist and reparative practices in their approach to planning the future of their communities. However, in this movement-allied form of planning, the role of professional planners again comes into question. While planning students, such as the UCLA Abolitionist Planning Group (2018), seek to forge new practices, questions remain about how radical or reform-oriented planners can be. For example, in Dozier's view,

> [a]bolition is not, nor ever will be, about "planners." It never has been. Instead, it is about practitioners of freedom dreams that occur outside of planning education and profession. Contributing to these movements and redistributing resources to them is a step in what "planners" can do. (Dozier 2018, para. 9)

There are many freedom dreamers in the EJ movement working alongside many other allies, including professional researchers, to experiment with this form of radical planning. An example can be seen in the Renewable Rikers Plan (Bratspies 2020). The Rikers Island prison complex occupies hundreds of acres in New York City and is one of the country's most notorious penal colonies. The Renewable Rikers Plan connects the current crises of mass incarceration, toxic prisons, and environmental racism to a vision grounded in restorative justice and reparations for the people and land harmed by the legacy of colonialism, incarceration, environmental injustice, and racism. This campaign is led by a coalition of organizations including the New York City Environmental Justice Alliance, New York Lawyers for the Public Interest, Urban Justice Center, NRDC, and A More Just NYC. Together these groups convened legislators, legal advocates, and activists across a range of social movements to conceive of a campaign not only to shut down the notorious prison, but to replace it with reparative projects that give opportunities to formerly incarcerated people, as well as residents of EJ communities, to produce renewable energy, grow food, and treat wastewater. In February 2021, the New York City Council (2021) passed three bills transferring Rikers Island from the Department of Corrections to other agencies for sustainability and resiliency purposes. The laws also require a feasibility study for renewable energy production and storage as well as wastewater treatment. An advisory committee will guide the process with survivors of Rikers and residents of EJ communities.

Another emergent CER practice in the EJ movement is planning focused on just transitions. Just transitions is both a concept and a process by which society shifts from an extractive, exploitative economy to a regenerative economic system by making connections between workers and community issues, organizing, and movement building (Córdova, Bravo, and Acosta-Córdova 2022). One example of this type of planning was developed by the Climate Justice Alliance (CJA) with Movement Generation in their guide to just transition planning (Gonzalez 2021), which details how communities can craft and lead their own vision for achieving a just transition. This guide provides insights into the role of planning, as well as curriculum and other tools to lead planning processes grounded in frontline community experiences. Some of the key roles of planning that the guide highlights are (1) activating cultural wealth and community assets, and practicing accountability to community vision and values; (2) alignment among key players in moving a just transition strategy; (3) advocacy and organizing that is responsive to community priorities; and (4) activating community capacity to take over public planning processes. EJ organizations—such as PUSH Buffalo (2017), Kentuckians for the Commonwealth (2013), and the Indigenous Environmental Network (2021)—along with public agencies and even some in the private sector, are increasingly developing similar plans that articulate their visions and strategies for shifting to a pollution-free and more just set of economic and social systems around which to build their communities. As the climate crisis deepens in the decades ahead, the ability for EJ communities to plan for and implement transformative change will be critical to their survival and resurgence. These emergent approaches to planning can break open a radical reimagining of future possibilities, allowing EJ communities to research, reclaim, restore, and remake their communities and the world through acts of reparations, freedom, and placemaking (Gilmore 2017).

12

Conservation

Ashwin J. Ravikumar, Deniss Martinez, Jeanyna Garcia, Malaya Jules,
Chad Raphael, and Martha Matsuoka

Advancing environmental justice in conservation requires undoing colonial relationships, centering traditional ecological knowledge and sovereignty in research that informs policy and practice, and shifting decision-making power to Indigenous and other communities so that they can thrive on their lands. In this chapter, we critique the history of conservation science and policy, and reflect on how Indigenous and other marginalized communities have reclaimed research to conserve nature on their own terms. We show how a small but growing body of community-engaged research (CER) has provided an alternative understanding of conservation of forests, freshwater and marine ecosystems, and wildlife in places such as the Putumayo watershed in the Amazon, and the Klamath Basin and the Great Bear Rainforest on the Pacific coast of North America. We offer guidance on how to navigate the fraught relationships between conservation and environmental justice (EJ) by presenting key lessons from these case studies.

Throughout the chapter, we foreground the role of CER that involves Indigenous-led research and that centers traditional ecological knowledge, for several reasons. Indigenous peoples have been harmed most powerfully by conservation policies that have removed or restricted people's access to land and their self-determination. Indigenous nations and tribes are also crucial contributors to conservation because around 80 percent of the planet's remaining biodiversity resides on Indigenous lands, covering over 20 percent of the world's land surface (Whyte 2021). In addition, because many Indigenous peoples' identities and livelihoods are inextricably rooted in their ancestral lands, focusing on the impact of conservation policies on Indigenous communities highlights most clearly how access to healthy land is central to peoples' cultural and economic well-being. Indigenous conservation also holds expansive views of intergenerational and interspecies justice, which

TABLE 12.1. CER for EJ in Conservation

Dimension of Justice	In CER for EJ in Conservation
Distribution *Who ought to get what?*	Devoting research resources to conserving and restoring access to land for Indigenous cultural, spiritual, and economic sustenance, and healing nature Funding Indigenous and community-led researchers and initiatives directly
Procedure *Who ought to decide?*	Exercising Indigenous self-determination and other affected communities' rights to influence conservation research and policies Promoting Indigenous knowledge sovereignty and control over data gathered on their ancestral lands
Recognition *Who ought to be respected and valued?*	Centering traditional ecological knowledge Recognizing responsibilities to past and future generations to care for land Recognizing reciprocal kinship relationships to nature
Transformation *What ought to change, and how?*	Decolonizing knowledge, institutions, and systems in conservation science to restore nature and self-determination to Indigenous peoples

include obligations to past and future generations of humans, and to the Earth, to care for lands and species in reciprocal kinship relations. Moreover, the historic exclusion of Indigenous ecological knowledges from Western science, as well as their complex rapprochement in some current conservation science, points to the importance and challenges of reconciling local knowledges with dominant forms of expertise. Table 12.1 summarizes how the main issues discussed in this chapter relate to the dimensions of justice common to CER and EJ.

THE LEGACY OF FORTRESS CONSERVATION

Historically, the conservation movement in the United States and around the world has often worked against the interests of marginalized people. Conservation science and policy were developed in the 19th and 20th centuries by people who saw human activity as largely incompatible with environmental conservation (Cronon 1996). Racism was often central to this project. Conservation policy was built to protect nature for the enjoyment of wealthy white settlers, to the exclusion of Indigenous people, people of color, and poor white people (Jacoby 2014). John Muir, the founder of the Sierra Club and an early "preservationist" and advocate for national parks, viewed Indigenous North Americans as nuisances to be removed so that landscapes might thrive. Muir described Indigenous Californians in the Yosemite Valley region as "mostly ugly, and some . . . altogether hideous" people who "seemed [to have] no right place in the landscape" and complained that he could not feel the "solemn calm" of wilderness when he was in their presence

(Spence 1999, 109). The other major stream of environmental ideology during this period, the "conservationist" movement, viewed nature as useful insofar as it delivered goods that would feed the engines of the growing capitalist economy. For example, Gifford Pinchot, the first head of the U.S. Forest Service, sought to manage and conserve the forests of the United States not for their beauty, spiritual value, biological diversity, or cultural value, but to maximize the production of timber—and to ensure that business interests could continue to profit from its availability (Rinfret and Pautz 2014).

While their objectives differed, neither Muir's preservationists nor Pinchot's conservationists were interested in learning from the traditional ecological knowledge of Indigenous North Americans, nor in sharing the benefits of nature with poor people of any race. Although early preservationists and conservationists helped pass policies to conserve some important ecosystems, these movements marginalized and removed Indigenous people from the lands they had managed for centuries, to preserve a mythologized "pristine" nature. They replaced Indigenous land management practices, including the strategic use of fire to maintain healthy mixed-aged forest stands that allow for high biodiversity and promote multiple ecological functions, with Western "scientific" management that focused solely on producing timber reliably. Similarly, the conservation science of white settlers ignored the deep connections that Black people in the U.S. had to nature, even as they became integrally involved in the work of building national parks, farming, and managing land, both as enslaved people and as legally freed folks (Finney 2014; Taylor 2016).

The United States exported this model, known as fortress conservation, to the rest of the world (Baletti 2011; Brockington 2002). Following this logic, countries in the Global South moved in the latter half of the 20th century to establish protected areas by displacing local people who had historical claims to these lands. From Southeast Asia, to the Congo Basin, to the Amazon, environmental nonprofits based in the United States often abetted these conservation schemes (Hance 2016; Myers and Muhajir 2015). These initiatives were ostensibly undergirded by science: in particular, ecologists from or trained in the Global North would prioritize regions for conservation based on biodiversity indicators. For much of the 20th century, and into the 21st century, research on how people used natural resources was absent from conservation science, and the preferences of local people were sublimated to the dogma of conserving biodiversity by removing people from the land.

Between 1970 and 2010, countries in the Global South would also create environmental ministries tasked with establishing and overseeing protected areas, enforcing pollution standards, and regulating industries through environmental permitting (Busch and Jörgens 2005). The World Bank conditioned loans to developing and newly decolonized countries upon their having national environmental protection strategies (Busch and Jörgens 2005). In the 1990s and 2000s, large environmental nonprofits grew to wield great influence over the conservation policies of the Global South (Hance 2016).

At the same time, a new movement emerged that some scholars call "neoliberal environmentalism." Neoliberal environmentalism eschewed top-down regulations on industry in favor of consumer action and market-based solutions, such as payments for ecosystem services (Clark 2015). International agencies, including the United Nations and the World Bank, have pushed this approach to tropical forest conservation through the REDD+ program (*r*educing *e*missions from *d*eforestation and forest *d*egradation + enhancing forest carbon stocks), which aims to conserve tropical forests by paying their owners to leave them standing. To date, the vast majority of funds for tropical forest conservation have been channeled through environmental nonprofits into local projects, without yielding major reductions in tropical deforestation (Angelsen et al. 2018). Indigenous communities have in many instances opposed REDD+ and market-based conservation programs, calling instead for non-conditional funding to support Indigenous priorities and cosmovisions (Osborne 2015).

COMMUNITY-ENGAGED CONSERVATION RESEARCH ACROSS CONTEXTS

Some researchers recognize an obligation to use their platforms and resources to support Indigenous-led movements for conservation around the world. Taking a community-engaged approach to this research can make an especially valuable contribution to decolonizing knowledge and building conservation policy that centers and supports Indigenous communities and other people who steward important ecosystems, while repairing historical harm done by states and the environmental movement.

Indigenous and allied scholars have created important scaffolding for researchers to understand how Indigenous cosmologies—including kinship relationships with land (Goeman 2015; Whyte 2021) and animals (Hessami et al. 2021; Todd 2014)—differ dramatically from more narrow and anthropocentric Western conceptions of "natural resource management" and "wildlife conservation." These scholars have also chronicled histories of resistance and environmental activism (Gilio-Whitaker 2019), innovative land stewardship and governance (Carroll 2015), and ethical research and data collaborations (Carroll, Rodriguez-Lonebear, and Martinez 2019; Smith 2021). These works lay out theoretical frameworks for understanding and carrying out decolonial research in the context of campaigns led by Indigenous communities, and for finding common policy ground among Western-trained and Indigenous conservationists.

As this body of work underscores, research is not confined to studies conceptualized and funded by universities and other formal institutions. We understand research to encompass the sum of ways that people systematically and intentionally gather information and disseminate knowledge. Through this lens, research includes activists and organizers collecting information to support their campaigns.

It also includes Indigenous people experimenting with horticultural, fishing, and farming techniques and passing this knowledge on to children who accompany adults while they work. In some cases, these communities may not need the sort of research produced by formal scientific institutions at all. While formal research has not always been beneficial to communities who live in and manage ecosystems, a growing body of CER has helped support conservation that empowers communities and uplifts their agendas in a variety of ecosystems that humans use.

Some of this research has addressed the struggles of forest-dwelling communities. For example, Fisher (2021) collaborated with farmers, youth, local village planners, and others in the Kajang community to analyze how they became the first Indigenous people to gain recognition of their land rights from Indonesia's forest authorities. Demeulenaere (2021) integrated ethnographic methods and participatory action research with CHamoru people to document their efforts to preserve access to their forested terraces, medicinal plants, and sacred sites threatened by construction of a U.S. Navy firing range in Guam/Guahan. Kuan (2021) examined the Tayal people's use of community mapping and dialogue with state agencies to integrate Indigenous agroforestry and state-sponsored land management strategies in Taiwan. Varese (2006) and Chirif and Hierro (2007) recount the history of social science as a tool for securing land rights for Indigenous people in the Peruvian Amazon. Lake and Long (2014) describe collaborations between Native American tribal governments and the U.S. Forest Service to apply Indigenous fire stewardship for social and ecological resilience.

CER has also focused on freshwater and marine ecosystems. Ayre, Wallis, and Daniell (2018) draw recommendations for conducting ethical and impactful CER on freshwater conservation from the literature on Indigenous community-based natural resource management and estuary management in Australia, management of flood and drought risks in Bulgaria, and climate resilience and water management in the Pacific. Ban and Frid (2018) examine relational dynamics and tensions among Indigenous peoples and other researchers involved in the creation and management of marine protected areas in Canada, Australia, Vanuatu, the Cook Islands, Palau, Hawai'i, and Samoa. The authors found that the majority of successful collaborations emphasized cultural and social benefits more than ecological ones. McGreavy et al. (2021) summarized insights from multiple participatory projects on forest conservation, river restoration, and co-management of fisheries by an interdisciplinary team of Native and White settler scholars with the Penobscot Nation, including recommendations for addressing tensions between Indigenous cultures and Western science and academic cultures.

Across ecosystems, CER has also begun to contribute to studies of climate justice. For example, Work et al. (2021) collaborated with local justice advocates and residents to analyze "green grabbing" of Indigenous land in Cambodia for climate mitigation projects. This is but one example of the growing problem of large environmental NGOs and governments using the urgent need to respond to the climate crisis as a new rationale for denying Indigenous peoples' rights to participate in decision making and access their ancestral lands (Whyte 2020). More hopefully, Manning and Reed (2019) review the process by which the Yurok Tribe made one of the largest tribal conservation land acquisitions in the U.S., funded in part by carbon offsets and accomplished through a web of partnerships with tribal, conservation, private, and public agencies. This was also a victory for tribal sovereignty, as the Yurok expanded recognition of Indigenous values and rights in California's natural resources policy, and engaged in diplomacy with Indigenous nations in other states that may adopt carbon cap-and-trade policies like California's. The Yurok's land management is informed in part through their rich history of CER on conservation issues, including forest management (Marks-Block, Lake, and Curran 2019), food sovereignty (Sowerwine, Mucioki, et al. 2019; Sowerwine, Sarna-Wojcicki, et al. 2019), and remediating river water contamination (Middleton et al. 2019).

These and other conservation studies increasingly advocate for "biocultural" approaches to conservation that put the well-being of communities, as defined by those communities themselves, at the core of conservation research (Sterling et al. 2017). While many researchers who are not from these communities have been working to center their values, priorities, and knowledge, there is still a long way to go. Researchers from the Global North, postcolonial governments, and nonprofit organizations still too often set research agendas, with community "participation" only rising to the level of a second-order consideration (Sterling et al. 2017). We argue that researchers should take further steps towards community-engaged EJ research that defers to the political aspirations of communities, centers and uplifts Indigenous knowledge, and builds real power for communities with the most at stake in conservation. Fully adopting a decolonizing approach to research is especially important.

DECOLONIZING CER FOR INDIGENOUS-LED CONSERVATION

Decolonization is not a metaphor—it is not a matter of changing language and attitudes, but one of shifting resources and power to Indigenous people (Tuck and Yang 2012). Decolonizing the academy is not just about bringing in Indigenous knowledge, but also about bringing the power of the academy to Indigenous communities themselves, and transforming academic structures to support respect and reciprocity with Indigenous partners. As climate change continues to threaten the well-being of Indigenous peoples, it is ever more important to mobilize the resources, capacity, and finances of academic institutions to solve environmental problems with communities, while finding ways to turn over power and land (Smith 2021). This orientation towards decolonization is explicitly counter to what some academics view as the role of the academy: namely, that of an "unbiased" and "apolitical" scientific force—a view that has long been critiqued by feminist scholars and political ecology (Rocheleau, Thomas-Slayter, and Wangari 2013). Instead, decolonization requires that academics work in support of Indigenous campaigns, carry out applied research that uplifts Indigenous knowledge systems, and explicitly acknowledge researchers' commitments and loyalties (Estes 2019).

In practice, decolonizing research involves several characteristic arrangements. Research partners often develop Indigenous research advisory boards and review systems, share co-authorship, create copyright agreements, and institute data-sharing agreements that allow for Indigenous communities to retain the rights to their contributions in a way that uplifts their cultural sovereignty (see chapters 4 and 5). These practical steps stem from an underlying commitment to respecting knowledge sovereignty.

Knowledge Sovereignty

Across biomes, Indigenous sovereignty over knowledge is central to solidarity research for conservation. Knowledge sovereignty is the ability for communities to meaningfully control the production, interpretation, use, and distribution of information that pertains to their territories (Norgaard 2014). Community-engaged researchers have made efforts to work with, rather than suppress, Indigenous knowledge. One of the concepts that has emerged from these efforts is traditional ecological knowledge (TEK). This term is used to describe the deep ecological and geographic knowledge woven throughout Indigenous peoples' culture, governance, and practice. *TEK* describes the vast and expansive knowledge Indigenous people across the world have formed about their respective homelands. It is also a useful term when describing these knowledge systems at a large scale and when uniting groups working on the resurgence and reclamation of Indigenous culture, practice, and land stewardship in different parts of the world. At the same time, it is important to acknowledge Indigenous science's distinct place- and culture-based contexts, as well as its dynamic and relational nature (Wyndham 2017). These are important tensions that can often come up in natural resource stewardship collaborations (Nadasdy 1999).

TEK is rooted in concepts of land, which is central to Indigenous identity, culture, and social movements (Goeman 2015). Indigenous knowledge of the flora, fauna, and ecosystem dynamics present in their homelands is a powerful toolbox that can support environmental decision making. However, this knowledge can only be successfully implemented by including Indigenous knowledge keepers as leaders, not merely as consultants (Norgaard 2014). For this reason, shared decision making and knowledge sovereignty are key to any collaboration, and are important for subverting settler colonialism (Gilio-Whitaker 2019). Collaborations with Indigenous people, organizations, and tribal governments can be experiments in decolonizing knowledge to the degree that they subvert knowledge hierarchies that privilege Western science and, instead, return power and resources to Indigenous people (Neale and Smith 2019). Several additional conceptual tools can help advance knowledge sovereignty.

Two-Eyed Seeing

Diverse Indigenous communities in the Global North and the Global South have found ways to produce knowledge that align with their own culture and values, often without any need for outside assistance. However, in some instances scholars from outside of the community can provide helpful support. Just as non-Indigenous research institutions have strict guidelines for how legitimate knowledge should be created, Indigenous communities often have expectations about knowledge production (Batz 2018). Reconciling both sets of expectations, worldviews, and knowledge systems can be a challenge.

One framework that can support collaborations attempting to include multiple knowledge systems is "Two-Eyed Seeing," a Mi'kmaw concept taught by elder Dr. Albert Marshall (Reid et al. 2021). It encourages "learning to see from one eye with the strengths of Indigenous knowledges and ways of knowing, and from the other eye with the strengths of mainstream knowledges and ways of knowing, and to use both these eyes together, for the benefit of all" (Barlett et al. 2015, quoted in Reid et al. 2021, 245).

Whereas many Western scientists have sought to "incorporate" Indigenous knowledge into their research to some degree, Two-Eyed Seeing calls upon them to defer to Indigenous knowledge by treating it as an equal or greater way of knowing. This provides a means to dismantle the unequal power dynamics that pervade conventional Western conservation science. When Western scientists seek only to incorporate and integrate Indigenous knowledge into non-Indigenous systems, they assume that there are parts of Indigenous knowledge that fit their aims and other parts that may not. Subsequently, this can lead non-Indigenous researchers to compartmentalize or selectively tap Indigenous knowledge systems to fit within colonial ways of organizing knowledge (Nadasdy 1999). Two-Eyed Seeing reminds non-Indigenous researchers that they are likely to be novices at a significant portion of the collaborative work they undertake with Indigenous partners, and need to honor these partners' expertise.

In addition, while scholars have organized to increase open access to data and research, many Indigenous communities hold their knowledge collectively and govern it with their own organizations. To respect knowledge sovereignty, researchers should defer to Indigenous authorities with respect to data management, and clarify with Indigenous authorities which knowledge must be kept confidential and which data might need to be made public based on the rules and regulations of funders and non-Native collaborators. Ensuring that Indigenous organizations and nations are making decisions around the collection and dispersal of data is critical to knowledge sovereignty (Carroll, Rodriguez-Lonebear, and Martinez 2019).

TABLE 12.2. The CARE Principles as a Guide for CER

CARE Principles	Description	Evaluative Questions
Collective benefit	Data ecosystems shall be designed and function in ways that enable Indigenous peoples to derive benefit from the data	Do communities' political and policy agendas drive research design and implementation? What policy or political agenda does the research support? How does it impact access to land, resources, funding, and political power?
Authority to control data and knowledge	Indigenous peoples' rights and interests in Indigenous data must be recognized and their authority to control such data be empowered	Who controls existing data? Who will collect new data? What form do the data take? Who can access the data and how? Are there any limits to how people could access data?
Responsibility	Those working with Indigenous data have a responsibility to share how those data are used to support Indigenous peoples' self-determination and collective benefit	What do researchers do to demonstrate that their work delivers on promises, provides benefits, etc.? What steps do researchers take to be accountable to Indigenous communities and to convey the story of this work to a wider audience?
Ethics	Indigenous peoples' rights and well-being should be the primary concern at all stages of the data life cycle and across the data ecosystem	Do researchers understand that the well-being of communities is paramount? Are any outside stakeholders bringing in priorities that are in tension with community goals?

To this end, researchers at the Global Indigenous Data Alliance created the CARE Principles for Indigenous Data Governance (Research Data Alliance International Indigenous Data Sovereignty Interest Group 2019). Building on earlier work (Wilkinson et al. 2016), the CARE framework encompasses the principles of *c*ollective benefit, *a*uthority to control data and knowledge, *r*esponsibility, and *e*thics. The principles emphasize justice, Indigenous data for governance and governance of data, capacity building, and minimizing harm. In table 12.2 we build on the CARE principles and present several specific questions that researchers should ask themselves as they approach conservation work in places where Indigenous people live.

CASES IN DECOLONIZING CONSERVATION RESEARCH

Academic research in ecology, and the policy and social sciences, has overwhelmingly prioritized scholarly publication and "scientific objectivity" over transferring resources to support the political priorities of Indigenous organizations. In contrast, some scholars have looked to re-orient their research and deploy their platforms and resources in the service of Indigenous campaigns and decolonial projects. Here, we offer three examples of how CER has served to empower Indigenous people around protected areas and supported grassroots Indigenous movements. We chose the Putumayo and Klamath Dam case studies from our firsthand experience carrying out CER in the regions where the research occurred. We added the Great Bear Rainforest case as an emerging example of strong collaboration between Indigenous and non-Indigenous researchers in North America.

Protected Areas in the Putumayo Watershed

Background. Since 1999, more than 10 million hectares of tropical forest land in the Peruvian Amazon have been legally protected (Wali et al. 2017). Many of these protected areas were supported by Indigenous organizations. Protected areas have colonial roots, and have historically been deployed to exclude rather than empower Indigenous communities (Spence 1999). Throughout the 20th century, Peru was no exception to this global pattern (Orihuela 2020). Despite this, in recent decades researchers have worked with Indigenous communities to advocate for community interests through collaboratively managed protected areas (Wali et al. 2017). While many communities have gained more rights to land and resources by collaborating with researchers and the government, some Indigenous groups, including some Wampis and Awajún communities, resist collaboration with the state and pursue alternative legal pathways to greater autonomy (Gómez Perochena 2019).

In this context, Indigenous communities have worked with researchers to support their demands for cultural autonomy, land rights, and economic resources. Here we describe the case of the Putumayo watershed, where Indigenous organizations have advanced their goals by strategically enlisting the help of environmental nonprofits, research institutions from Peru and the United States, and international environmental foundations.

The presidency of Juan Velasco Alvarado in Peru (1968–1975) saw a significant land reform and, for the first time, collective land titles for Indigenous communities (Varese 2006). In the 1970s, Amazonian communities in Peru began organizing themselves into watershed level federations and regional organizations in order to fight for land rights and resources from the state. In the Putumayo watershed (see map 12.1), regional conservation areas and Yaguas National Park have been created since 2005 as a result of advocacy by Indigenous organizations and allied environmental groups. The regional conservation areas are collaboratively managed and used by communities, while Yaguas National Park (shown in dark green in map 12.1) has more restrictive legal uses.

Four major Indigenous federations led the charge to establish the park in 2018: the Federation of Native Peoples of the Putumayo Frontier (FECONAFROPU, for its initials in Spanish), the Federation of Native Peoples of the Ampiyacu-Apayacu Basin (FECONA), the Federation of Native Peoples of the Lower Putumayo (FECOIBAP), and the Federation of Native Peoples of the Maijuna Ethnicity (FECONAMAI). Of these organizations, FECONAMAI and FECONA both co-manage regional protected areas and built their constituents' interest in protected areas through these experiences (Pitman et al. 2016).

In 2021, Liz Chicaje Churay was awarded the Goldman Environmental Prize for her efforts to establish collaboratively managed protected areas in the region (Praeli 2021). She and other Indigenous leaders have for many years taken a strategic approach to working with outside researchers and organizations. They recognized early on that the titles that their communities held were not adequate to protect the lands that they actually used, valued, and cherished from extractive interests of loggers, gold miners, and large agribusinesses, among others. In this context, they needed to convince the government not only that these extended territories needed protection, but that they ought to be collaboratively managed by the Indigenous communities who had in fact steered them for generations.

Approach and Participants. To gather the information that they needed to make the case for Indigenous-led conservation in the region, the Indigenous organizations worked with national and international partner organizations to carry out "Rapid Social and Biological Inventories." These rapid inventories are intensive interdisciplinary data collection campaigns that bring Indigenous experts and Western scientists together to build a common understanding of the landscape, a shared vision for its future, and a strategy to advocate for this vision.

For support in these campaigns, Indigenous groups looked to organizations including the Peruvian nonprofit Instituto del Bien Común; the Field Museum of Natural History based in Chicago, IL; several national and regional government agencies; the Colombian nonprofit Foundation for Conservation and Sustainable Development; the National University of San Marcos based in Lima; and the National University of the Peruvian Amazon. Crucially, the Instituto del Bien Común had built long-standing relationships with Indigenous communities in the region, and elsewhere in the Amazon, by supporting their campaigns to title lands.

To collect data, a team of biologists led by the Field Museum and bolstered by Indigenous experts and Peruvian scientists carried out rapid field assessments of flora and fauna in key locations in the forest identified by communities. Meanwhile, a team of social scientists led by the elected leader of the Indigenous federation carried out a rapid social inventory. These social inventories involved the following elements: documenting stories and legends from elders; participatory mapping of natural resources use with focus groups of men, women, and youth; interactive exercises to visually depict the relationships between the community and state agencies; household economic surveys focused on the economic value that people derive from forest products and natural resources; interviews with knowledgeable community members to catalog key plant and animal species that they use; visits to horticultural plots to describe agricultural practices; semi-structured interviews with villagers to describe their concerns

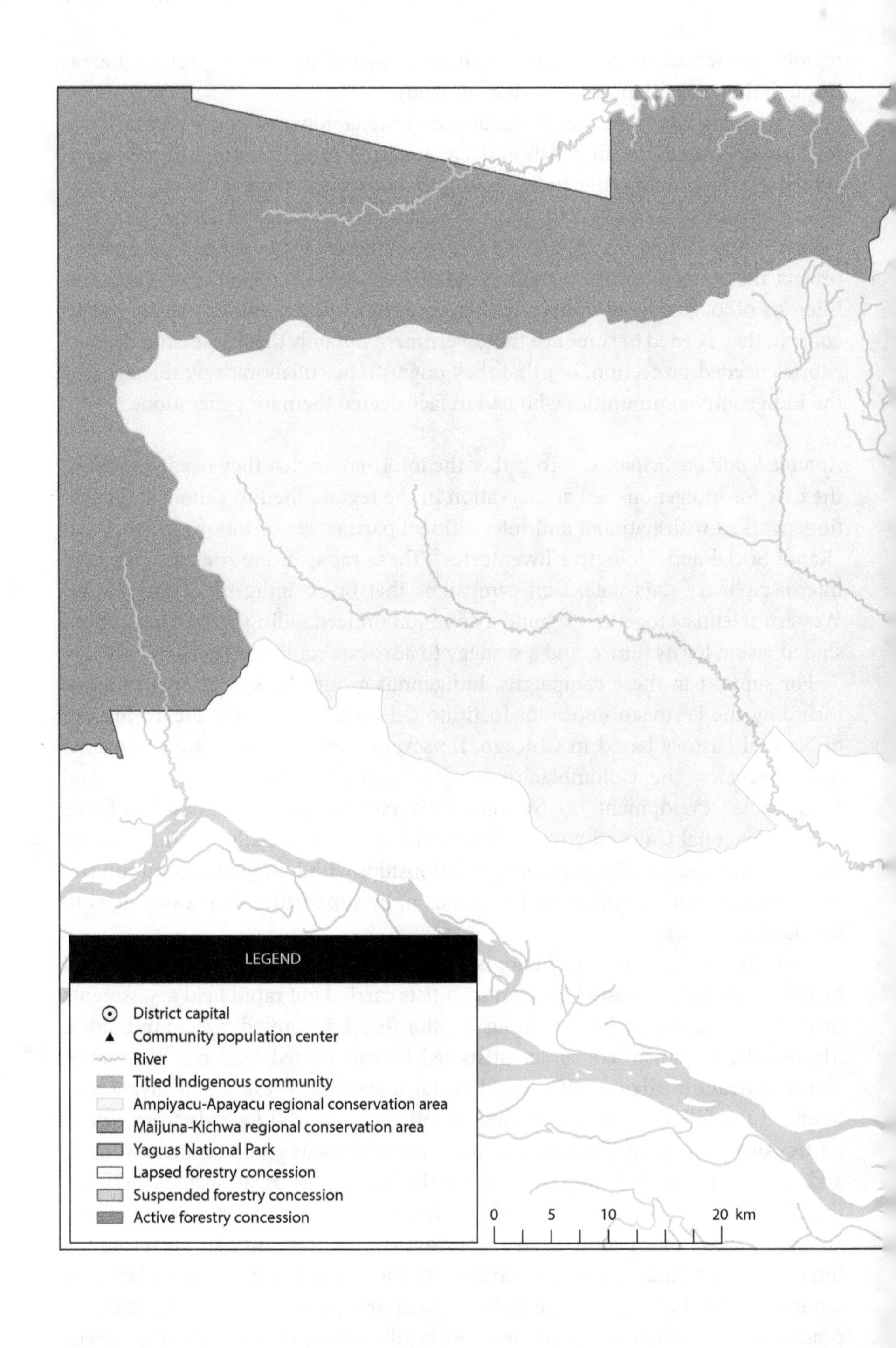
LEGEND
District capital
Community population center
River
Titled Indigenous community
Ampiyacu-Apayacu regional conservation area
Maijuna-Kichwa regional conservation area
Yaguas National Park
Lapsed forestry concession
Suspended forestry concession
Active forestry concession
0
5
10
20 km

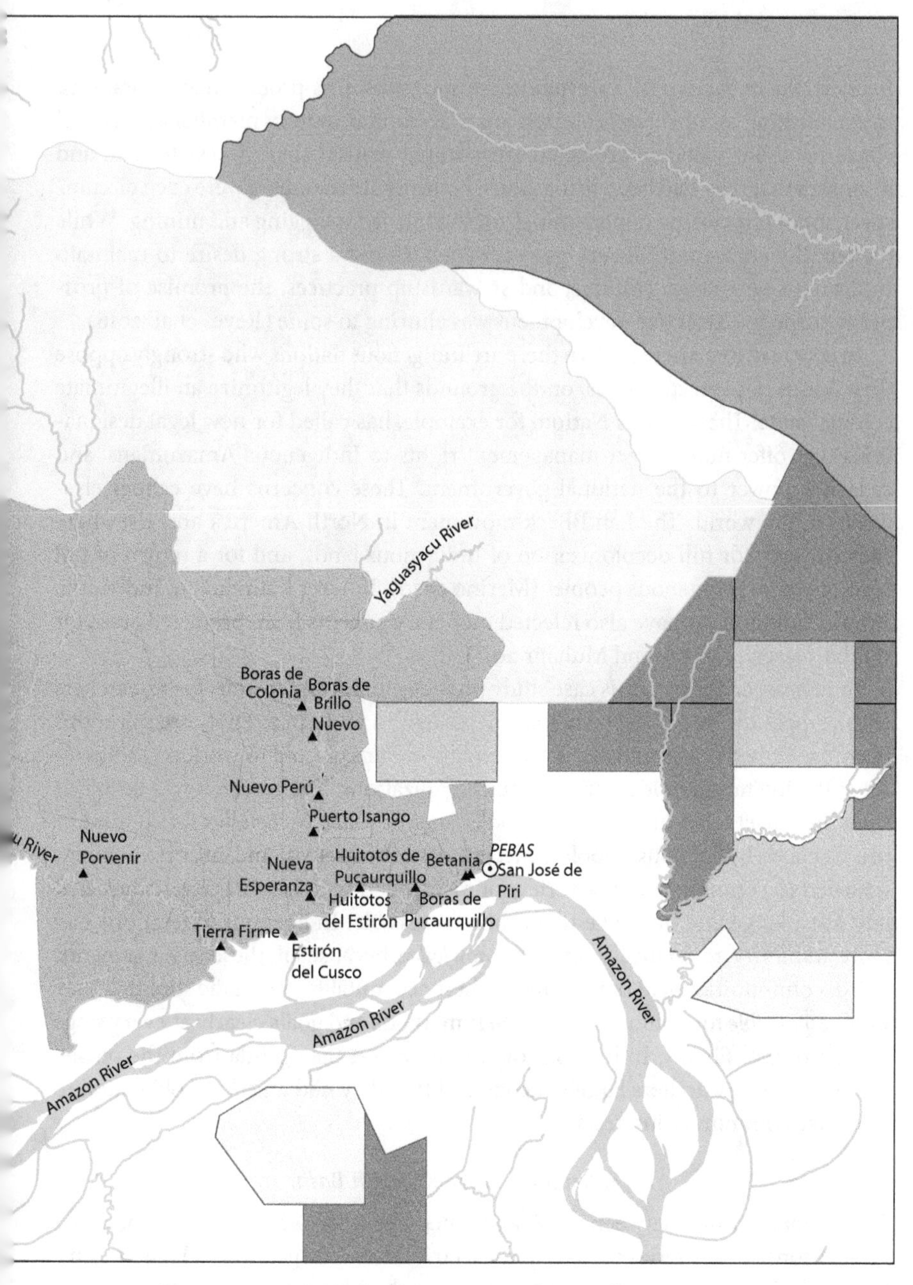

MAP 12.1. Conservation areas in the Putumayo Corridor, Northern Peruvian Amazon.
Map created by Jose Luis Jibaja-Aspajo.

and their vision for the future; and participant observation during hunting and fishing expeditions.

Implications and Lessons. Despite these successes, this process had limitations. In establishing all of these protected areas, some community members expressed concerns about whether protected areas might restrict their access to land and resources that they had been using. Some community members were even circumspect about foreclosing opportunities for income from logging and mining. While the lengthy community meetings generally surfaced a strong desire to maintain Indigenous languages, cultures, and stewardship practices, the promise of prosperity through extractive development was alluring to some (Reyes et al. 2016).

In contrast, elsewhere in Peru there are Indigenous nations who strongly oppose these kinds of protected areas, on the grounds that they legitimize an illegitimate colonial state. The Wampis Nation, for example, has called for new legal designations that offer more direct management rights to Indigenous Amazonians, and cede less power to the national government. These concerns have echoes elsewhere in the world. The Land Back movement in North America and elsewhere calls strongly for full decolonization of Indigenous lands, and for a return of full sovereignty to Indigenous peoples (Merino 2020). In West Kalimantan, Indonesia, Dayak communities have also rejected monetary benefits from protected areas for similar reasons (Myers and Muhajir 2015).

In this larger context, this case study provides important lessons for researchers with respect to the CARE principles described in table 12.2. First, organizations from the United States and the urban centers of Peru elected to work on this project at the invitation of local Indigenous organizations. The research was designed from the outset to secure ecological and economic *collective benefits* for communities. Second, Indigenous people collected data themselves, and information was returned to communities in a variety of media, and with key messages translated into local languages, to make them more accessible. Indigenous federations had more *authority to control data and knowledge* because of these arrangements. Third, communities held outside researchers accountable and made sure that they were *responsible* for communicating their methods and goals clearly at every stage of the process. Finally, Indigenous organizations set the agenda from the outset, meaning that researchers largely recognized that they had an *ethical* obligation to prioritize community interests.

Dam Removal in the Klamath Basin

Background. Built between 1908 and 1964, the Klamath River Hydroelectric Project consists of a series of four hydroelectric dams (Norgaard 2019). These dams have had severe impacts on salmon fisheries in the Klamath Basin, as the dams do not have fish passages and salmon cannot access over 150 miles of spawning and rearing habitat (Norgaard 2019). Salmon are central to culture, sustenance, and identity of Indigenous people in the area. The three major tribes along the Klamath—the Yurok, Hoopa, and Karuk—all depended on fish for sustenance, and the fish provided a source of wealth and well-being. In fact, the Klamath was once the third most abundant salmon-producing river in the lower 48 states (Gosnell and Kelly 2010). In 2001, the U.S. Fish and Wildlife Service and the National Marine Fisheries Service issued biological opinions that required higher water levels for endangered sucker fish in the upper basin and higher in-stream flow levels for the coho salmon. This caused a curtailment of water for irrigators, which led to losses between $37.5 and $54 million in gross crop revenues (Gosnell and Kelly 2010). In response to activism against these regulations, and a National Research Council report criticizing the science behind the 2001 biological opinions, the Bureau of Reclamation released a new management plan that provided a long-term irrigation water allotment. The fall of 2002 brought about a fish kill involving 33,000 adult salmon (Gosnell and Kelly 2010).

This fish kill was devastating given the importance of salmon for Klamath Basin tribes. For example, Karuk tribal members were once able to harvest 450 pounds per person per year (Reed and Norgaard 2010). Now salmon consumption has dropped to less than 5 pounds of salmon per person per year (Norgaard 2019). This event and a decline of other traditional food and fiber plants via fire suppression (Lake and Long 2014) have led to a drastic change in diet for Karuk tribal members, which comes with significant health and cultural implications, given salmon's centrality to Karuk identity and health. Activist and traditional dip net fisherman Ron Reed knew this, and when PacifiCorp filed to renew their dam license in 2004, he made every effort to voice his concerns (Norgaard 2019).

Approach and Participants. One of those efforts included a collaborative report with Dr. Kari Norgaard. The project consisted of surveys and interviews of Karuk tribal members about their health and their fish consumption. Interview and survey questions were informed by and developed in collaboration with Karuk tribal members. The research found that loss of access to traditional food was increasing diabetes rates for Karuk people to nearly four times the national average. The dramatic decline in eel and salmon populations, which provide essential nutrients important for the prevention of diabetes, happened within the lifetime of most Karuk adults alive at the time of the report (Norgaard 2004). These essential proteins once made up half the Karuk diet, and while diabetes was nearly unheard of prior to 1950, it became more common by the 1970s (Norgaard 2019). This report was groundbreaking in that it was the first time that a tribe had named diabetes as an impact of a dam in a federal process (Norgaard 2019). In 2008, an agreement was reached to remove the four dams along the Klamath in 2020, though that was delayed and is now slated for 2023 (Bacher 2021). Through tribal leadership, direct action, collaboration, and research partnerships that demonstrated the negative impacts of dams, tribes were able to change the political dynamics of dams. Tribes and advocates continue to advocate for the dams to come down without further delay (www.californiasalmon.org).

Implications and Lessons. With respect to the CARE principles, research documenting the impact of salmon loss on Karuk health and culture helped make the tribe's case for the *collective benefit* of dam removal. Tribal members retained *authority to control the data* gathered by co-conducting the community survey, documenting their own situation in multiple media, and retaining copyright over their academic research partner's resulting book. The tribe ensured that the research practiced *responsibility* to their interests and *ethics* by executing a contract with Norgaard to do the work, and through tribal review of and participation in the research.

The Karuk Tribe continues to be a leader in making the connections between health and the environment. The Karuk Department of Natural Resources has been a strong advocate for food and cultural sovereignty through tribal stewardship of forest, wildlife, and watersheds. Research coming from Karuk country benefits from careful scrutiny by community members via a process called "Practicing Pikyav," meaning "to fix it." Created in collaboration with researchers at University of California, Berkeley, this process was an effort to begin to fix the long history of harm done by researchers. The document outlines expectations and requirements for researchers that have created a strong body of research based in, led by, and relevant to the community. Requirements include a review by the Karuk Resources Advisory Board, an established team of local mentors, and use of community-based research, as well as a list of required research principles that protect Indigenous intellectual property, confidentiality, and self-determination. These protections ensure that tribal members can continue to leverage research for their decision making. Having a formal process also supports researchers who now have guidelines for how to engage the Karuk Tribe, as well as a touch point for guidance and support. The document can be used as a model or conversation starter in other collaborations that might not have a formalized process, helping to set clear expectations, boundaries, and goals for researchers and tribes.

Land Governance in the Great Bear Rainforest

Background. In the Great Bear Rainforest in British Columbia, Indigenous communities worked with nonprofits and independent researchers from universities, including the University of Victoria, to secure legal protection and resources for the forests that these communities traditionally stewarded. In 1997, the Supreme Court of Canada ruled that First Nations hold the rights to vast swathes of land and resources in British Columbia (Esbjorn-Hargens and Zimmerman 2009). First Nations worked with the local government, with large environmental nonprofits, including Greenpeace, the Sierra Club, World Wildlife Fund, Nature United, and the Nature Conservancy, and with independent scientists to campaign for restrictions on logging and a higher share of profits from any logging that does happen on Indigenous land. The Indigenous-led groups leveraged the support of the Nature Conservancy and Nature United to access funding for Indigenous-led conservation projects and investments in local businesses. More importantly, Indigenous groups demanded a right to co-manage their land, with the new agreement ensuring Indigenous rights in the context of the newly protected Great Bear Rainforest (Gaworecki 2016).

Approach and Participants. An Indigenous-led organization known as Coastal First Nations was established as a Great Bear initiative. Prior to mass organizing around Great Bear conservation, Indigenous tribes operated independently of one another due to the physical distance and cultural differences amongst them. However, the collaboration around Great Bear conservation inspired an alliance of the Wuikinuxv Nation, Heiltsuk, Kitasoo/Xai'xais, Nuxalk Nation, Gitga'at, Metlakatla, Old Massett, Skidegate, and Council of the Haida Nation (Low and Shaw 2011). Collectively, these nations held much more power than before. Nonprofits and independent researchers carried out ecological surveys to catalog ecosystem functions and traditional uses of the land in order to advocate for its protection under the leadership of the Coastal First Nations (Low and Shaw 2011).

The Coastal First Nations were crucial in the legal negotiations that led to the development of the Conservation Investments and Incentives Initiative. This initiative established financial support for the First Nations in their creation of a conservation-based coastal economy. This $120 million investment signaled an important shift in the definition of conservation (Low and Shaw 2011). For large environmental groups and researchers focused on environmental protection, conservation had been limited to the preservation of the natural environment. First Nations in Great Bear challenged this definition of conservation, expanding it to include the well-being of the Indigenous communities who lived in the rainforest (Low and Shaw 2011). Therefore, the fund allowed First Nations to manage and invest in sustainable business initiatives directly led by Indigenous groups to support the communities in the rainforest.

Indigenous communities in Great Bear also established a new category of protected areas called conservancies, which allowed Indigenous groups to insert themselves into the governing practices of these lands, whereas they had been excluded from governance of other land designations. The new designation enabled Indigenous groups to establish the management plan for conservancies, and empowered First Nations in each specific conservancy to serve as co-developers (Low and Shaw 2011).

Implications and Lessons. With respect to the CARE principles, environmental groups that had fought for the conservation of Great Bear Rainforest since the

1990s shifted to emphasize First Nations' demands for economic support and other material *collective benefits*, which ensured that community interests were at the center of environmental advocacy. By creating a coalition, the Coastal First Nations gained more *authority to control* information and how it was used—namely, by directing science towards their campaign objectives. Through direct actions and coordinated social movements, First Nations held nonprofits accountable to their principles, and outside researchers took *responsibility* for publishing information that advanced the campaign. Nonprofits recognized their *ethical* obligation to center the interests of First Nations. Over the course of the 1990s and 2000s they began to recognize this obligation by using their research to advocate not only for the protection of the ecosystem, but for the Coastal First Nations' vision of the future.

CONCLUSION

Fundamentally, Indigenous knowledge comes from people with ancestral connections to the places where they live. Researchers from non-Indigenous institutions—such as universities, government agencies, NGOs, and foundations that do not have these personal connections to the places where they work—must make Indigenous data and knowledge sovereignty a core priority. Conservation science continues to be dominated by organizations and researchers from non-Indigenous communities in the Global North. At the same time, Indigenous organizers and researchers have made impressive steps to reorient conservation science towards Indigenous policy demands, and to decolonize conservation.

Future CER that aims to be comprehensive must recognize how diverse Indigenous peoples relate differently to land, and consider all communities who live in and depend on these ecosystems, not just those with collective land rights or those postcolonial states officially recognize as Indigenous (Cossío et al. 2014). Indigenous communities are diverse in how they relate to states, nature, and other communities. In addition, in many landscapes where Indigenous people live, people who are not legally considered Indigenous and/or who do not self-identify as Indigenous often live as smallholder producers. These people are often refugees or migrants from elsewhere, and in many cases they have ecological knowledge that they use to care for and value tropical forests too. These peoples are also important actors in these ecosystems, frequently sharing histories of colonization and marginalization with legally and self-identified Indigenous people, and yet often ignored in international conservation discussions.

The principles outlined in this chapter provide a road map for researchers to support movements to decolonize conservation among Indigenous peoples and their neighbors and allies. The case studies provide insights into the complexities of carrying out conservation research in solidarity with communities who are most impacted by conservation policy. While these cases offer examples of what individual researchers and teams have done to support Indigenous movements, many institutional changes need to be made to improve relationships with Indigenous communities and to increase structural support for CER (see chapter 5). Researchers engaging in this type of work must not forget to continue opening spaces for others to join. By creating opportunities for students and trainees, and holding their institutions accountable to ethical and reciprocal relationships with Indigenous groups, professional researchers can continue to make this work possible for themselves and others.

REFERENCES

10EPHSFITF (10 Essential Public Health Services Futures Initiative Task Force). 2020. "10 Essential Public Health Services." Public Health National Center for Innovations. https://phnci.org/uploads/resource-files/EPHS-English.pdf.

Abelson, Julia, Erika Blacksher, Kathy Li, Sarah Boesveld, and Susan Goold. 2013. "Public Deliberation in Health Policy and Bioethics: Mapping an Emerging, Interdisciplinary Field." *Journal of Public Deliberation* 9 (1): 5.

Acabado, Stephen. 2018. "Zones of Refuge: Resisting Conquest in the Northern Philippine Highlands Through Environmental Practice." *Journal of Anthropological Archaeology* 52: 180–95.

Acabado, Stephen, and Da-Wei Kuan. 2021. "Indigenous Peoples: Heritage and Landscape in the Asia Pacific." In *Indigenous Peoples, Heritage and Landscape in the Asia Pacific: Knowledge Co-Production and Empowerment*, edited by Stephen Acabado and Da-Wei Kuan, 1–15. New York: Routledge.

———, eds. 2021. *Indigenous Peoples, Heritage and Landscape in the Asia Pacific: Knowledge Co-Production and Empowerment*. New York: Routledge.

Adams, Alison E., Thomas E. Shriver, Anne Saville, and Gary Webb. 2018. "Forty Years on the Fenceline: Community, Memory, and Chronic Contamination." *Environmental Sociology* 4 (2): 210–20.

Adams, Crystal, Phil Brown, Rachel Morello-Frosch, Julia Green Brody, Ruthann Rudel, Ami Zota, Sarah Dunagan, Jessica Tovar, and and Sharyle Patton. 2011. "Disentangling the Exposure Experience: The Roles of Community Context and Report-Back of Environmental Exposure Data." *Journal of Health and Social Behavior* 52 (2): 180–96.

Adkins-Jackson, Paris B., Tongtan Chantarat, Zinzi D. Bailey, and Ninez A. Ponce. 2022. "Measuring Structural Racism: A Guide for Epidemiologists and Other Health Researchers." *American Journal of Epidemiology* 191 (4): 539–47.

Affolderbach, Julia, and Rob Krueger. 2017. "'Just' Ecopreneurs: Re-Conceptualising Green Transitions and Entrepreneurship." *Local Environment* 22 (4): 410–23.

Agénor, Madina. 2020. "Future Directions for Incorporating Intersectionality into Quantitative Population Health Research." *American Journal of Public Health* 110 (6): 803–6.

Agénor, Madina, Carly Perkins, Catherine Stamoulis, Rahsaan D. Hall, Mihail Samnaliev, Stephanie Berland, and S. Bryn Austin. 2021. "Developing a Database of Structural Racism-Related State Laws for Health Equity Research and Practice in the United States." *Public Health Reports* 136 (4): 428–40.

Agyeman, Julian, Robert D. Bullard, and Bob Evans, eds. 2003. *Just Sustainabilities: Development in an Unequal World*. Cambridge: MIT Press.

Agyeman, Julian, David Schlosberg, Luke Craven, and Caitlin Matthews. 2016. "Trends and Directions in Environmental Justice: From Inequity to Everyday Life, Community, and Just Sustainabilities." *Annual Review of Environment and Resources* 41 (1): 321–40.

Ahmed, Syed M., and Ann-Gel S. Palermo. 2010. "Community Engagement in Research: Frameworks for Education and Peer Review." *American Journal of Public Health* 100 (8): 1380–87.

Akchurin, Maria. 2015. "Constructing the Rights of Nature: Constitutional Reform, Mobilization, and Environmental Protection in Ecuador." *Law & Social Inquiry* 40 (4): 937–68.

Akuno, Kali. 2017. "Build and Fight: The Program and Strategy of Cooperation Jackson." In *Jackson Rising: The Struggle for Economic Democracy and Black Self-Determination in Jackson, Mississippi*, edited by Kali Akuno and Ajamu Nangwaya, 3–42. Wakefield, Canada: Daraja Press.

Alang, Sirry, Rachel Hardeman, J'Mag Karbeah, Odichinma Akosionu, Cydney McGuire, Hamdi Abdi, and Donna McAlpine. 2021. "White Supremacy and the Core Functions of Public Health." *American Journal of Public Health* 111 (5): 815–19.

Alexeeff, Stacey E., Ananya Roy, Jun Shan, Xi Liu, Kyle Messier, Joshua S. Apte, Christopher Portier, Stephen Sidney, and Stephen K. Van Den Eeden. 2018. "High-Resolution Mapping of Traffic Related Air Pollution with Google Street View Cars and Incidence of Cardiovascular Events Within Neighborhoods in Oakland, CA." *Environmental Health* 17: 38.

Alkon, Alison Hope. 2018. "Food Justice: An Environmental Justice Approach to Food and Agriculture." In *The Routledge Handbook of Environmental Justice*, edited by Ryan Holifield, Jayajit Chakraborty, and Gordon Walker, 412–24. New York: Routledge.

Alkon, Alison Hope, and Julian Agyeman, eds. 2011. *Cultivating Food Justice: Race, Class, and Sustainability*. Cambridge: MIT Press.

Allen, Douglas, Mary Lawhon, and Joseph Pierce. 2019. "Placing Race: On the Resonance of Place with Black Geographies." *Progress in Human Geography* 43 (6): 1001–19.

Allen, Patricia, and Carolyn Sachs. 2012. "Women and Food Chains: The Gendered Politics of Food." In *Taking Food Public: Redefining Foodways in a Changing World*, edited by Psyche Williams-Forson and Carole Counihan, 23–40. New York: Routledge.

Alvarado Garcia, Adriana, Alyson L. Young, and Lynn Dombrowski. 2017. "On Making Data Actionable: How Activists Use Imperfect Data to Foster Social Change for Human Rights Violations in Mexico." *Proceedings of the ACM on Human-Computer Interaction* 1, CSCW: 19.

Alvarez, Camila H., and Clare Rosenfeld Evans. 2021. "Intersectional Environmental Justice and Population Health Inequalities: A Novel Approach." *Social Science & Medicine* 269: 113559.

American Institute of Certified Planners. 2021. "AICP Code of Ethics and Professional Conduct." Last modified November 2021. https://www.planning.org/ethics/ethicscode/.

Amiri, Azita, and Shuang Zhao. 2019. "Working with an Environmental Justice Community: Nurse Observation, Assessment, and Intervention." *Nursing Forum* 54 (2): 270–79.

Ammons, Shorlette, Sarah Blacklin, Dara Bloom, Shironda Brown, Marcello Cappellazzi, Nancy Creamer, Angel Cruz, et al. 2021. "A Collaborative Approach to COVID-19 Response." *Journal of Agriculture, Food Systems, and Community Development* 10 (2): 297–302.

Ammons, Shorlette, Nancy Creamer, Paul B. Thompson, Hunter Francis, Joanna Friesner, Casey Hoy, Tom Kelly, Christine M. Porter, and Thomas P. Tomich. 2018. "A Deeper Challenge of Change: The Role of Land-Grant Universities in Assessing and Ending Structural Racism in the U.S. Food System." Davis: Inter-Institutional Network for Food, Agriculture, and Sustainability. https://asi.ucdavis.edu/programs/infas/a-deeper-challenge-of-change-the-role-of-land-grant-universities-in-assessing-and-ending-structural-racism-in-the-us-food-system.

Anderson, Elijah. 2015. "The White Space." *Sociology of Race and Ethnicity* 1 (1): 10–21.

Andress, Lauri, Tristen Hall, Sheila Davis, Judith Levine, Kimberly Cripps, and Dominique Guinn. 2020. "Addressing Power Dynamics in Community-Engaged Research Partnerships." *Journal of Patient-Reported Outcomes* 4: 24.

Aneesh, Patnaik, Jiahn Son, Alice Feng, and Crystal Ade. 2020. "Racial Disparities and Climate Change." *Princeton Student Climate Initiative* (blog). August 15, 2020. https://psci.princeton.edu/tips/2020/8/15/racial-disparities-and-climate-change.

Angelsen, Arild, Christopher Martius, Veronique De Sy, Amy E. Duchelle, Anne M. Larson, and Pham Thu Thuy, eds. 2018. *Transforming REDD+: Lessons and New Directions*. Bogor, Indonesia: Center for International Forestry Research.

Anguelovski, Isabelle. 2013. "New Directions in Urban Environmental Justice: Rebuilding Community, Addressing Trauma, and Remaking Place." *Journal of Planning Education and Research* 33 (2): 160–75.

———. 2015. "From Toxic Sites to Parks as (Green) LULUs? New Challenges of Inequity, Privilege, Gentrification, and Exclusion for Urban Environmental Justice." *Journal of Planning Literature* 31 (1): 23–36.

Anguelovski, Isabelle, Anna Livia Brand, Eric Chu, and Kian Goh. 2018. "Urban Planning, Community (Re)Development, and Environmental Gentrification: Emerging Challenges for Green and Equitable Neighborhoods." In *The Routledge Handbook of Environmental Justice*, edited by Ryan Holifield, Jayajit Chakraborty, and Gordon Walker, 449–62. New York: Routledge.

Anguelovski, Isabelle, James JT Connolly, Melissa Garcia-Lamarca, Helen Cole, and Hamil Pearsall. 2019. "New Scholarly Pathways on Green Gentrification: What Does the Urban 'Green Turn' Mean and Where Is It Going?" *Progress in Human Geography* 43 (6): 1064–86.

Antonellis, Paul James, and Gwendolyn Berry. 2017. "Practical Steps for the Utilization of Action Research in Your Organization: A Qualitative Approach for Non-Academic Research." *International Journal of Human Resource Studies* 7 (2): 41–59.

Apostolopoulou, Elia, and Jose A. Cortes-Vazquez, eds. 2018. *The Right to Nature: Social Movements, Environmental Justice and Neoliberal Natures*. New York: Routledge.

Appalachian Land Ownership Task Force. 1983. *Who Owns Appalachia? Landownership and Its Impact*. Lexington: University Press of Kentucky.

Appe, Susan, Nadia Rubaii, Sebastian Líppez-De Castro, and Stephen Capobianco. 2017. "The Concept and Context of the Engaged University in the Global South: Lessons from Latin America to Guide a Research Agenda." *Journal of Higher Education Outreach and Engagement* 21 (2): 7–36.

Arcaya, Mariana C., Reginald D. Tucker-Seeley, Rockli Kim, Alina Schnake-Mahl, Marvin So, and S. V. Subramanian. 2016. "Research on Neighborhood Effects on Health in the United States: A Systematic Review of Study Characteristics." *Social Science & Medicine* 168: 16–29.

Arizona Board of Regents. 2018. "Policy 1–118 Tribal Consultation." Last modified September 2018. https://public.azregents.edu/Policy%20Manual/1-118-Tribal%20Consultation.pdf.

Armiero, Marco, Thanos Andritsos, Stefania Barca, Rita Brás, Sergio Ruiz Cauyela, Çağdaş Dedeoğlu, Marica Di Pierri, Lúcia de Oliveira Fernandes, Filippo Gravagno, and Laura Greco. 2019. "Toxic Bios: Toxic Autobiographies—A Public Environmental Humanities Project." *Environmental Justice* 12 (1): 7–11.

Armstrong, Rebecca, Elizabeth Waters, Maureen Dobbins, Laurie Anderson, Laurence Moore, Mark Petticrew, Rachel Clark, et al. 2013. "Knowledge Translation Strategies to Improve the Use of Evidence in Public Health Decision Making in Local Government: Intervention Design and Implementation Plan." *Implementation Science* 8: 121.

Arnstein, Sherry R. 1969. "A Ladder of Citizen Participation." *Journal of the American Institute of Planners* 35 (4): 216–24.

Arquette, Mary, Maxine Cole, Katsi Cook, Brenda LaFrance, Margaret Peters, James Ransom, Elvera Sargent, Vivian Smoke, and Arlene Stairs. 2002. "Holistic Risk-Based Environmental Decision Making: A Native Perspective." *Environmental Health Perspectives* 110 (S2): 259–64.

Arredondo, Elva, Kristin Mueller, Elizabeth Mejia, Tanya Rovira-Oswalder, Dana Richardson, and Tracy Hoos. 2013. "Advocating for Environmental Changes to Increase Access to Parks: Engaging Promotoras and Youth Leaders." *Health Promotion Practice* 14 (5): 759–66.

Asad, Asad L., and Matthew Clair. 2018. "Racialized Legal Status as a Social Determinant of Health." *Social Science & Medicine* 199: 19–28.

Atalay, Sonya. 2012. *Community-Based Archaeology: Research with, by, and for Indigenous and Local Communities*. Berkeley: University of California Press.

Atapattu, Sumudu A., Carmen G. Gonzalez, and Sara L. Seck, eds. 2021. *The Cambridge Handbook of Environmental Justice and Sustainable Development*. Cambridge: Cambridge University Press.

Axel-Lute, Miriam. 2021. "Understanding Community Land Trusts." *Shelterforce*. https://shelterforce.org/2021/07/12/understanding-community-land-trusts/.

Axner, Myra. n.d. "Working Together for Racial Justice and Inclusion: Understanding Culture and Diversity in Building Communities." Community Tool Box. Accessed November 14, 2021. https://ctb.ku.edu/en/table-of-contents/culture/cultural-competence/culture-and-diversity/main.

Ayre, Margaret L., Philip J. Wallis, and Katherine A. Daniell. 2018. "Learning from Collaborative Research on Sustainably Managing Fresh Water: Implications for Ethical Research-Practice Engagement." *Ecology and Society* 23 (1): 6.

Babich, Adam. 2013. "Twenty Questions (and Answers) About Environmental Law School Clinics." *The Professional Lawyer* 22 (1): 45–53.

Bacher, Dan. 2021. "Spring 2023 the Target for Removing Four NorCal Dams to Free 400 Miles of Rivers and Tributaries." *Sacramento News & Review*. December 10, 2021. https://sacramento.newsreview.com/2021/12/10/spring-2023-the-target-for-removing-four-nor-cal-dams-to-free-400-miles-of-rivers-and-tributaries/.

Backlund Jarquín, Paige. 2012. "Data Sharing: Creating Agreements—In Support of Community-Academic Partnerships." Aurora: Colorado Clinical and Translational Sciences Institute and Rocky Mountain Prevention Research Center. http://trailhead.institute/wp-content/uploads/2017/04/tips_for_creating_data_sharing_agreements_for_partnerships.pdf.

Bacon, Christopher, Saneta deVuono-Powell, Mary Louise Frampton, Tony LoPresti, and Camille Pannu. 2013. "Introduction to Empowered Partnerships: Community-Based Participatory Action Research for Environmental Justice." *Environmental Justice* 6 (1): 1–8.

Bacon, Christopher M., William A. Sundstrom, and María Eugenia Flores Gómez Gómez. 2014. "Explaining the 'Hungry Farmer Paradox': Smallholders and Fair Trade Cooperatives Navigate Seasonality and Change in Nicaragua's Corn and Coffee Markets." *Global Environmental Change* 25: 133–49.

Bacon, Christopher M., William A. Sundstrom, Iris T. Stewart, Ed Maurer, and Lisa C. Kelley. 2021. "Towards Smallholder Food and Water Security: Climate Variability in the Context of Multiple Livelihood Hazards in Nicaragua." *World Development* 143: 105468.

Bagchee, Nandini. 2019. "Building a Transition City: The Ewing Street Eco-Village Coop Pilot Project." New York: Advanced Design Studio, City College of New York. https://www.scribd.com/document/431893329/Building-a-Transition-City-Landscape-Online-Version#.

Bailey, Kerry A. 2020. "Salmon and Acorns Feed Our People: Colonialism, Nature, and Social Action." *Ethnic and Racial Studies* 43 (13): 2442–44.

Bailey, Zinzi D., Nancy Krieger, Madina Agénor, Jasmine Graves, Natalia Linos, and Mary T. Bassett. 2017. "Structural Racism and Health Inequities in the USA: Evidence and Interventions." *The Lancet* 389 (10077): 1453–63.

Baiocchi, Gianpaolo, Patrick Heller, and Marcelo Silva. 2011. *Bootstrapping Democracy: Transforming Local Governance and Civil Society in Brazil*. Redwood City: Stanford University Press.

Baker, Elijah, Cambria Wilson, Fabiana Lake, and David N. Pellow. 2020. "Environmental Justice Struggles in Prisons and Jails Around the World: The 2020 Annual Report of the Prison Environmental Justice Project." Santa Barbara: Global Environmental Justice Project. https://www.es.ucsb.edu/es-news-2020/around-es/prison-ecology.

Balazs, Carolina L., and Rachel Morello-Frosch. 2013. "The Three R's: How Community Based Participatory Research Strengthens the Rigor, Relevance and Reach of Science." *Environmental Justice* 6 (1): 9–16.

Baletti, Brenda. 2011. "Saving the Amazon? Land Grabs and 'Sustainable Soy' as the New Logic of Conservation." *Proceedings of the 1st International Conference of Global Land Grabbing*. Brighton: Land Deal Politics Initiative. https://www.future-agricultures.org/wp-content/uploads/pdf-archive/Brenda%20Baletti.pdf.

BAMCO and UFW (Bon Appétit Management Company Foundation and United Farm Workers of America). 2011. "The Inventory of Farmworker Issues and Protections in the United States." Palo Alto: Bon Appétit Management Company Foundation and United Farm Workers. https://www.bamco.com/timeline/farmworker-inventory/.

Ban, Natalie C., and Alejandro Frid. 2018. "Indigenous Peoples' Rights and Marine Protected Areas." *Marine Policy* 87: 180–85.

Ban, Natalie C., Alejandro Frid, Mike Reid, Barry Edgar, Danielle Shaw, and Peter Siwallace. 2018. "Incorporate Indigenous Perspectives for Impactful Research and Effective Management." *Nature Ecology & Evolution* 2 (11): 1680–83.

Bang, Megan, and Shirin Vossoughi. 2016. "Participatory Design Research and Educational Justice: Studying Learning and Relations Within Social Change Making." *Cognition and Instruction* 34 (3): 173–93.

Banks, Sarah, and Mary Brydon-Miller, eds. 2018. *Ethics in Participatory Research for Health and Social Well-Being*. New York: Routledge.

Baptista, Ana. 2021. "Zoning, Land Use, and Local Policies for Environmental Justice." *Zoning Practice* March: 1–8. Chicago: American Planning Association. https://www.planning.org/publications/document/9212433/.

Baptista, Ana Isabel, Amanda Sachs, and Claudia Rot. 2019. "Local Policies for Environmental Justice: A National Scan." New York: Tishman Environment and Design Center. https://www.nrdc.org/sites/default/files/local-policies-environmental-justice-national-scan-tishman-201902.pdf.

Barge, J. Kevin. 2016. "Crossing Boundaries Between Communication Activism Research and Applied Communication Research Discourses." *International Journal of Communication* 10: 4000–8.

Barnett, Steven, Jessica Cuculick, Lori Dewindt, Kelly Matthews, and Erika Sutter. 2017. "National Center for Deaf Health Research: CBPR with Deaf Communities." In *Community-Based Participatory Research for Health: Advancing Social and Health Equity*, edited by Nina Wallerstein, Bonnie Duran, John G. Oetzel, and Meredith Minkler, 3rd ed., 157–74. San Francisco: Jossey-Bass.

Baron, Sherry, Raymond Sinclair, Devon Payne-Sturges, Jerry Phelps, Harold Zenick, Gwen W. Collman, and Liam R. O'Fallon. 2009. "Partnerships for Environmental and Occupational Justice: Contributions to Research, Capacity and Public Health." *American Journal of Public Health* 99 (S3): 517–25.

Barrett, Meredith, Veronica Combs, Jason G. Su, Kelly Henderson, and Michael Tuffli. 2018. "AIR Louisville: Addressing Asthma with Technology, Crowdsourcing, Cross-Sector Collaboration, and Policy." *Health Affairs* 37 (4): 525–34.

Bartlett, Cheryl, Murdena Marshall, and Albert Marshall. 2012. "Two-Eyed Seeing and Other Lessons Learned Within a Co-Learning Journey of Bringing Together Indigenous and Mainstream Knowledges and Ways of Knowing." *Journal of Environmental Studies and Sciences* 2: 331–40.

Battaglia, Tracy A., Jennifer Pamphile, Sharon Bak, Nikki Spencer, and Christine Gunn. 2019. "Connecting Community to Research: A Training Program to Increase Community Engagement in Research." *Progress in Community Health Partnerships* 13 (2): 209–17.

Batz, Giovanni. 2018. "The Ixil University and the Decolonization of Knowledge." In *Indigenous and Decolonizing Studies in Education: Mapping the Long View*, edited by Linda Tuhiwai Smith, Eve Tuck, and K. Wayne Yang, 103–15. New York: Routledge.

Bauer, Katherine W., Rachel Widome, John H. Himes, Mary Smyth, Bonnie Holy Rock, Peter J. Hannan, and Mary Story. 2012. "High Food Insecurity and Its Correlates Among Families Living on a Rural American Indian Reservation." *American Journal of Public Health* 102 (7): 1346–52.

Bayuo, Blaise Booponoyeng, Cristina Chaminade, and Bo Göransson. 2020. "Unpacking the Role of Universities in the Emergence, Development and Impact of Social Innovations: A Systematic Review of the Literature." *Technological Forecasting and Social Change* 155: 120030.

Beans, Julie A., Bobby Saunkeah, R. Brian Woodbury, Terry S. Ketchum, Paul G. Spicer, and Vanessa Y. Hiratsuka. 2019. "Community Protections in American Indian and Alaska Native Participatory Research: A Scoping Review." *Social Sciences* 8 (4): 127.

Bedoya, Roberto. 2013. "Placemaking and the Politics of Belonging and Dis-Belonging." *GIA Reader* 24 (1). https://www.giarts.org/article/placemaking-and-politics-belonging-and-dis-belonging.

Beebeejaun, Yasminah, Catherine Durose, James Rees, Joanna Richardson, and Liz Richardson. 2014. "'Beyond Text': Exploring Ethos and Method in Co-Producing Research with Communities." *Community Development Journal* 49 (1): 37–53.

Bell, Derek, and Jayne Carrick. 2018. "Procedural Environmental Justice." In *The Routledge Handbook of Environmental Justice*, edited by Ryan Holifield, Jayajit Chakraborty, and Gordon Walker, 101–12. New York: Routledge.

Bell, Sarah, Cassandra Phoenix, Rebecca Lovell, and Benedict Wheeler. 2015. "Using GPS and Geo-Narratives: A Methodological Approach for Understanding and Situating Everyday Green Space Encounters." *Area* 47 (1): 88–96.

Bell, Shannon Elizabeth. 2015. "Bridging Activism and the Academy: Exposing Environmental Injustices Through the Feminist Ethnographic Method of Photovoice." *Human Ecology Review* 21 (1): 27–58.

Benford, Robert. 2005. "The Half-Life of the Environmental Justice Frame: Innovation, Diffusion, and Stagnation." In *Power, Justice, and the Environment: A Critical Appraisal of the Environmental Justice Movement*, edited by David N. Pellow, and Robert J. Brulle, 37–53. Cambridge: MIT Press.

Bennett, Gavin, and Nasreen Jessani, eds. 2011. *The Knowledge Translation Toolkit: Bridging the Know-Do Gap: A Resource for Researchers*. Thousand Oaks: SAGE Publications and International Development Research Centre.

Berkes, Fikret, Johan Colding, and Carl Folke. 2000. "Rediscovery of Traditional Ecological Knowledge as Adaptive Management." *Ecological Applications* 10 (5): 1251–62.

Bhatia, Rajiv. 2011. "Health Impact Assessment: A Guide for Practice." Oakland: Human Impact Partners. https://www.pewtrusts.org/~/media/assets/2011/01/01/bhatia_2011_hia_guide_for_practice.pdf.

Bickerstaff, Karen. 2018. "Justice in Energy System Transitions: A Synthesis and Agenda." In *The Routledge Handbook of Environmental Justice*, edited by Ryan Holifield, Jayajit Chakraborty, and Gordon Walker, 388–99. New York: Routledge.

Black, Kristin Z., Christina Yongue Hardy, Molly De Marco, Alice Ammerman, Giselle Corbie-Smith, Barbara Council, Danny Ellis, et al. 2013. "Beyond Incentives for Involvement to Compensation for Consultants: Increasing Equity in CBPR Approaches." *Progress in Community Health Partnerships* 7 (3): 263–70.

BlackSpace. n.d. "BlackSpace: Home." Accessed December 21, 2021. https://www.blackspace.org.

Blumenthal, Daniel S., Ralph J. DiClemente, Ronald L. Braithwaite, and Selina A. Smith, eds. 2013. *Community-Based Participatory Health Research: Issues, Methods, and Translation to Practice*. 2nd ed. New York: Springer.

Blumenthal, Daniel S., Ernest Hopkins, and Elleen Yancey. 2013. "Community-Based Participatory Research: An Introduction." In *Community-Based Participatory Health Research: Issues, Methods, and Translation to Practice*, edited by Daniel S. Blumenthal, Ralph J. DiClemente, Ronald L. Braithwaite, and Selina A. Smith, 2nd ed., 1–18. New York: Springer.

Blumgart, Jake. 2015. "Affordable Housing's Forever Solution." *Next City* (blog), August 10, 2015. https://nextcity.org/features/affordable-housings-forever-solution.

Boll-Bosse, Amber J., and Katherine B. Hankins. 2018. "'These Maps Talk for Us': Participatory Action Mapping as Civic Engagement Practice." *The Professional Geographer* 70 (2): 319–26.

Bor, Jacob, Gregory H. Cohen, and Sandro Galea. 2017. "Population Health in an Era of Rising Income Inequality: USA, 1980–2015." *The Lancet* 389 (10077): 1475–90.

Bor, Jacob, Atheendar S. Venkataramani, David R. Williams, and Alexander C. Tsai. 2018. "Police Killings and Their Spillover Effects on the Mental Health of Black Americans: A Population-Based, Quasi-Experimental Study." *The Lancet* 392 (10144): 302–10.

Borrell, Carme, Laia Palència, Carles Muntaner, Marcelo Urquía, Davide Malmusi, and Patricia O'Campo. 2014. "Influence of Macrosocial Policies on Women's Health and Gender Inequalities in Health." *Epidemiologic Reviews* 36 (1): 31–48.

Bowen, Sarah, Tannis Erickson, Patricia Martens, and Susan Crockett. 2009. "More Than 'Using Research': The Real Challenges in Promoting Evidence-Informed Decision -Making." *Healthcare Policy* 4 (3): 87–102.

Bowleg, Lisa. 2012. "The Problem with the Phrase 'Women and Minorities: Intersectionality'—An Important Theoretical Framework for Public Health." *American Journal of Public Health* 102 (7): 1267–73.

Bowman, Nicole. 2020. "Nation-to-Nation in Evaluation: Utilizing an Indigenous Evaluation Model to Frame Systems and Government Evaluations." *New Directions for Evaluation* 166: 101–18.

Boydell, Katherine M., Michael Hodgins, Brenda M. Gladstone, Elaine Stasiulis, George Belliveau, Hoi Cheu, Pia Kontos, and Janet Parsons. 2016. "Arts-Based Health Research and Academic Legitimacy: Transcending Hegemonic Conventions." *Qualitative Research* 16 (6): 681–700.

Boyer, Ernest L. 1996. "The Scholarship of Engagement." *Journal of Public Service and Outreach* 1 (1): 11–20.

Brabham, Daren C., Kurt M. Ribisl, Thomas R. Kirchner, and Jay M. Bernhardt. 2014. "Crowdsourcing Applications for Public Health." *American Journal of Preventive Medicine* 46 (2): 179–87.

Bradbury, Hilary, ed. 2015. *The SAGE Handbook of Action Research*. 3rd ed. Los Angeles: SAGE Publications.

Bradley, Katherine, and Hank Herrera. 2016. "Decolonizing Food Justice: Naming, Resisting, and Researching Colonizing Forces in the Movement." *Antipode* 48 (1): 97–114.

Bratspies, Rebecca M. 2020. "Renewable Rikers: A Plan for Restorative Environmental Justice." *Loyola Law Review* 66: 371–400.

Breckwich Vásquez, Victoria, Dana Lanza, Susana Hennessey-Lavery, Shelley Facente, Helen Ann Halpin, and Meredith Minkler. 2007. "Addressing Food Security Through

Public Policy Action in a Community-Based Participatory Research Partnership." *Health Promotion Practice* 8 (4): 342–49.

Brenner, Neil. 2015. "Is 'Tactical Urbanism' an Alternative to Neoliberal Urbanism?" *POST* (blog), March 24, 2015. https://teputahi.org.nz/wp-content/uploads/2015/08/Is-%E2%80%9CTactical-Urbanism%E2%80%9D-an-Alternative-to-Neoliberal-Urbanism.pdf.

Brockington, Dan. 2002. *Fortress Conservation: The Preservation of the Mkomazi Game Reserve, Tanzania.* Bloomington and Dar es Salaam: Indiana University Press and International African Institute.

brown, adrienne maree. 2017. *Emergent Strategy.* Chico: AK Press.

Brown, Phil. 1992. "Popular Epidemiology and Toxic Waste Contamination: Lay and Professional Ways of Knowing." *Journal of Health and Social Behavior* 33 (3): 267–81.

———. 1993. "When the Public Knows Better: Popular Epidemiology Challenges the System." *Environment: Science and Policy for Sustainable Development* 35 (8): 16–41.

———. 2007. *Toxic Exposures: Contested Illnesses and the Environmental Health Movement.* New York: Columbia University Press.

Brown, Phil, Vanessa de la Rosa, and Alissa Cordner. 2020. "Toxic Trespass: Science, Activism, and Policy Concerning Chemicals in Our Bodies." In *Toxic Truths: Environmental Justice and Citizen Science in a Post-Truth Age*, edited by Thom Davies and Alice Mah, 34–58. Manchester: Manchester University Press.

Brown, Phil, and Edwin J. Mikkelsen. 1997. *No Safe Place: Toxic Waste, Leukemia, and Community Action.* Berkeley: University of California Press.

Brown, Sandy, and Christy Getz. 2011. "Farmworker Food Insecurity and the Production of Hunger in California." In *Cultivating Food Justice: Race, Class, and Sustainability*, edited by Alison Hope Alkon and Julian Agyeman, 121–46. Cambridge: MIT Press.

Brownson, Ross C., Amy A. Eyler, Jenine K. Harris, Justin B. Moore, and Rachel G. Tabak. 2018. "Getting the Word Out: New Approaches for Disseminating Public Health Science." *Journal of Public Health Management and Practice* 24 (2): 102–11.

Brownson, Ross C., Jonathan E. Fielding, and Christopher M. Maylahn. 2009. "Evidence-Based Public Health: A Fundamental Concept for Public Health Practice." *Annual Review of Public Health* 30: 175–201.

Brugge, Doug, and Rob Goble. 2002. "The History of Uranium Mining and the Navajo People." *American Journal of Public Health* 92 (9): 1410–19.

Brydon-Miller, Mary, Davydd Greenwood, and Patricia Maguire. 2003. "Why Action Research?" *Action Research* 1 (1): 9–28.

Bua, Adrian, Myfanwy Taylor, and Sara Gonzalez. 2018. *Measuring the Value of Traditional Retail Markets.* London: New Economics Foundation. https://neweconomics.org/2018/11/measuring-the-value-of-traditional-retail-markets.

Bullard, Robert D. 1983. "Solid Waste Sites and the Black Houston Community." *Sociological Inquiry* 53 (2–3): 273–88.

———. 1990. *Dumping in Dixie: Race, Class, and Environmental Quality.* Boulder: Westview Press.

Bullard, Robert D., Paul Mohai, Robin Saha, and Beverly Wright. 2008. "Toxic Wastes and Race at Twenty: Why Race Still Matters After All of These Years." *Environmental Law* 38 (2): 371–411.

Bullard, Robert D., and Beverly Wright. 2012. *The Wrong Complexion for Protection: How the Government Response to Disaster Endangers African American Communities.* New York: NYU Press.

Burke, Jessica G., Sally Hess, Kamden Hoffmann, Lisa Guizzetti, Ellyn Loy, Andrea Gielen, Maryanne Bailey, Adrienne Walnoha, Genevieve Barbee, and Michael Yonas. 2013. "Translating Community-Based Participatory Research (CBPR) Principles into Practice: Building a Research Agenda to Reduce Intimate Partner Violence." *Progress in Community Health Partnerships* 7 (2): 115–22.

Burke, Michael P., Sonya J. Jones, Edward A. Frongillo, Maryah S. Fram, Christine E. Blake, and Darcy A. Freedman. 2018. "Severity of Household Food Insecurity and Lifetime Racial Discrimination Among African-American Households in South Carolina." *Ethnicity & Health* 23 (3): 276–92.

Burris, Scott, Marice Ashe, Donna Levin, Matthew Penn, and Michelle Larkin. 2016. "A Transdisciplinary Approach to Public Health Law: The Emerging Practice of Legal Epidemiology." *Annual Review of Public Health* 37: 135–48.

Busch, Per-Olof, and Helge Jörgens. 2005. "The International Sources of Policy Convergence: Explaining the Spread of Environmental Policy Innovations." *Journal of European Public Policy* 12 (5): 860–84.

Buse, Chris G., Valerie Lai, Katie Cornish, and Margot W. Parkes. 2019. "Towards Environmental Health Equity in Health Impact Assessment: Innovations and Opportunities." *International Journal of Public Health* 64 (1): 15–26.

Butts, Shannon, and Madison Jones. 2021. "Deep Mapping for Environmental Communication Design." *Communication Design Quarterly* 9 (1): 4–19.

Buytaert, Wouter, Art Dewulf, Bert De Bièvre, Julian Clark, and David M. Hannah. 2016. "Citizen Science for Water Resources Management: Toward Polycentric Monitoring and Governance?" *Journal of Water Resources Planning and Management* 142 (4): 01816002.

Buzzelli, Michael. 2018. "Air Pollution and Respiratory Health: Does Better Evidence Lead to Policy Paralysis?" In *The Routledge Handbook of Environmental Justice*, edited by Ryan Holifield, Jayajit Chakraborty, and Gordon Walker, 327–37. New York: Routledge.

Byrne, Jason. 2018. "Urban Parks, Gardens and Greenspace." In *The Routledge Handbook of Environmental Justice*, edited by Ryan Holifield, Jayajit Chakraborty, and Gordon Walker, 437–48. New York: Routledge.

Cable, Sherry. 2012. "Confessions of the Parasitic Researchers to the Man in the Cowboy Hat." In *Confronting Ecological Crisis in Appalachia and the South: University and Community Partnerships*, edited by Stephanie McSpirit, Lynne Faltraco, and Connor Bailey, 21–38. Lexington: University Press of Kentucky.

Cable, Sherry, Tamara Mix, and Donald Hastings. 2005. "Mission Impossible? Environmental Justice Activists' Collaborations with Professional Environmentalists and with Academics." In *Power, Justice, and the Environment: A Critical Appraisal of the Environmental Justice Movement*, edited by David N. Pellow and Robert J. Brulle, 55–75. Cambridge: MIT Press.

Cacari-Stone, Lisa, Nina Wallerstein, Analilia P. Garcia, and Meredith Minkler. 2014. "The Promise of Community-Based Participatory Research for Health Equity: A Conceptual Model for Bridging Evidence with Policy." *American Journal of Public Health* 104 (9): 1615–23.

Cachelin, Adrienne, Jeff Rose, and Danya Lee Rumore. 2016. "Leveraging Place for Critical Sustainability Education: The Promise of Participatory Action Research." *Journal of Sustainability Education* 11. http://www.jsedimensions.org/wordpress/wp-content/uploads/2016/03/Cachelin-Rose-Rumore-JSE-February-2016-Place-Issue-PDF-Ready.pdf.

Cagney, Kathleen A., Erin York Cornwell, Alyssa W. Goldman, and Liang Cai. 2020. "Urban Mobility and Activity Space." *Annual Review of Sociology* 46: 623–48.

Cahalan, Margaret W., Laura W. Perna, Marisha Addison, Murray Chelsea, Pooja R. Patel, and Nathan Jiang. 2020. *Indicators of Higher Education Equity in the United States: 2020 Historical Trend Report*. Washington, DC: The Pell Institute for the Study of Opportunity in Higher Education and Alliance for Higher Education and Democracy. https://eric.ed.gov/?id=ED606010.

Cahill, Caitlin, and Rachel Pain. 2019. "Representing Slow Violence and Resistance: On Hiding and Seeing." *ACME: An International Journal for Critical Geographies* 18 (5): 1054–65.

Cairns, Kate. 2018. "Youth, Temporality, and Territorial Stigma: Finding Good in Camden, New Jersey." *Antipode* 50 (5): 1224–43.

Caldwell, Wilma Brakefield, Angela G. Reyes, Zachary Rowe, Julia Weinert, and Barbara A. Israel. 2015. "Community Partner Perspectives on Benefits, Challenges, Facilitating Factors, and Lessons Learned from Community-Based Participatory Research Partnerships in Detroit." *Progress in Community Health Partnerships: Research, Education, and Action* 9 (2): 299–311.

California Environmental Justice Alliance. 2011. "Green Zones for Economic and Environmental Sustainability: A Concept Paper from the California Environmental Justice Alliance." Oakland and Huntington Park: CEJA. https://perma.cc/DNE2-WFMF.

Campbell, Scott. 1996. "Green Cities, Growing Cities, Just Cities? Urban Planning and the Contradictions of Sustainable Development." *Journal of the American Planning Association* 62 (3): 296–312.

Canaan, Joyce E., and Wesley Shumar, eds. 2008. *Structure and Agency in the Neoliberal University*. New York: Routledge.

Candipan, Jennifer, Nolan Edward Phillips, Robert J. Sampson, and Mario Small. 2021. "From Residence to Movement: The Nature of Racial Segregation in Everyday Urban Mobility." *Urban Studies* 58 (15): 3095–117.

Cannon, Clare E. B. 2020. "Towards Convergence: How to Do Transdisciplinary Environmental Health Disparities Research." *International Journal of Environmental Research and Public Health* 17 (7): 2303.

Capeheart, Loretta, and Dragan Milovanovic. 2020. *Social Justice: Theories, Issues, and Movements*. 2nd ed. New Brunswick: Rutgers University Press.

Čapek, Stella M. 1993. "The 'Environmental Justice' Frame: A Conceptual Discussion and an Application." *Social Problems* 40 (1): 5–24.

Carlson, Barbara A., Diane Neal, and Gayenell Magwood. 2006. "A Community-Based Participatory Health Information Needs Assessment to Help Eliminate Diabetes Information Disparities." *Health Promotion Practice* 7 (S3): 213–22.

Carmin, JoAnn, and Julian Agyeman. 2011. *Environmental Inequalities Beyond Borders: Local Perspectives on Global Injustices*. Cambridge: MIT Press.

Carnegie Classification of Institutions of Higher Education. n.d. "The Elective Classification for Community Engagement." Accessed December 31, 2022. https://carnegieelectiveclassifications.org/the-2024-elective-classification-for-community-engagement/.

Carpenter, Deborah, Veronica Nieva, Tarek Albaghal, and Joann Sorra. 2005. "Development of a Planning Tool to Guide Research Dissemination." In *Advances in Patient Safety*, edited by Kerm Henriksen, James B. Battles, Eric S. Marks, and David I. Lewin, vol. 4, 83–92. Rockville: Agency for Healthcare Research and Quality.

Carrel, Margaret, Sean G. Young, and Eric Tate. 2016. "Pigs in Space: Determining the Environmental Justice Landscape of Swine Concentrated Animal Feeding Operations (CAFOs) in Iowa." *International Journal of Environmental Research and Public Health* 13 (9): 849.

Carroll, Clint. 2015. *Roots of Our Renewal: Ethnobotany and Cherokee Environmental Governance*. Minneapolis: University of Minnesota Press.

Carroll, Stephanie Russo, Desi Rodriguez-Lonebear, and Andrew Martinez. 2019. "Indigenous Data Governance: Strategies from United States Native Nations." *Data Science Journal* 18 (1): 31.

Carruthers, David V., ed. 2008. *Environmental Justice in Latin America: Problems, Promise, and Practice*. Cambridge: MIT Press.

Catalani, Caricia, and Meredith Minkler. 2010. "Photovoice: A Review of the Literature in Health and Public Health." *Health Education & Behavior*, 37 (3): 424–51.

Casey, Joan A., Peter James, Lara Cushing, Bill M. Jesdale, and Rachel Morello-Frosch. 2017. "Race, Ethnicity, Income Concentration and 10-Year Change in Urban Greenness in the United States." *International Journal of Environmental Research and Public Health* 14 (12): 1546.

Casey, Joan A., Rachel Morello-Frosch, Daniel J. Mennitt, Kurt Fristrup, Elizabeth L. Ogburn, and Peter James. 2017. "Race/Ethnicity, Socioeconomic Status, Residential Segregation, and Spatial Variation in Noise Exposure in the Contiguous United States." *Environmental Health Perspectives* 125 (7): 077017.

Cashman, Suzanne B., Sarah Adeky, Alex J. Allen III, Jason Corburn, Barbara A. Israel, Jaime Montaño, Alvin Rafelito, et al. 2008. "The Power and the Promise: Working with Communities to Analyze Data, Interpret Findings, and Get to Outcomes." *American Journal of Public Health* 98 (8): 1407–17.

Castillo, Federico, Ana M. Mora, Georgia L. Kayser, Jennifer Vanos, Carly Hyland, Audrey R. Yang, and Brenda Eskenazi. 2021. "Environmental Health Threats to Latino Migrant Farmworkers." *Annual Review of Public Health* 42: 257–76.

Castleden, Heather, Ella Bennett, Diana Lewis, and Debbie Martin. 2017. "'Put It Near the Indians': Indigenous Perspectives on Pulp Mill Contaminants in Their Traditional Territories (Pictou Landing First Nation, Canada)." *Progress in Community Health Partnerships: Research, Education, and Action* 11 (1): 25–33.

Castleden, Heather, Theresa Garvin, and Huu-ay-aht First Nation. 2009. "'Hishuk Tsawak' (Everything Is One/Connected): A Huu-ay-aht Worldview for Seeing Forestry in British Columbia, Canada." *Society & Natural Resources* 22 (9): 789–804.

Castro, Fábio de, Barbara Hogenboom, and Michiel Baud, eds. 2016. *Environmental Governance in Latin America*. New York: Palgrave Macmillan.

Cato, Molly Scott. 2006. *Market, Schmarket: Building the Post-Capitalist Economy*. Cheltenham: New Clarion Press.

CBDIO (Binational Center for Indigenous Oaxacan Community Development), Vista Community Clinic / FarmWorker CARE Coalition, California Institute for Rural Studies, and the COVID-19 Farmworker Study Collective. 2021. "Experts in their Fields: Contributions and Realities of Indigenous Campesinos in California during COVID-19." https://cirsinc.org/wp-content/uploads/2021/10/COFS_Experts-in-Their-Fields_10.18.21_Final.pdf.

Center for Story-Based Strategy. 2017. "Story-Based Strategy 101." https://www.storybasedstrategy.org/tools-and-resources.

Center for Urban Pedagogy. n.d. "CUP: Home." Accessed December 21, 2021. http://welcometocup.org/.

Cernea, Michael M., ed. 1985. *Putting People First: Sociological Variables in Rural Development*. New York: World Bank.

Cha, J. Mijin, Manuel Pastor, Madeline Wander, James Sadd, and Rachel Morello-Frosch. 2019. "A Roadmap to an Equitable Low-Carbon Future: Four Pillars for a Just Transition." Climate Equity Network. http://dornsifelive.usc.edu/pere/roadmap-equitable-low-carbon-future/.

Chakraborty, Jayajit. 2018. "Spatial Representation and Estimation of Environmental Risk: A Review of Analytic Approaches." In *The Routledge Handbook of Environmental Justice*, edited by Ryan Holifield, Jayajit Chakraborty, and Gordon Walker, 175–89. New York: Routledge.

Chakraborty, Jayajit, Timothy W. Collins, and Sara E. Grineski. 2016. "Environmental Justice Research: Contemporary Issues and Emerging Topics." *International Journal of Environmental Research and Public Health* 13 (11): 1072.

Chalupka, Stephanie, Laura Anderko, and Emma Pennea. 2020. "Climate Change, Climate Justice, and Children's Mental Health: A Generation at Risk?" *Environmental Justice* 13 (1): 10–14.

Chambers, Robert. 1997. *Whose Reality Counts? Putting the First Last*. London: Intermediate Technology Publications.

Chang, David, and Taylor Morton. 2016. "WE ACT's Environmental Health and Justice Leadership Training (EHJLT) Program: A Tool to Help Disseminate Research to Community Members Who Live/Work in Northern Manhattan (NYC)." Boston: UMass Chan Medical School. http://hdl.handle.net/20.500.14038/26653.

Chang, Vera L. 2020. "After #MeToo, This Group Has Nearly Erased Sexual Harassment in Farm Fields." *Civil Eats*, March 9, 2020. https://civileats.com/2020/03/09/after-metoo-this-group-has-nearly-erased-sexual-harassment-in-farm-fields/.

Chávez, Vivian, Bonnie Duran, Quinton E. Baker, Magdalena Avila, and Nina Wallerstein. 2008. "The Dance of Race and Privilege in CBPR." In *Community-Based Participatory Research for Health: From Process to Outcomes*, edited by Meredith Minkler and Nina Wallerstein, 2nd ed., 91–106. San Francisco: Jossey-Bass.

Chen, Peggy G., Nitza Diaz, Georgina Lucas, and Marjorie S. Rosenthal. 2010. "Dissemination of Results in Community-Based Participatory Research." *American Journal of Preventive Medicine* 39 (4): 372–78.

Chevalier, Jacques M., and Daniel J. Buckles. 2019. *Participatory Action Research: Theory and Methods for Engaged Inquiry*. 2nd ed. New York: Routledge.

Chirif, Alberto, and Pedro García Hierro. 2007. *Marcando Territorio: Progresos y Limitaciones de La Titulación de Territorios Indígenas En La Amazonía*. Copenhagen: International Work Group for Indigenous Affairs.

Chorus Foundation. 2014. "The Navajo Green Economy." *Chorus Foundation* (blog). December 1, 2014. https://chorusfoundation.org/the-navajo-green-economy/.

Chu, Eric, Isabelle Anguelovski, and JoAnn Carmin. 2016. "Inclusive Approaches to Urban Climate Adaptation Planning and Implementation in the Global South." *Climate Policy* 16 (3): 372–92.

Chung-Do, Jane J., Ilima Ho-Lastimosa, Samantha Keaulana, Kenneth Ho, Phoebe W. Hwang, Theodore Radovich, Luana Albinio, et al. 2019. "Waimānalo Pono Research Hui:

A Community-Academic Partnership to Promote Native Hawaiian Wellness through Culturally Grounded and Community-Driven Research and Programming." *American Journal of Community Psychology* 64 (1–2): 107–17.

Ciplet, David, J. Timmons Roberts, and Mizan R. Khan. 2015. *Power in a Warming World: The New Global Politics of Climate Change and the Remaking of Environmental Inequality*. Cambridge: MIT Press.

City of Los Angeles. 2016. "Los Angeles Municipal Code, Ordinance No. 184246." https://planning.lacity.org/ordinances/docs/cugu/184246.pdf.

Clark, Katherine Ann. 2015. "Grasping Green Electrons: Power in Markets or the Structure of Power?" PhD diss., University of Colorado at Boulder.

Claudio, Luz. 2007. "Standing on Principle: The Global Push for Environmental Justice." *Environmental Health Perspectives* 115 (10): 500–3.

Clinical and Translational Science Awards Consortium. 2011. *Principles of Community Engagement*. 2nd ed. Washington, DC: National Institutes of Health. https://www.atsdr.cdc.gov/communityengagement/.

Coelho, Vera Schattan Ruas Pereira, and Laura Waisbich. 2016. "Participatory Mechanisms and Inequality Reduction: Searching for Plausible Relations." *Journal of Public Deliberation* 12 (2): 13.

Coemans, Sara, and Karin Hannes. 2017. "Researchers under the Spell of the Arts: Two Decades of Using Arts-Based Methods in Community-Based Inquiry with Vulnerable Populations." *Educational Research Review* 22: 34–49.

Cohen, Alison Klebanoff, Andrea Lopez, Nile Malloy, and Rachel Morello-Frosch. 2016. "Surveying for Environmental Health Justice: Community Organizing Applications of Community-Based Participatory Research." *Environmental Justice* 9 (5): 129–36.

Cohen, Alison, Andrea Lopez, Nile Malloy, and Rachel Morello-Frosch. 2012. "Our Environment, Our Health: A Community-Based Participatory Environmental Health Survey in Richmond, California." *Health Education & Behavior* 39 (2): 198–209.

Cohen, Alison K., Travis Richards, Barbara L. Allen, Yolaine Ferrier, Johanna Lees, and Louisa H. Smith. 2018. "Health Issues in the Industrial Port Zone of Marseille, France: The Fos EPSEAL Community-Based Cross-Sectional Survey." *Journal of Public Health* 26 (2): 235–43.

Cole, Brian L., Kara E. MacLeod, and Raenita Spriggs. 2019. "Health Impact Assessment of Transportation Projects and Policies: Living Up to Aims of Advancing Population Health and Health Equity?" *Annual Review of Public Health* 40: 305–18.

Cole, Luke W. 1992. "Empowerment as the Key to Environmental Protection: The Need for Environmental Poverty Law." *Ecology Law Quarterly* 19 (4): 619–83.

———. 1995. "Macho Law Brains, Public Citizens, and Grassroots Activists: Three Models of Environmental Advocacy." *Virginia Environmental Law Journal* 14 (4): 687–710.

Cole, Luke W., and Sheila R. Foster. 2001. *From the Ground Up: Environmental Racism and the Rise of the Environmental Justice Movement*. New York: NYU Press.

Coleman-Jensen, Alisha, Matthew P. Rabbitt, Christian A. Gregory, and Anita Singh. 2021. "Household Food Security in the United States in 2020." Washington, DC: U.S. Department of Agriculture, Economic Research Service. https://www.ers.usda.gov/publications/pub-details/?pubid=102075#.

Collective for Community, Culture, and Environment. 2019. "Sunset Park Green Resilient Industrial District." New York: UPROSE and Protect Our Working Waterfront Alliance.

https://static1.squarespace.com/static/581b72c32e69cfaa445932df/t/5d7fa701be8e5528f6becen/1568646954044/GRID_UPROSE+EDITS.pdf.

———. 2021. "Collective for Community, Culture and Environment: Start." http://collectiveforcce.com/.

Collet, Bruce. 2008. "Confronting the Insider-Outsider Polemic in Conducting Research with Diasporic Communities: Towards a Community-Based Approach." *Refuge* 25 (1): 77–83.

Collins, Susan E., Seema L. Clifasefi, Joey Stanton, Kee J. E. Straits, Eleanor Gil-Kashiwabara, Patricia Rodriguez Espinosa, Andel V. Nicasio, et al. 2018. "Community-Based Participatory Research (CBPR): Towards Equitable Involvement of Community in Psychology Research." *American Psychologist* 73 (7): 884–98.

Commission for Racial Justice. 1987. "Toxic Wastes and Race in the United States: A National Report on the Racial and Socio-Economic Characteristics of Communities with Hazardous Waste Sites." New York: United Church of Christ. https://www.nrc.gov/docs/ML1310/ML13109A339.pdf.

Commodore, Adwoa, Sacoby Wilson, Omar Muhammad, Erik Svendsen, and John Pearce. 2017. "Community-Based Participatory Research for the Study of Air Pollution: A Review of Motivations, Approaches, and Outcomes." *Environmental Monitoring and Assessment* 189 (8): 378.

Cooke, Bill, and Uma Kothari, eds. 2001. *Participation: The New Tyranny?* London: Zed Books.

Coolsaet, Brendan, and Pierre-Yves Néron. 2021. "Recognition and Environmental Justice." In *Environmental Justice: Key Issues*, edited by Brendan Coolsaet, 52–63. New York: Routledge.

Coombe, Chris M., Amy J. Schulz, Lello Guluma, Alex J. Allen, Carol Gray, Wilma Brakefield Caldwell, J. Ricardo Guzman, et al. 2020. "Enhancing Capacity of Community-Academic Partnerships to Achieve Health Equity: Results from the CBPR Partnership Academy." *Health Promotion Practice* 21 (4): 552–63.

Cooper, Caren B., Chris L. Hawn, Lincoln R. Larson, Julia K. Parrish, Gillian Bowser, Darlene Cavalier, Robert R. Dunn, et al. 2021. "Inclusion in Citizen Science: The Conundrum of Rebranding." *Science* 372 (6549): 1386–88.

Cooperation Jackson. 2019. "2019 Year in Review: Resistance and Reflection." *Cooperation Jackson* (blog). December 19, 2019. https://cooperationjackson.org/announcementsblog/2019yearinreview.

Corburn, Jason. 2002. "Environmental Justice, Local Knowledge, and Risk: The Discourse of a Community-Based Cumulative Exposure Assessment." *Environmental Management* 29 (4): 451–66.

———. 2005. *Street Science: Community Knowledge and Environmental Health Justice*. Cambridge: MIT Press.

———. 2009. *Toward the Healthy City: People, Places, and the Politics of Urban Planning*. Cambridge: MIT Press.

———. 2017. "Concepts for Studying Urban Environmental Justice." *Current Environmental Health Reports* 4 (1): 61–67.

Corburn, Jason, Ives Rocha, Alexei A. Dunaway, and Jack Makau. 2017. "Global Health Policy: Slum Settlement Mapping in Nairobi and Rio De Janeiro." In *Community-Based Participatory Research for Health: Advancing Social and Health Equity*, edited by Nina

Wallerstein, Bonnie Duran, John G. Oetzel, and Meredith Minkler, 3rd ed., 321–36. San Francisco: Jossey-Bass.

Córdova, Teresa. 2002. "Grassroots Mobilizations in the Southwest for Environmental and Economic Justice." *International Journal of Public Administration* 25 (2–3): 333–49.

Córdova, Teresa, José Bravo, and José Miguel Acosta-Córdova. 2022. "Environmental Justice and The Alliance for a Just Transition: Grist for Climate Justice Planning." *Journal of Planning Literature*. https://doi.org/10.1177/08854122221121120.

Córdova, Teresa, Jose T. Bravo, Jeanne Gauna, Richard Moore, and Ruben Solis. 2000. "Building Networks to Tackle Global Restructuring: The Environmental and Economic Justice Movement." In *The Collaborative City: Opportunities and Struggles for Blacks and Latinos in U.S. Cities*, edited by John Bentancur and Douglas Gills, 177–96. New York: Routledge.

Cossío, Rosa, Mary Menton, Peter Cronkleton, and Anne Larson. 2014. "Community Forest Management in the Peruvian Amazon: A Literature Review." Working Paper 136. Bogor, Indonesia: Center for International Forestry Research.

Coulthard, Glen Sean. 2014. *Red Skin, White Masks: Rejecting the Colonial Politics of Recognition*. Minneapolis: University of Minnesota Press.

Coventry, Philip, and Chukwumerije Okereke. 2018. "Climate Change and Environmental Justice." In *The Routledge Handbook of Environmental Justice*, edited by Ryan Holifield, Jayajit Chakraborty, and Gordon Walker, 362–73. New York: Routledge.

Crenshaw, Kimberlé. 1989. "Demarginalizing the Intersection of Race and Sex: A Black Feminist Critique of Antidiscrimination Doctrine, Feminist Theory and Antiracist Politics." *University of Chicago Legal Forum* 1: 139–67.

Crimmins, Gail, ed. 2019. *Strategies for Resisting Sexism in the Academy: Higher Education, Gender and Intersectionality*. 1st ed. New York: Palgrave Macmillan.

Cronon, William. 1996. "The Trouble with Wilderness: Or, Getting Back to the Wrong Nature." *Environmental History* 1 (1): 7–28.

Crosby, Richard, and Seth M. Noar. 2011. "What Is a Planning Model? An Introduction to PRECEDE-PROCEED." *Journal of Public Health Dentistry* 71 (S1): 7–15.

Cummins, Crescentia, John Doyle, Larry Kindness, Myra J. Lefthand, Urban J. Bear, Ada L. Bends, Susan C. Broadaway, et al. 2010. "Community-Based Participatory Research in Indian Country: Improving Health Through Water Quality Research and Awareness." *Family and Community Health* 33 (3): 166–74.

Curley, Andrew. 2018. "A Failed Green Future: Navajo Green Jobs and Energy 'Transition' in the Navajo Nation." *Geoforum* 88: 57–65.

Cushing, Lara, John Faust, Laura Meehan August, Rose Cendak, Walker Wieland, and George Alexeeff. 2015. "Racial/Ethnic Disparities in Cumulative Environmental Health Impacts in California: Evidence from a Statewide Environmental Justice Screening Tool (CalEnviroScreen 1.1)." *American Journal of Public Health* 105 (11): 2341–48.

Cyril, Sheila, Ben J. Smith, Alphia Possamai-Inesedy, and Andre M. N. Renzaho. 2015. "Exploring the Role of Community Engagement in Improving the Health of Disadvantaged Populations: A Systematic Review." *Global Health Action* 8 (1): 29842.

Dalton, Craig M., and Tim Stallmann. 2018. "Counter-Mapping Data Science." *The Canadian Geographer / Le Géographe Canadien* 62 (1): 93–101.

D'Amore, Chiara, Clare Hintz, Cirien Saadeh, and Jeremy Solin, eds. 2016. "Place and Resilience in Sustainability Education." Special issue, *Journal of Sustainability Education*. April. www.susted.com/wordpress/content/place-and-resilience_2016_04/.

Dandekar, Hemalata C. 2001. "Rural Planning: General." In *International Encyclopedia of the Social and Behavioral Sciences*, 13425–29. Elsevier Science.

DataCenter. 2015a. "An Introduction to Research Justice." Oakland: DataCenter. http://www.datacenter.org/wp-content/uploads/Intro_Research_Justice_Toolkit_FINAL1.pdf.

———. 2015b. "Campaign Research: A Toolkit for Grassroots Organizing." Oakland: DataCenter. http://www.datacenter.org/wp-content/uploads/CampaignResearch_FINAL1.pdf.

———. 2015c. "Our Voices, Our Lands: A Guide to Community-Based Strategies for Mapping Indigenous Stories." Oakland: DataCenter. http://www.datacenter.org/wp-content/uploads/OurVoicesOurLand_Final1.pdf.

Davidoff, Paul. 1965. "Advocacy and Pluralism in Planning." *Journal of the American Institute of Planners* 31 (4): 331–38.

Davies, Thom, and Alice Mah. 2020a. "Introduction: Tackling Environmental Injustice in a Post-Truth Age." In *Toxic Truths: Environmental Justice and Citizen Science in a Post-Truth Age*, edited by Thom Davies and Alice Mah, 1–28. Manchester: Manchester University Press.

Davies, Thom, and Alice Mah, eds. 2020b. *Toxic Truths: Environmental Justice and Citizen Science in a Post-Truth Age*. Manchester: Manchester University Press.

Davis, Leona F., and Mónica D. Ramírez-Andreotta. 2021. "Participatory Research for Environmental Justice: A Critical Interpretive Synthesis." *Environmental Health Perspectives* 129 (2): 026001.

Davis, Meghan F., Nora Pisanic, Sarah M. Rhodes, Alexis Brown, Haley Keller, Maya Nadimpalli, Andrea Christ, et al. 2018. "Occurrence of Staphylococcus Aureus in Swine and Swine Workplace Environments on Industrial and Antibiotic-Free Hog Operations in North Carolina, USA: A One Health Pilot Study." *Environmental Research* 163: 88–96.

De Filippo, Daniela, Nuria Bautista-Puig, Elba Mauleón, and Elías Sanz-Casado. 2018. "A Bridge Between Society and Universities: A Documentary Analysis of Science Shops." *Publications* 6 (3): 36.

De Lacy-Vawdon, Cassandra, and Charles Livingstone. 2020. "Defining the Commercial Determinants of Health: A Systematic Review." *BMC Public Health* 20: 1022.

De Lara, Juan D. 2018. *Inland Shift: Race, Space, and Capital in Southern California*. Oakland: University of California Press.

De Lara, Juan D., Ellen R. Reese, and Jason Struna. 2016. "Organizing Temporary, Subcontracted, and Immigrant Workers: Lessons from Change to Win's Warehouse Workers United Campaign." *Labor Studies Journal* 41 (4): 309–32.

De las Nueces, Denise, Karen Hacker, Ann DiGirolamo, and LeRoi S. Hicks. 2012. "A Systematic Review of Community-Based Participatory Research to Enhance Clinical Trials in Racial and Ethnic Minority Groups." *Health Services Research* 47 (3): 1363–86.

DeCarlo, Matt, Cory Cummings, and Kate Agnelli. 2021. *Graduate Research Methods in Social Work: A Project-Based Approach*. Black Rock City: Open Social Work.

Deeb-Sossa, Natalia, ed. 2019. *Community-Based Participatory Research: Testimonios from Chicana/o Studies*. Tucson: University of Arizona Press.

DeFilippis, James. 2012. "Community Control and Development: The Long View." In *The Community Development Reader*, edited by James DeFilippis and Susan Saegert, 2nd ed., 30–37. New York: Routledge.

Dehar, Mary-Anne, Sally Casswell, and Paul Duignan. 1993. "Formative and Process Evaluation of Health Promotion and Disease Prevention Programs." *Evaluation Review* 17 (2): 204–20.

Delgado, Richard. 1989. "Storytelling for Oppositionists and Others: A Plea for Narrative." *Michigan Law Review* 87 (8): 2411–41.

Demeulenaere, Else. 2021. "Prutehi Litekyan: A Social Movement to Protect Biocultural Diversity and Restore Indigenous Land Sovereignty on Guåhan." In *Indigenous Peoples, Heritage and Landscape in the Asia Pacific: Knowledge Co-Production and Empowerment*, edited by Stephen Acabado and Da-Wei Kuan, 54–71. New York: Routledge.

Democracy Collaborative. 2019. "Linking Anchor Institutions to Outcomes for Families, Children, and Communities." Community-Wealth.org. https://community-wealth.org/indicators.

Dennis, Samuel F., Suzanne Gaulocher, Richard M. Carpiano, and David Brown. 2009. "Participatory Photo Mapping (PPM): Exploring an Integrated Method for Health and Place Research with Young People." *Health & Place* 15 (2): 466–73.

Denyer, David, and David Tranfield. 2006. "Using Qualitative Research Synthesis to Build an Actionable Knowledge Base." *Management Decision* 44 (2): 213–27.

Denzin, Norman K., Yvonna S. Lincoln, and Linda Tuhiwai Smith, eds. 2008. *Handbook of Critical and Indigenous Methodologies*. Los Angeles: SAGE Publications.

Denzongpa, Kunga, Tracy Nichols, and Sharon D. Morrison. 2020. "Situating Positionality and Power in CBPR Conducted with a Refugee Community: Benefits of a Co-Learning Reflective Model." *Reflective Practice* 21 (2): 237–50.

Derose, Kathryn P., Malcolm V. Williams, Cheryl A. Branch, Karen R. Flórez, Jennifer Hawes-Dawson, Michael A. Mata, Clyde W. Oden, and Eunice C. Wong. 2019. "A Community-Partnered Approach to Developing Church-Based Interventions to Reduce Health Disparities Among African-Americans and Latinos." *Journal of Racial and Ethnic Health Disparities* 6 (2): 254–64.

Detroit Black Community Food Security Network. n.d. "Detroit Black Community Food Security Network: Home." Accessed December 21, 2021. https://www.dbcfsn.org.

———. n.d. "D-Town Farm." Accessed December 21, 2021. https://www.dbcfsn.org/dtownfarm2022.

Dewey, John. 1916. *Democracy and Education*. New York: Macmillan.

———. 1938. *Experience and Education*. New York: Macmillan.

Dhaliwal, Kanwarpal, Jill Casey, Kimberly Aceves-Iñiguez, and Jara Dean-Coffey. 2020. "Radical Inquiry—Liberatory Praxis for Research and Evaluation." *New Directions for Evaluation* 166: 49–64.

Di Chiro, Giovanna. 2021. "Mobilizing 'Intersectionality' in Environmental Justice Research and Action in a Time of Crisis." In *Environmental Justice: Key Issues*, edited by Brendan Coolsaet, 316–33. New York: Routledge.

Di Chiro, Giovanna, and Laura Rigell. 2018. "Situating Sustainability Against Displacement: Building Campus-Community Collaboratives for Environmental Justice from the Ground Up." In *Sustainability: Approaches to Environmental Justice and Social Power*, edited by Julie Sze, 76–101. New York: NYU Press.

DiAngelo, Robin J. 2018. *White Fragility: Why It's So Hard for White People to Talk About Racism*. Boston: Beacon Press.

Díaz Ríos, Claudia, Michelle L. Dion, and Kelsey Leonard. 2020. "Institutional Logics and Indigenous Research Sovereignty in Canada, the United States, Australia, and New Zealand." *Studies in Higher Education* 45 (2): 403–15.

D'Ignazio, Catherine, and Rahul Bhargava. 2015. "Approaches to Building Big Data Literacy." *Proceedings of the 2015 Bloomberg Data for Good Exchange Conference*. New York: MIT Media Lab. https://www.media.mit.edu/publications/approaches-to-building-big-data-literacy/.

D'Ignazio, Catherine, Jeffrey Warren, and Don Blair. 2014. "The Role of Small Data for Governance in the 21st Century." In *Governança Digital*, edited by Marcelo Soares Pimenta and Diego Rafael Canabarro, 115–129. Porto Alegre, Brazil: Universidade Federal Rio Grande do Sul.

Dittmer, Kristofer. 2013. "Local Currencies for Purposive Degrowth? A Quality Check of Some Proposals for Changing Money-as-Usual." *Journal of Cleaner Production* 54: 3–13.

Dittmer, Livia, Frank Mugagga, Alexander Metternich, Petra Schweizer-Ries, George Asiimwe, and Manuel Riemer. 2018. "'We Can Keep the Fire Burning': Building Action Competence through Environmental Justice Education in Uganda and Germany." *Local Environment*, 23 (2): 144–57.

Doberneck, Diane M., and John H. Schweitzer. 2017. "Disciplinary Variations in Publicly Engaged Scholarship: An Analysis Using the Biglan Classification of Academic Disciplines." *Journal of Higher Education Outreach and Engagement* 21 (1): 78–103.

Dodson, Robin E., Katherine E. Boronow, Herbert Susmann, Julia O. Udesky, Kathryn M. Rodgers, David Weller, Million Woudneh, Julia Green Brody, and Ruthann A. Rudel. 2020. "Consumer Behavior and Exposure to Parabens, Bisphenols, Triclosan, Dichlorophenols, and Benzophenone-3: Results from a Crowdsourced Biomonitoring Study." *International Journal of Hygiene and Environmental Health* 230: 113624.

Downing, Rupert. 2009. "Lessons and Opportunities from Canada's Social Economy Research Program." In *Solidarity Economy I: Building Alternatives for People and Planet; Papers and Reports from the 2009 U.S. Forum on the Solidarity Economy*, edited by Emily Kawano, Jonathan Teller-Elsberg, and Thomas Neal Materson, 277–86. Amherst: Center for Popular Economics.

Dozier, Deshonay. 2018. "A Response to Abolitionist Planning: There Is No Room for 'Planners' in the Movement for Abolition." *Planners Network* (blog), August 9, 2018. https://www.plannersnetwork.org/2018/08/response-to-abolitionist-planning/.

Drahota, Amy, Rosemary D. Meza, Brigitte Brikho, Meghan Naaf, Jasper A. Estabillo, Emily D. Gomez, Sarah F. Vejnoska, Sarah Dufek, and Aubyn C. Stahmer. 2016. "Community-Academic Partnerships: A Systematic Review of the State of the Literature and Recommendations for Future Research." *Milbank Quarterly* 94 (1): 163–214.

Draus, Paul, Dagmar Haase, Jacob Napieralski, Juliette Roddy, and Salman Qureshi. 2019. "Wounds, Ghosts and Gardens: Historical Trauma and Green Reparations in Berlin and Detroit." *Cities* 93 (8): 153–63.

Dubb, Steve, and Ted Howard. 2012. "Leveraging Anchor Institutions for Local Job Creation and Wealth Building." Shaker Heights: The Democracy Collaborative. https://thedemocracycollaborative.com/learn/publication/leveraging-anchor-institutions-local-job-creation-and-wealth-building-0.

Ducre, Kishi Animashaun. 2018. "The Black Feminist Spatial Imagination and an Intersectional Environmental Justice." *Environmental Sociology* 4 (1): 22–35.

Duran, Bonnie, John Oetzel, Maya Magarati, Myra Parker, Chuan Zhou, Yvette Roubideaux, Michael Muhammad, et al. 2019. "Toward Health Equity: A National Study of Promising

Practices in Community-Based Participatory Research." *Progress in Community Health Partnerships* 13 (4): 337–52.

Dutta, Mohan J. 2015. "Decolonizing Communication for Social Change: A Culture-Centered Approach." *Communication Theory* 25 (2): 123–43.

Dyck, Erika. 2021. "Doing History That Matters: Going Public and Activating Voices as a Form of Historical Activism." *Journal of the History of the Behavioral Sciences* 57: 75–86.

East Yard Communities for Environmental Justice. n.d. "I-710 Corridor." Accessed January 2, 2022. http://eycej.org/campaigns/i-710/.

Echo-Hawk, Abigail. 2019. "Indigenous Health Equity." Seattle: Urban Indian Health Institute. https://www.uihi.org/resources/indigenous-health-equity/.

El Puente. n.d. "Our Air! / ¡Nuestro Aire!" Accessed January 25, 2022. https://elpuente.us/our-air-nuestro-aire.

Ellis, Galen, and Sheryl Walton. 2012. "Building Partnerships Between Local Health Departments and Communities: Case Studies in Capacity Building and Cultural Humility." In *Community Organizing and Community Building for Health and Welfare*, edited by Meredith Minkler, 3rd ed., 130–48. New Brunswick: Rutgers University Press.

Elwood, William N., James G. Corrigan, and Kathryn A. Morris. 2019. "NIH-Funded CBPR: Self-Reported Community Partner and Investigator Perspectives." *Journal of Community Health* 44 (4): 740–48.

Emmett Environmental Law and Policy Clinic. 2017. "A Manual for Citizen Scientists Starting or Participating in Data Collection and Environmental Monitoring Projects." Harvard Law School. http://clinics.law.harvard.edu/environment/files/2017/09/HLS-Env-Clinic-Citizen-Science-Manual-Sept-2017-FULL.pdf.

Eng, Eugenia, Jennifer Schaal, Stephanie Baker, Kristin Black, Samuel Cykert, Nora Jones, Alexandra Lightfoot, et al. 2017. "Partnership, Transparency, and Accountability: Changing Systems to Enhance Racial Equity in Cancer Care and Outcomes." In *Community-Based Participatory Research for Health: Advancing Social and Health Equity*, edited by Nina Wallerstein, Bonnie Duran, John G. Oetzel, and Meredith Minkler, 3rd ed., 107–22. San Francisco: Jossey-Bass.

Eng, Eugenia, Karen Strazza, Scott D. Rhodes, Derek Griffith, Kate Shirah, and Elvira Mebane. 2013. "Insiders and Outsiders Assess Who Is the Community: Participant Observation, Key Informant Interview, Focus Group Interview, and Community Forum." In *Methods for Community-Based Participatory Research for Health*, edited by Barbara A. Israel, Eugenia Eng, Amy J. Schulz, and Edith A. Parker, 2nd ed., 133–60. San Francisco: Jossey-Bass.

Eng, Tiffany, Amy Vanderwarker, and Marybelle Nzegwu. 2018. "CalEnviroScreen: A Critical Tool for Achieving Environmental Justice in California." Oakland and Huntington Park: California Environmental Justice Alliance. https://caleja.org/wp-content/uploads/2018/08/CEJA-CES-Report-2018_web.pdf.

Engle, Elyzabeth W. 2019. "'Coal Is in Our Food, Coal Is in Our Blood': Everyday Environmental Injustices of Rural Community Gardening in Central Appalachia." *Local Environment* 24 (8): 746–61.

English, Paul B., Maxwell J. Richardson, and Catalina Garzón-Galvis. 2018. "From Crowdsourcing to Extreme Citizen Science: Participatory Research for Environmental Health." *Annual Review of Public Health* 39: 335–50.

Environmental Health Coalition. n.d. "Healthy Communities." San Diego: Environmental Health Coalition. Accessed December 21, 2021. https://www.environmentalhealth.org/index.php/en/what-we-do/toxic-free-neighborhoods/community-land-use-planning.

———. 2004. "Globalization at the Crossroads: Ten Years of NAFTA at the San Diego/Tijuana Border Region." San Diego: Environmental Health Coalition. http://www.geographycantakeyouthere.weebly.com/uploads/5/7/7/7/57778601/globalizationfnlrel_10_18_04.pdf.

Environmental Justice and the Common Good Initiative. 2020. "Racial and Environmental Justice." Santa Clara: Santa Clara University. www.scu.edu/ej/resources/standing-up-for-ej/.

———. 2021. "A Food Justice Response to Covid-19." Santa Clara: Santa Clara University. https://www.scu.edu/ej/events/ej-and-cg-initiative-news/a-food-justice-response-to-covid-19/.

Esbjorn-Hargens, Sean, and Michael E. Zimmerman. 2009. *Integral Ecology: Uniting Multiple Perspectives on the Natural World*. Boston: Integral Books.

Eskenazi, Brenda, Asa Bradman, and Rosemary Castorina. 1999. "Exposures of Children to Organophosphate Pesticides and Their Potential Adverse Health Effects." *Environmental Health Perspectives* 107 (S3): 409–19.

Eskenazi, Brenda, Jonathan Chevrier, Stephen A. Rauch, Katherine Kogut, Kim G. Harley, Caroline Johnson, Celina Trujillo, Andreas Sjödin, and Asa Bradman. 2013. "In Utero and Childhood Polybrominated Diphenyl Ether (PBDE) Exposures and Neurodevelopment in the CHAMACOS Study." *Environmental Health Perspectives* 121 (2): 257–62.

Eskenazi, Brenda, Kim Harley, Asa Bradman, Erin Weltzien, Nicholas P. Jewell, Dana B. Barr, Clement E. Furlong, and Nina T. Holland. 2004. "Association of in Utero Organophosphate Pesticide Exposure and Fetal Growth and Length of Gestation in an Agricultural Population." *Environmental Health Perspectives* 112 (10): 1116–24.

Eskenazi, Brenda, Amy R. Marks, Asa Bradman, Kim Harley, Dana B. Barr, Caroline Johnson, Norma Morga, and Nicholas P. Jewell. 2007. "Organophosphate Pesticide Exposure and Neurodevelopment in Young Mexican-American Children." *Environmental Health Perspectives* 115 (5): 792–98.

Espiritu, Belinda F. 2017. "The Lumad Struggle for Social and Environmental Justice: Alternative Media in a Socio-Environmental Movement in the Philippines." *Journal of Alternative & Community Media* 2 (1): 45–59.

Estes, Nick. 2019. *Our History Is the Future: Standing Rock Versus the Dakota Access Pipeline, and the Long Tradition of Indigenous Resistance*. New York: Verso Books.

Estes, Nick, and Jaskiran Dhillon. 2019. *Standing with Standing Rock: Voices from the #NoDAPL Movement*. Minneapolis: University of Minnesota Press.

Etieyibo, Edwin. 2017. "Ubuntu and the Environment." In *The Palgrave Handbook of African Philosophy*, edited by Adeshina Afolayan and Toyin Falola, 633–57. New York: Palgrave Macmillan.

Evans, Mei Mei. 2002. "Testimonies." In *The Environmental Justice Reader: Politics, Poetics, and Pedagogy*, edited by Joni Adamson, Mei Mei Evans, and Rachel Stein, 2nd ed., 29–43. Tucson: University of Arizona Press.

Evans-Agnew, Robin A., and Chris Eberhardt. 2019. "Uniting Action Research and Citizen Science: Examining the Opportunities for Mutual Benefit Between Two Movements Through a Woodsmoke Photovoice Study." *Action Research* 17 (3): 357–77.

Evans-Agnew, Robin A., and Marie-Anne S. Rosemberg. 2016. "Questioning Photovoice Research: Whose Voice?" *Qualitative Health Research* 26 (8): 1019–30.

Exley, Karen, Noemi Cano, Dominique Aerts, Pierre Biot, Ludwine Casteleyn, Marike Kolossa-Gehring, Gerda Schwedler, et al. 2015. "Communication in a Human Biomonitoring Study: Focus Group Work, Public Engagement and Lessons Learnt in 17 European Countries." *Environmental Research* 14: 31–41.

Eyben, Rosalind, Colette Harris, and Jethro Pettit. 2006. "Introduction: Exploring Power for Change." *IDS Bulletin* 37, no. 6.

Ezell, Jerel M., Samira Salari, Clinton Rooker, and Elizabeth C. Chase. 2021. "Intersectional Trauma: COVID-19, the Psychosocial Contract, and America's Racialized Public Health Lineage." *Traumatology* 27 (1): 78–85.

Faber, Daniel. 2018. "The Political Economy of Environmental Justice." In *The Routledge Handbook of Environmental Justice*, edited by Ryan Holifield, Jayajit Chakraborty, and Gordon Walker, 61–73. New York: Routledge.

Fagundes, Colton, Lorette Picciano, Willard Tillman, Jennifer Mleczko, Stephanie Schwier, Garrett Graddy-Lovelace, Felicia Hall, and Tracy Watson. 2020. "Ecological Costs of Discrimination: Racism, Red Cedar and Resilience in Farm Bill Conservation Policy in Oklahoma." *Renewable Agriculture and Food Systems* 35 (4): 420–34.

Fainstein, Susan S. 2000. "New Directions in Planning Theory." *Urban Affairs Review* 35 (4): 451–78.

Fainstein, Susan S., and James DeFilippis, eds. 2015. *Readings in Planning Theory*. 4th ed. West Sussex: John Wiley & Sons.

Fallman, Daniel. 2008. "The Interaction Design Research Triangle of Design Practice, Design Studies, and Design Exploration." *Design Issues* 24 (3): 4–18.

Fals Borda, Orlando. 1987. "The Application of Participatory Action-Research in Latin America." *International Sociology* 2 (4): 329–47.

———. 2006. "The North-South Convergence: A 30-Year First-Person Assessment of PAR." *Action Research* 4 (3): 351–58.

Farhang, Sean, and Ira Katznelson. 2005. "The Southern Imposition: Congress and Labor in the New Deal and Fair Deal." *Studies in American Political Development* 19: 1–30.

Farquhar, John W., Stephen P. Fortmann, Nathan Maccoby, William L. Haskell, Paul T. Williams, June A. Flora, C. Barr Taylor, Byron W. Brown, Douglas S. Solomon, and Stephen B. Hulley. 1985. "The Stanford Five-City Project: Design and Methods." *American Journal of Epidemiology* 122 (2): 323–34.

Ferman, Barbara, Miriam Greenberg, Thao Le, and Steven C. McKay. 2021. "The Right to the City and to the University: Forging Solidarity Beyond the Town/Gown Divide." *The Assembly* 3 (1): 10–34.

Fernández, Jesica Siham. 2021. *Growing Up Latinx: Coming of Age in a Time of Contested Citizenship*. New York: NYU Press.

Fernandez, Maria E., Natalia I. Heredia, Lorna H. McNeill, Maria Eugenia Fernandez-Esquer, Yen-Chi L. Le, and Kelly G. McGauhey. 2017. "Special Issues in Conducting Community-Based Participatory Research Studies with Ethnic and Racial Minorities." In *Handbook of Community-Based Participatory Research*, edited by Steven S. Coughlin, Selina A. Smith, and Maria E. Fernandez, 81–114. Oxford: Oxford University Press.

Ferrera, Maria J., Tina K. Sacks, Miriam Perez, John P. Nixon, Dale Asis, and Walter L. Coleman. 2015. "Empowering Immigrant Youth in Chicago: Utilizing CBPR to Document the Impact of a Youth Health Service Corps Program." *Family & Community Health* 38 (1): 12–21.

Fine, Michelle. 1994. "Dis-Stance and Other Stances: Negotiations of Power Inside Feminist Research." In *Power and Method: Political Activism and Educational Research*, edited by Andrew Gitlin, 13–35. New York: Routledge.

Finn, Symma, Mose Herne, and Dorothy Castille. 2017. "The Value of Traditional Ecological Knowledge for the Environmental Health Sciences and Biomedical Research." *Environmental Health Perspectives* 125 (8): 085006.

Finney, Carolyn. 2014. *Black Faces, White Spaces: Reimagining the Relationship of African Americans to the Great Outdoors*. Chapel Hill: University of North Carolina Press.

Fiorino, Daniel J. 2017. "Green Economy: Reframing Ecology, Economics, and Equity." In *Conceptual Innovation in Environmental Policy*, edited by James Meadowcroft and Daniel J. Fiorino, 233–57. Cambridge: MIT Press.

First National People of Color Environmental Leadership Summit. 1991. "Principles of Environmental Justice." Accessed February 10, 2022. https://www.ejnet.org/ej/principles .html.

———. 1992. "A Call to Action: National Conference Gives Birth to a New Environmental Movement." Accessed February 10, 2022. https://www.threecircles.org/wp-content /uploads/2015/12/1992JMEE.pdf.

Fischer, Frank. 2017. *Climate Crisis and the Democratic Prospect: Participatory Governance in Sustainable Communities*. Oxford: Oxford University Press.

Fisher, Micah R. 2021. "Engaging Voices in the Landscape: Participatory Geography in Indigenous Land Rights Recognition." In *Indigenous Peoples, Heritage and Landscape in the Asia Pacific: Knowledge Co-Production and Empowerment*, edited by Stephen Acabado and Da-Wei Kuan, 31–53. New York: Routledge.

Fleckman, Julia M., Mark Dal Corso, Shokufeh Ramirez, Maya Begalieva, and Carolyn C. Johnson. 2015. "Intercultural Competency in Public Health: A Call for Action to Incorporate Training into Public Health Education." *Frontiers in Public Health* 3: 210.

Fletcher, Fay, Brent Hammer, and Alicia Hibbert. 2014. "'We Know We Are Doing Something Good, but What Is It?': The Challenge of Negotiating Between Service Delivery and Research in a CBPR Project." *Journal of Community Engagement and Scholarship* 7 (2): 19–30.

Flicker, Sarah, Adrian Guta, and Robb Travers. 2017. "Everyday Challenges in the Life Cycle of CBPR: Broadening Our Bandwidth on Ethics." In *Community-Based Participatory Research for Health: Advancing Social and Health Equity*, edited by Nina Wallerstein, Bonnie Duran, John G. Oetzel, and Meredith Minkler, 3rd ed., 227–36. San Francisco: Jossey-Bass.

Foell, Andrew, Jason Q. Purnell, Rachel Barth, Michelle Witthaus, Timetria Murphy-Watson, Sal Martinez, and Mike Foley. 2020. "Resident-Led Neighborhood Development to Support Health: Identifying Strategies Using CBPR." *American Journal of Community Psychology* 66 (3–4): 404–16.

Ford, Chandra L., and Collins O. Airhihenbuwa. 2010. "The Public Health Critical Race Methodology: Praxis for Antiracism Research." *Social Science & Medicine* 71 (8): 1390–98.

Ford-Thompson, Adriana E. S., Carolyn Snell, Glen Saunders, and Piran C. L. White. 2012. "Stakeholder Participation in Management of Invasive Vertebrates." *Conservation Biology* 26 (2): 345–56.

Forester, John. 1988. *Planning in the Face of Power*. Berkeley: University of California Press.

———. 1999. *The Deliberative Practitioner: Encouraging Participatory Planning Processes*. Cambridge: MIT Press.

Foronda, Cynthia, Diana-Lyn Baptiste, Maren M. Reinholdt, and Kevin Ousman. 2016. "Cultural Humility: A Concept Analysis." *Journal of Transcultural Nursing* 27 (3): 210–17.

Forum for Food Sovereignty. 2007. "Declaration of the Forum for Food Sovereignty, Nyéléni." *Nyéléni Forum* (blog). February 27, 2007. https://nyeleni.org/spip.php?article290.

Foster, Sheila R. 2018. "Vulnerability, Equality and Environmental Justice: The Potential and Limits of Law." In *The Routledge Handbook of Environmental Justice*, edited by Ryan Holifield, Jayajit Chakraborty, and Gordon Walker, 136–48. New York: Routledge.

Fox, Carly, Rebecca Fuentes, Fabiola Ortiz Valdez, Gretchen Purser, and Kathleen Sexsmith. 2017. *Milked: Immigrant Dairy Farmworkers in New York State*. New York: Workers' Center of Central New York and Worker Justice Center. http://www.iwj.org/resources/milked-immigrant-dairy-farmworkers-in-new-york-state.

Freese, Bill, and Ashley Lukens. 2015. *Pesticides in Paradise: Hawai'i's Health and Environment at Risk*. Hawai'i Center for Food Safety. https://www.centerforfoodsafety.org/reports/3901/pesticides-in-paradise-hawaiis-health-and-environment-at-risk.

Freire, Paulo. 1970. *The Pedagogy of the Oppressed*. New York: Seabury Press.

———. 1982. "Creating Alternative Research Methods: Learning to Do It by Doing It." In *Creating Knowledge: A Monopoly?* edited by Budd Hall, Arthur Gillette, and Rajesh Tandon, 29–37. Participatory Research 1. New Delhi: Society for Participatory Research in Asia.

Frickel, Scott, Sahra Gibbon, Jeff Howard, Joanna Kempner, Gwen Ottinger, and David J. Hess. 2010. "Undone Science: Charting Social Movement and Civil Society Challenges to Research Agenda Setting." *Science, Technology, & Human Values* 35 (4): 444–73.

Friendly, Abigail. 2022. "Insurgent Planning in Pandemic Times: The Case of Rio de Janeiro." *International Journal of Urban and Regional Research* 46 (1): 115–25.

Fujishiro, Kaori, Emily Q. Ahonen, David Gimeno Ruiz de Porras, I-Chen Chen, and Fernando G. Benavides. 2021. "Sociopolitical Values and Social Institutions: Studying Work and Health Equity Through the Lens of Political Economy." *SSM—Population Health* 14: 100787.

Funk, Julia. 2021. "Energy Independence in Puerto Rico: On Fighting Disaster Capitalism with Community-Based Solar Projects." ArcGIS StoryMaps. April 26, 2021. https://storymaps.arcgis.com/stories/4b19a10915cc409db4bb21b033ab36e5.

Gaard, Greta. 2017. *Critical Ecofeminism*. Lanham: Lexington Books.

———. 2018. "Feminism and Environmental Justice." In *The Routledge Handbook of Environmental Justice*, edited by Ryan Holifield, Jayajit Chakraborty, and Gordon Walker, 74–88. New York: Routledge.

Gachupin, Francine C., and Fatima Molina. 2019. "How to Build and Sustain a Tribal IRB." Tucson: University of Arizona Department of Family and Community Medicine. https://in.nau.edu/wp-content/uploads/sites/142/2019/05/IRB-1.pdf.

Galea, Sandro, and Roger D. Vaughan. 2019. "Public Health, Politics, and the Creation of Meaning: A Public Health of Consequence, July 2019." *American Journal of Public Health* 109 (7): 966–68.

Ganz, Marshall. 2011. "Public Narrative, Collective Action, and Power." In *Accountability through Public Opinion: From Inertia to Public Action*, edited by Sina Odugbemi and Taeku Lee, 273–89. Washington, DC: World Bank.

Garcia, Analilia P., Nina Wallerstein, Andrea Hricko, Jesse N. Marquez, Angelo Logan, Elina Green Nasser, and Meredith Minkler. 2013. "THE (Trade, Health, Environment) Impact Project: A Community-Based Participatory Research Environmental Justice Case Study." *Environmental Justice* 6 (1): 17–26.

Garth, Hanna, and Ashanté M. Reese, eds. 2020. *Black Food Matters: Racial Justice in the Wake of Food Justice*. Minneapolis: University of Minnesota Press.

Garzón, Catalina, Brian Beveridge, Margaret Gordon, Cassandra Martin, Eyal Matalon, and Eli Moore. 2013. "Power, Privilege, and the Process of Community-Based Participatory Research: Critical Reflections on Forging an Empowered Partnership for Environmental Justice in West Oakland, California." *Environmental Justice* 6 (2): 71–78.

Gassett, Parker Randall, Katie O'Brien-Clayton, Carolina Bastidas, Jennie E. Rheuban, Christopher W. Hunt, Elizabeth Turner, Matthew Liebman, et al. 2021. "Community Science for Coastal Acidification Monitoring and Research." *Coastal Management* 49 (5): 510–31.

Gaudry, Adam. 2018. "Next Steps in Indigenous Community-Engaged Research: Supporting Research Self-Sufficiency in Indigenous Communities." In *Towards a New Ethnohistory: Community-Engaged Scholarship Among the People of the River*, edited by Keith Thor Carlson, John Sutton Lutz, David M. Schaepe, and Naxaxalhts'i, 254–58. Winnipeg: University of Manitoba Press.

Gaworecki, Mike. 2016. "The Inside Story of How Great Bear Rainforest Went from a 'War in the Woods' to an Unprecedented Environmental and Human Rights Agreement." *Mongabay Environmental News*, February. https://news.mongabay.com/2016/02/the-inside-story-of-how-great-bear-rainforest-went-from-a-war-in-the-woods-to-an-unprecedented-environmental-and-human-rights-agreement/.

Gee, Gilbert C., Katrina M. Walsemann, and Elizabeth Brondolo. 2012. "A Life Course Perspective on How Racism May Be Related to Health Inequities." *American Journal of Public Health* 102 (5): 967–74.

Geiser, Ken, Joel Tickner, Sally Edwards, and Mark Rossi. 2015. "The Architecture of Chemical Alternatives Assessment." *Risk Analysis* 35 (12): 2152–61.

Geronimus, Arline T., Margaret Hicken, Danya Keene, and John Bound. 2006. "'Weathering' and Age Patterns of Allostatic Load Scores Among Blacks and Whites in the United States." *American Journal of Public Health* 96 (5): 826–33.

Giacomini, Terran, and Terisa Turner. 2015. "The 2014 People's Climate March and Flood Wall Street Civil Disobedience: Making the Transition to a Post-Fossil Capitalist, Commoning Civilization." *Capitalism Nature Socialism* 26 (2): 27–45.

Gibbs, Jenna L., Michael G. Yost, Maria Negrete, and Richard A. Fenske. 2017. "Passive Sampling for Indoor and Outdoor Exposures to Chlorpyrifos, Azinphos-Methyl, and Oxygen Analogs in a Rural Agricultural Community." *Environmental Health Perspectives* 125 (3): 333–41.

Gilio-Whitaker, Dina. 2019. *As Long as Grass Grows: The Indigenous Fight for Environmental Justice, from Colonization to Standing Rock*. Boston: Beacon Press.

Gilmore, Jalisa, Priya Mulgaonkar, Tok Michelle Oyewole, and Michael Heimbinder. 2021. "CAMP-EJ: Findings and Recommendations Report." New York City Environmental Justice Alliance. https://www.nyc-eja.org/wp-content/uploads/2021/02/CAMP-EJ-2020-Report-Final-021821-Reduced.pdf.

Gilmore, Ruth Wilson. 2017. "Abolition Geography and the Problem of Innocence." In *Futures of Black Radicalism*, edited by Gaye Theresa Johnson and Alex Lubin, 225–40. London: Verso Books.

Gislason, Maya K., Vanessa Sloan Morgan, Kendra Mitchell-Foster, and Margot W. Parkes. 2018. "Voices from the Landscape: Storytelling as Emergent Counter-Narratives and Collective Action from Northern BC Watersheds." *Health & Place* 54: 191–99.

Glass, Ronald David, Jennifer M. Morton, Joyce E. King, Patricia Krueger-Henney, Michele S. Moses, Sheeva Sabati, and Troy Richardson. 2018. "The Ethical Stakes of Collaborative Community-Based Social Science Research." *Urban Education* 53 (4): 503–31.

Godden, Naomi Joy. 2017. "The Participation Imperative in Co-operative Inquiry: Personal Reflections of an Initiating Researcher." *Systemic Practice and Action Research* 30 (1): 1–18.

Goeman, Mishuana. 2015. "Land as Life: Unsettling the Logics of Containment." In *Native Studies Keywords*, edited by Michelle Raheja, Andrea Smith, and Stephanie Nohelani Teves, 71–89. Tucson: University of Arizona Press.

Goldberg, Hanna, and Laura-Anne Minkoff-Zern. 2021. "Teaching Labor in Food Studies: Challenging Consumer-Based Approaches to Social Change Through Student Research Community Partnerships." *Food, Culture & Society* 25 (2): 328–44.

Gómez Perochena, Tania Daniela. 2019. "Reinventar El Estado Desde La Autonomía Indígena: Un Rastreo Del Proceso De La Articulación Del Gobierno Territorial Autónomo De La Nación Wampis (GTANW)." PhD diss., Pontificia Universidad Católica del Perú.

Gonsalves, Kavita, Marcus Foth, Glenda Caldwell, and Waldemar Jenek. 2020. "Radical Placemaking: Immersive, Experiential and Activist Approaches for Marginalised Communities." In *Connections: Exploring Heritage, Architecture, Cities, Art, Media*, 20.1: 237–52. Canterbury: Architecture, Media, Politics, Society.

González, Erualdo Romero, Raul P. Lejano, Guadalupe Vidales, Ross F. Conner, Yuki Kidokoro, Bahram Fazeli, and Robert Cabrales. 2007. "Participatory Action Research for Environmental Health: Encountering Freire in the Urban Barrio." *Journal of Urban Affairs* 29 (1): 77–100.

Gonzalez, Rosa. 2017. "Community-Driven Climate Resilience Planning: A Framework (Version 2.0)." National Association of Climate Resilience Planners. https://kresge.org/sites/default/files/library/community_drive_resilience_planning_from_movement_strategy_center.pdf.

———. 2021. "A Guide to Community-Driven Just Transition Planning." Climate Justice Alliance. https://climatejusticealliance.org/wp-content/uploads/2021/03/CJA_Curriculum_C_FinalWeb.pdf.

Gosnell, Hannah, and Erin Kelly. 2010. "Peace on the River? Social-Ecological Restoration and Large Dam Removal in the Klamath Basin, USA." *Water Alternatives* 3 (2): 362–83.

Gottlieb, Robert. 2005. *Forcing the Spring: The Transformation of the American Environmental Movement*. Washington, DC: Island Press.

Gottlieb, Robert, and Anupama Joshi. 2010. *Food Justice*. Cambridge: MIT Press.

Graham, Louis F., Mark B. Padilla, William D. Lopez, Alexandra M. Stern, Jerry Peterson, and Danya E. Keene. 2016. "Spatial Stigma and Health in Postindustrial Detroit." *International Quarterly of Community Health Education* 36 (2): 105–13.

Gray, Kathleen M. 2018. "From Content Knowledge to Community Change: A Review of Representations of Environmental Health Literacy." *International Journal of Environmental Research and Public Health* 15 (3): 466.

Gray, Margaret. 2013. *Labor and the Locavore*. Berkeley: University of California Press.

Grayson, Shira, Megan Doerr, and Joon-Ho Yu. 2020. "Developing Pathways for Community-Led Research with Big Data: A Content Analysis of Stakeholder Interviews." *Health Research Policy and Systems* 18: 76.

Green, Lawrence W. 2001. "From Research to 'Best Practices' in Other Settings and Populations." *American Journal of Health Behavior* 25 (3): 165–78.

Green, Lawrence W., Russell E. Glasgow, David Atkins, and Kurt Stange. 2009. "Making Evidence from Research More Relevant, Useful, and Actionable in Policy, Program Planning, and Practice." *American Journal of Preventive Medicine* 37 (6): 187–91.

Green, Lawrence W., Michel O'Neill, Maria Faria Westphal, and Donald E. Morisky. 1996. "The Challenges of Participatory Action Research for Health Promotion." *Promotion & Education* 3 (4): 3–5.

Greenberg, Miriam. 2014. "The Disaster inside the Disaster: Hurricane Sandy and Post-Crisis Redevelopment." *New Labor Forum* 23 (1): 44–52.

Greenwood, Davydd J. 2015. "An Analysis of the Theory/Concept Entries in the SAGE Encyclopedia of Action Research: What We Can Learn About Action Research in General from the Encyclopedia." *Action Research* 13 (2): 198–213.

Griffin, Toni L., Ariella Cohen, and David Maddox, eds. 2015. "The Just City Essays: 26 Visions for Urban Equity, Inclusion and Opportunity." New York: J. Max Bond Center on Design for the Just City. https://thenatureofcities.com/thejustcity/TheJustCityEssays.pdf.

Griffiths, Tom. 2007. "The Humanities and an Environmentally Sustainable Australia." *Australian Humanities Review* 43. http://australianhumanitiesreview.org/2007/03/01/the-humanities-and-an-environmentally-sustainable-australia/.

Guajardo, Andrea D., Grisel M. Robles-Schrader, Lisa Aponte-Soto, and Leah C. Neubauer. 2020. "LatCrit Theory as a Framework for Social Justice Evaluation: Considerations for Evaluation and Evaluators." *New Directions for Evaluation* 166: 65–75.

Gubrium, Aline, and Krista Harper. 2013. *Participatory Visual and Digital Methods*. New York: Routledge.

Guidry, Virginia T., Sarah M. Rhodes, Courtney G. Woods, Devon J. Hall, and Jessica L. Rinsky. 2018. "Connecting Environmental Justice and Community Health: Effects of Hog Production in North Carolina." *North Carolina Medical Journal* 79 (5): 324–28.

Gutberlet, Jutta. 2008. "Empowering Collective Recycling Initiatives: Video Documentation and Action Research with a Recycling Co-Op in Brazil." *Resources Conservation and Recycling* 52 (4): 659–70.

Gutberlet, Jutta, and Bruno de Oliveira Jayme. 2010. "The Story of My Face: How Environmental Stewards Perceive Stigmatization (Re)Produced by Discourse." *Sustainability* 2 (11): 3339–53.

HabitatMap. 2021. "Health." Accessed December 21, 2021. https://www.habitatmap.org/blog/categories/health.

Haklay, Muki. 2013. "Citizen Science and Volunteered Geographic Information: Overview and Typology of Participation." In *Crowdsourcing Geographic Knowledge: Volunteered Geographic Information (VGI) in Theory and Practice*, edited by Daniel Sui, Sarah Elwood, and Michael Goodchild, 105–22. Berlin: Springer.

Haklay, Muki, and Louise Francis. 2018. "Participatory GIS and Community-Based Citizen Science for Environmental Justice Action." In *The Routledge Handbook of Environmental*

Justice, edited by Ryan Holifield, Jayajit Chakraborty, and Gordon Walker, 297–308. New York: Routledge.

Hale, Charles R., ed. 2008. *Engaging Contradictions: Theory, Politics, and Methods of Activist Scholarship*. Berkeley: University of California Press.

Hall, Budd, Rajesh Tandon, and Crystal Tremblay, eds. 2015. *Strengthening Community University Research Partnerships: Global Perspectives*. Victoria: University of Victoria and PRIA.

Hall, Melvin E. 2020. "Blest Be the Tie That Binds." *New Directions for Evaluation*, 166: 13–22.

Halliday, Emma, Jennie Popay, Rachel Anderson de Cuevas, and Paula Wheeler. 2020. "The Elephant in the Room? Why Spatial Stigma Does Not Receive the Public Health Attention It Deserves." *Journal of Public Health* 42 (1): 38–43.

Hammond, Chad, Wendy Gifford, Roanne Thomas, Seham Rabaa, Ovini Thomas, and Marie-Cécile Domecq. 2018. "Arts-Based Research Methods with Indigenous Peoples: An International Scoping Review." *AlterNative: An International Journal of Indigenous Peoples* 14(3): 260–76.

Hance, Jeremy. 2016. "Has Big Conservation Gone Astray?" *Mongabay Latam*. April 26, 2016. https://news.mongabay.com/2016/04/big-conservation-gone-astray/.

Hancock, Trevor, and Meredith Minkler. 2012. "Community Health Assessment or Healthy Community Assessment: Whose Community? Whose Health? Whose Assessment?" In *Community Organizing and Community Building for Health and Welfare*, edited by Meredith Minkler, 3rd ed., 153–70. Brunswick: Rutgers University Press.

Hardeman, Rachel R., Patricia A. Homan, Tongtan Chantarat, Brigette A. Davis, and Tyson H. Brown. 2022. "Improving the measurement of structural racism to achieve antiracist health policy." *Health Affairs* 41 (2): 179–86.

Harding, Anna, Barbara Harper, Dave Stone, Catherine O'Neill, Patricia Berger, Stuart Harris, and Jamie Donatuto. 2012. "Conducting Research with Tribal Communities: Sovereignty, Ethics, and Data-Sharing Issues." *Environmental Health Perspectives* 120 (1): 6–10.

Harley, Anne, and Eurig Scandrett, eds. 2019. *Environmental Justice, Popular Struggle and Community Development*. Bristol: Policy Press.

Harley, Kim G., Robert Gunier, Asa Bradman, Chloe Lessard, James Nolan, Kim Anderson, Carolyn Poutasse, et al. 2018. "The COSECHA Study: Youth Peer-to-Peer Research on Pesticide Exposure in a Farmworker Community." *ISES-IEE Conference Abstracts*. https://ehp.niehs.nih.gov/doi/abs/10.1289/isesisee.2018.S03.02.14.

Harley, Kim G., Kimberly L. Parra, Jose Camacho, Asa Bradman, James E. S. Nolan, Chloe Lessard, Kim A. Anderson, et al. 2019. "Determinants of Pesticide Concentrations in Silicone Wristbands Worn by Latina Adolescent Girls in a California Farmworker Community: The COSECHA Youth Participatory Action Study." *Science of the Total Environment* 652: 1022–29.

Harmsworth, Sally, and Sarah Turpin. 2000. "Creating an Effective Dissemination Strategy: An Expanded Interactive Workbook for Educational Development Projects." TQEF National Co-Ordination Team. https://ajpp-online.org/resources/downloads/06-CreatingAnEffectiveDisseminationStrategy-AnExpandedWorkbook.pdf.

Harris, Leila M., and Helen D. Hazen. 2005. "Power of Maps: (Counter) Mapping for Conservation." *ACME: An International Journal for Critical Geographies* 4 (1): 99–130.

Harris, Leila M., Scott McKenzie, Lucy Rodina, Sameer H. Shah, and Nicole J. Wilson. 2018. "Water Justice: Key Concepts, Debates and Research Agendas." In *The Routledge Handbook of Environmental Justice*, edited by Ryan Holifield, Jayajit Chakraborty, and Gordon Walker, 338–49. New York: Routledge.

Harris, Leslie M. 2020. "Higher Education's Reckoning with Slavery." *Academe*. https://www.aaup.org/article/higher-education%E2%80%99s-reckoning-slavery#.Y7SzxofMLyQ.

Harris-Roxas, Ben, and Elizabeth Harris. 2011. "Differing Forms, Differing Purposes: A Typology of Health Impact Assessment." *Environmental Impact Assessment Review*, 31 (4): 396–403.

Harrison, Jill L. 2011. *Pesticide Drift and the Pursuit of Environmental Justice*. Cambridge: MIT Press.

Hartmann, Betsy. 1995. *Reproductive Rights and Wrongs: The Global Politics of Population Control*. Boston: South End Press.

Harvey, David. 2005. *A Brief History of Neoliberalism*. Oxford: Oxford University Press.

———. 2008. "The Right to the City." In *The City Reader*, edited by Richard T. LeGates and Frederic Stout, 6th ed., 1: 23–40. New York: Routledge.

Harvey, Michael. 2021. "The Political Economy of Health: Revisiting Its Marxian Origins to Address 21st-Century Health Inequalities." *American Journal of Public Health* 111 (2): 293–300.

Haywood, Benjamin K., Julia K. Parrish, and Jane Dolliver. 2016. "Place-Based and Data-Rich Citizen Science as a Precursor for Conservation Action." *Conservation Biology* 30 (3): 476–86.

Healey, Patsy. 2012. "Communicative Planning: Practices, Concepts, and Rhetorics." In *Planning Ideas That Matter: Livability, Territoriality, Governance, and Reflective Practice*, edited by Bishwapriya Sanyal, Lawrence J. Vale, and Christina D. Rosan, 333–58. Cambridge: MIT Press.

Heaney, Christopher D., Kevin Myers, Steve Wing, Devon Hall, Dothula Baron, and Jill R. Stewart. 2015. "Source Tracking Swine Fecal Waste in Surface Water Proximal to Swine Concentrated Animal Feeding Operations." *Science of the Total Environment* 511: 676–83.

Heaney, Christopher D., Sacoby Wilson, Omega Wilson, John Cooper, Natasha Bumpass, and Marilyn Snipes. 2011. "Use of Community-Owned and -Managed Research to Assess the Vulnerability of Water and Sewer Services in Marginalized and Underserved Environmental Justice Communities." *Journal of Environmental Health* 74 (1): 8–17.

Heaney, Christopher D., Steve Wing, Sacoby M. Wilson, Robert L. Campbell, David Caldwell, Barbara Hopkins, Shannon O'Shea, and Karin Yeatts. 2013. "Public Infrastructure Disparities and the Microbiological and Chemical Safety of Drinking and Surface Water Supplies in a Community Bordering a Landfill." *Journal of Environmental Health* 75 (10): 24–36.

Heeks, Richard, and Satyarupa Shekhar. 2019. "Datafication, Development and Marginalised Urban Communities: An Applied Data Justice Framework." *Information, Communication & Society* 22 (7): 992–1011.

Heizmann, Helena, and Michael R. Olsson. 2015. "Power Matters: The Importance of Foucault's Power/Knowledge as a Conceptual Lens in KM Research and Practice." *Journal of Knowledge Management* 19 (4): 756–69.

Heller, Jonathan, Marjory L. Givens, Tina K. Yuen, Solange Gould, Maria Benkhalti Jandu, Emily Bourcier, and Tim Choi. 2014. "Advancing Efforts to Achieve Health Equity: Equity Metrics for Health Impact Assessment Practice." *International Journal of Environmental Research and Public Health* 11 (11): 11054–64.

Henke, Christopher. 2008. *Cultivating Science, Harvesting Power: Science and Industrial Agriculture in California*. Cambridge: MIT Press.

Hessami, Mateen A., Ella Bowles, Jesse N. Popp, and Adam T. Ford. 2021. "Indigenizing the North American Model of Wildlife Conservation." *FACETS* 6 (1): 1285–1306.

Heying, Charles, and Stephen Marotta. 2016. "Portland Made: Building Partnerships to Support the Local Artisan/Maker Community." In *University–Community Partnerships*, edited by B. D. Wortham-Galvin, Jennifer H. Allen, and Jacob D. B. Sherman, 112–22. London: Routledge.

Heynen, Nik, and Megan Ybarra. 2021. "On Abolition Ecologies and Making 'Freedom as a Place.'" *Antipode* 53 (1): 21–35.

Hickey, Sam, and Giles Mohan. 2005. "Relocating Participation Within a Radical Politics of Development." *Development and Change* 36 (2): 237–62.

Hicks, Sarah, Bonnie Duran, Nina Wallerstein, Magdalena Avila, Lorenda Belone, Julie Lucero, Maya Magarati, et al. 2021. "Evaluating Community-Based Participatory Research to Improve Community-Partnered Science and Community Health." *Progress in Community Health Partnerships: Research, Education, and Action* 6 (3): 289–99.

Hipp, John R., and Cynthia M. Lakon. 2010. "Social Disparities in Health: Disproportionate Toxicity Proximity in Minority Communities over a Decade." *Health & Place* 16 (4): 674–83.

Hiratsuka, Vanessa Y., Julie A. Beans, Renee F. Robinson, Jennifer L. Shaw, Ileen Sylvester, and Denise A. Dillard. 2017. "Self-Determination in Health Research: An Alaska Native Example of Tribal Ownership and Research Regulation." *International Journal of Environmental Research and Public Health* 14 (11): 1324.

Hirschman, Albert O. 1984. *Getting Ahead Collectively: Grassroots Experiences in Latin America*. New York: Pergamon Press.

Hodges, Rita Axelroth, and Steve Dubb. 2012. *The Road Half Traveled: University Engagement at a Crossroads*. Lansing: Michigan State University Press.

Hoeft, Theresa J., Wylie Burke, Scarlett E. Hopkins, Walkie Charles, Susan B. Trinidad, Rosalina D. James, and Bert B. Boyer. 2014. "Building Partnerships in Community-Based Participatory Research: Budgetary and Other Cost Considerations." *Health Promotion Practice* 15 (2): 263–70.

Hoffman, Andrew J. 2015. "Isolated Scholars: Making Bricks, Not Shaping Policy." *Chronicle of Higher Education*. February 9, 2015, A48.

Hoffman, Jeremy S. 2020. "Learn, Prepare, Act: 'Throwing Shade' on Climate Change." *Journal of Museum Education* 45 (1): 28–41.

Holifield, Ryan, Jayajit Chakraborty, and Gordon Walker, eds. 2018. *The Routledge Handbook of Environmental Justice*. New York: Routledge.

Holland, Breena. 2021. "Capabilities, Well-Being, and Justice." In *Environmental Justice: Key Issues*, edited by Brendan Coolsaet, 64–77. New York: Routledge.

Horton, Billy D. 1993. "The Appalachian Land Ownership Study: Research and Citizen Action in Appalachia." In *Voices of Change: Participatory Research in the United States and Canada*, edited by Peter Park, Mary Brydon-Miller, Budd Hall, and Ted Jackson, 85–102. Westport: Bergin and Garvey.

Horton, Myles, and Paulo Freire. 1990. *We Make the Road by Walking: Conversations on Education and Social Change*, edited by Brenda Bell, John Gaventa, and John Peters. Philadelphia: Temple University Press.

Horton, Rachel Avery, Steve Wing, Stephen W. Marshall, and Kimberly A. Brownley. 2009. "Malodor as a Trigger of Stress and Negative Mood in Neighbors of Industrial Hog Operations." *American Journal of Public Health* 99 (S3): 610–15.

Houston, Donna, and Pavithra Vasudevan. 2018. "Storytelling Environmental Justice: Cultural Studies Approaches." In *The Routledge Handbook of Environmental Justice*, edited by Ryan Holifield, Jayajit Chakraborty, and Gordon Walker, 241–51. New York: Routledge.

Howard, Phillip H. 2016. *Concentration and Power in the Food System: Who Controls What We Eat*. New York: Bloomsbury.

Howard, Ted, and Michelle Camou. 2018. "Economic Development as if Low-Income Communities Mattered." In *Community Development and Public Administration Theory*, edited by Ashley E. Nickels and Jason D. Rivera, 265–83. New York: Routledge.

Howell, Junia, and James R. Elliott. 2019. "Damages Done: The Longitudinal Impacts of Natural Hazards on Wealth Inequality in the United States." *Social Problems* 66 (3): 448–67.

Hricko, Andrea. 2008. "Global Trade Comes Home: Community Impacts of Goods Movement." *Environmental Health Perspectives* 116 (2): 78–81.

Huang, Ganlin, and Jonathan K. London. 2012. "Cumulative Environmental Vulnerability and Environmental Justice in California's San Joaquin Valley." *International Journal of Environmental Research and Public Health* 9 (5): 1593–1608.

———. 2016. "Mapping in and out of 'Messes': An Adaptive, Participatory, and Transdisciplinary Approach to Assessing Cumulative Environmental Justice Impacts." *Landscape and Urban Planning* 154: 57–67.

Hufschmidt, Maynard M. 1983. *Environment, Natural Systems, and Development: An Economic Valuation Guide*. Baltimore: Johns Hopkins University Press.

Human Impact Partners. 2011. "I-710 Corridor Project HIA." https://humanimpact.org/wp-content/uploads/2017/09/HIA-I710-Air-Quality-Plan.pdf.

Human Impact Partners, Ohio Justice and Policy Center, and Ohio Organizing Collaborative. 2015. *Stress on the Streets (SOS): Race, Policing, Health, and Increasing Trust Not Trauma*. Oakland: Human Impact Partners. https://humanimpact.org/hipprojects/trust-not-trauma.

Hunt, Dallas, and Shaun A. Stevenson. 2017. "Decolonizing Geographies of Power: Indigenous Digital Counter-Mapping Practices on Turtle Island." *Settler Colonial Studies* 7 (3): 372–92.

Hunt, Kathleen P. 2014. "'It's More Than Planting Trees, It's Planting Ideas': Ecofeminist Praxis in the Green Belt Movement." *Southern Communication Journal* 79 (3): 235–49.

Huntjens, Patrick, Jasper Eshuis, Catrien Termeer, and Arwin Van Buuren. 2014. "Forms and Foundations of Action Research." In *Action Research for Climate Change Adaptation: Developing and Applying Knowledge for Governance*, edited by Arwin Van Buuren, Jasper Eshuis, and Mathijs Van Vliet, 19–34. New York: Routledge.

Huq, Efadul. 2020. "Seeing the Insurgent in Transformative Planning Practices." *Planning Theory* 19 (4): 371–91.

Hyde, Cheryl A. 2017. "Challenging Ourselves: Critical Self-Reflection on Power and Privilege." In *Community-Based Participatory Research for Health: Advancing Social and Health Equity*, edited by Nina Wallerstein, Bonnie Duran, John Oetzel, and Meredith Minkler, 3rd ed., 337–44. San Francisco: Jossey-Bass.

IAP2 (International Association for Public Participation). 2018. "IAP2 Spectrum of Public Participation." https://iap2.org.au/wp-content/uploads/2020/01/2018_IAP2_Spectrum.pdf.

ILR Worker Institute. 2021. "NYS Passes Labor Standards on Clean Energy Work." www.ilr.cornell.edu/worker-institute/labor-leading-climate/nys-passes-labor-standards-clean-energy-work.

Indigenous Environmental Network. 2021. "Indigenous Principles of Just Transition." https://www.ienearth.org/justtransition/.

International Collaboration for Participatory Health Research. 2013. *What Is Participatory Health Research?* Berlin: International Collaboration for Participatory Health Research.

Ironbound Community Corporation. n.d. "Community Planning and Development: Envisioning Its Future." Accessed December 21, 2021. http://ironboundcc.org/community-planning/.

Irwin, Lauren N., and Zak Foste. 2021. "Service-Learning and Racial Capitalism: On the Commodification of People of Color for White Advancement." *Review of Higher Education* 44 (4): 419–46.

Islam, Nadia, Charlotte Yu-Ting Chang, Pam Tau Lee, and Chau Trinh-Shevrin. 2017. "CBPR in Asian American Communities." In *Community-Based Participatory Research for Health: Advancing Social and Health Equity*, edited by Nina Wallerstein, Bonnie Duran, John G. Oetzel, and Meredith Minkler, 3rd ed., 175–88. San Francisco: Jossey-Bass.

Israel, Barbara A., Barry Checkoway, Amy Schulz, and Marc Zimmerman. 1994. "Health Education and Community Empowerment: Conceptualizing and Measuring Perceptions of Individual, Organizational, and Community Control." *Health Education Quarterly* 21 (2): 149–70.

Israel, Barbara A., Eugenia Eng, Amy J. Schulz, and Edith A. Parker. 2013a. "Introduction to Methods for CBPR for Health." In *Methods for Community-Based Participatory Research for Health*, edited by Barbara A. Israel, Eugenia Eng, Amy J. Schulz, and Edith A. Parker, 2nd ed., 3–38. San Francisco: Jossey-Bass.

Israel, Barbara A., Eugenia Eng, Amy J. Schulz, and Edith A. Parker, eds. 2013b. *Methods for Community-Based Participatory Research for Health*. 2nd ed. San Francisco: Jossey-Bass.

Jacobs, Fayola. 2019. "Black Feminism and Radical Planning: New Directions for Disaster Planning Research." *Planning Theory* 18 (1): 24–39.

Jacoby, Karl. 2014. *Crimes Against Nature: Squatters, Poachers, Thieves, and the Hidden History of American Conservation*. Berkeley: University of California Press.

Jarratt-Snider, Karen, and Marianne O. Nielsen. 2020. *Indigenous Environmental Justice*. Tucson: University of Arizona Press.

Jasanoff, Sheila. 2017. "Virtual, Visible, and Actionable: Data Assemblages and the Sightlines of Justice." *Big Data & Society* 4 (2).

Jelks, Na'Taki Osborne, Timothy L. Hawthorne, Dajun Dai, Christina H. Fuller, and Christine Stauber. 2018. "Mapping the Hidden Hazards: Community-Led Spatial Data

Collection of Street-Level Environmental Stressors in a Degraded, Urban Watershed." *International Journal of Environmental Research and Public Health* 15 (4): 825.

Jenkins, Carolyn, Heather Bittner Fagan, Jennifer Passarella, Nicole Fournakis, and Dana Burshell. 2020. "Training Academic and Community Investigator Teams for Community-Engaged Research: Program Development, Implementation, Evaluation and Replication." *Progress in Community Health Partnerships* 14 (2): 229–42.

Jennings, Viniece, Richard Schulterbrandt Gragg, C. Perry Brown, Dudley Hartel, Eric Kuehler, Alex Sinykin, Elijah Johnson, and Michelle Kondo. 2019. "Structural Characteristics of Tree Cover and the Association with Cardiovascular and Respiratory Health in Tampa, FL." *Journal of Urban Health* 96 (5): 669–81.

Jernigan, Valarie Blue Bird, Kimberly R. Huyser, Jimmy Valdes, and Vanessa Watts Simonds. 2017. "Food Insecurity Among American Indians and Alaska Natives: A National Profile Using the Current Population Survey." *Journal of Hunger & Environmental Nutrition* 12 (1): 1–10.

Jernigan, Valarie Blue Bird, Alicia L. Salvatore, Dennis M. Styne, and Marilyn Winkleby. 2012. "Addressing Food Insecurity in a Native American Reservation Using Community-Based Participatory Research." *Health Education Research* 27 (4): 645–55.

Johansson, Patrik, and Kendra K. Schmid. 2015. "The Waponahki Tribal Health Assessment: Successfully Using CBPR to Conduct a Comprehensive and Baseline Health Assessment of Waponahki Tribal Members." *Journal of Health Care for the Poor and Underserved* 26 (3): 889–907.

Johnston, Jill, and Lara Cushing. 2020. "Chemical Exposures, Health, and Environmental Justice in Communities Living on the Fenceline of Industry." *Current Environmental Health Reports* 7 (1): 48–57.

Johnston, Jill E., Zully Juarez, Sandy Navarro, Ashley Hernandez, and Wendy Gutschow. 2020. "Youth Engaged Participatory Air Monitoring: A 'Day in the Life' in Urban Environmental Justice Communities." *International Journal of Environmental Research and Public Health* 17 (1): 93.

Jojola, Ted. 2008. "Indigenous Planning—An Emerging Context." *Canadian Journal of Urban Research* 17 (1): 37–47.

Jones, Miranda R., Ana V. Diez-Roux, Anjum Hajat, Kiarri N. Kershaw, Marie S. O'Neill, Eliseo Guallar, Wendy S. Post, Joel D. Kaufman, and Ana Navas-Acien. 2014. "Race/Ethnicity, Residential Segregation, and Exposure to Ambient Air Pollution: The Multi-Ethnic Study of Atherosclerosis (MESA)." *American Journal of Public Health* 104 (11): 2130–37.

Jones, Nancy L., Stephen E. Gilman, Tina L. Cheng, Stacy S. Drury, Carl V. Hill, and Arline T. Geronimus. 2019. "Life Course Approaches to the Causes of Health Disparities." *American Journal of Public Health* 109 (S1): 48–55.

Joosse, Sofie, Stina Powell, Hanna Bergeå, Steffen Böhm, Camilo Calderón, Elvira Caselunghe, Anke Fischer, et al. 2020. "Critical, Engaged and Change-Oriented Scholarship in Environmental Communication: Six Methodological Dilemmas to Think With." *Environmental Communication* 14 (6): 758–71.

Jordan, Catherine M., Yvonne A. Joosten, Rebecca C. Leugers, and Sharon L. Shields. 2009. "The Community-Engaged Scholarship Review, Promotion, and Tenure Package: A Guide for Faculty and Committee Members." *Metropolitan Universities* 20 (2): 66–86.

Jung, Jin-Kyu, and Sarah Elwood. 2010. "Extending the Qualitative Capabilities of GIS: Computer-Aided Qualitative GIS." *Transactions in GIS* 14 (1): 63–87.

Kang, Helen. 2009. "Pre-Litigation Considerations for Grassroots Groups." In *Environmental Justice: Law, Policy & Regulation*, edited by Clifford Rechtschaffen, Eileen Gauna, and Catherine A. O'Neill, 2nd ed. Durham: Carolina Academic Press.

Kano, Miria, Kelley P. Sawyer, and Cathleen E. Willging. 2017. "Guiding CBPR Principles: Fostering Equitable Health Care for LGBTQ+ People." In *Community-Based Participatory Research for Health: Advancing Social and Health Equity*, edited by Nina Wallerstein, Bonnie Duran, John Oetzel, and Meredith Minkler, 3rd ed., 345–50. San Francisco: Jossey-Bass.

Kanuha, Valli Kalei. 2000. "'Being' Native Versus 'Going Native': Conducting Social Work Research as an Insider." *Social Work* 45 (5): 439–47.

Karner, Alex, Aaron Golub, Karel Martens, and Glenn Robinson. 2018. "Transportation and Environmental Justice: History and Emerging Practice." In *The Routledge Handbook of Environmental Justice*, edited by Ryan Holifield, Jayajit Chakraborty, and Gordon Walker, 400–411. New York: Routledge.

Karpowitz, Christopher F., and Chad Raphael. 2014. *Deliberation, Democracy, and Civic Forums: Improving Equality and Publicity*. New York: Cambridge University Press.

Karuk-UCB Collaborative. n.d. "Practicing Píkyav: A Guiding Policy for Collaborative Projects and Research Initiatives with the Karuk Tribe." Accessed July 21, 2021. https://nature.berkeley.edu/karuk-collaborative/?page_id=165.

Karvonen, Andrew, and Ralf Brand. 2014. "Expertise: Specialized Knowledge in Environmental Politics and Sustainability." In *Routledge Handbook of Global Environmental Politics*, edited by Paul G. Harris, 215–30. New York: Routledge.

Kastelic, Sarah L., Nina Wallerstein, Bonnie Duran, and John G. Oetzel. 2017. "Socio-Ecologic Framework for CBPR." In *Community-Based Participatory Research for Health: Advancing Social and Health Equity*, edited by N. Wallerstein, Bonnie Duran, John G. Oetzel, and Meredith Minkler, 3rd ed., 77–94. San Francisco: Jossey-Bass.

Kaswan, Alice. 2021. "Distributive Environmental Justice." In *Environmental Justice: Key Issues*, edited by Brendan Coolsaet, 21–36. New York: Routledge.

Kaufman, Amanda, Ron Williams, Timothy Barzyk, Molly Greenberg, Marie O'Shea, Patricia Sheridan, Anhthu Hoang, et al. 2017. "A Citizen Science and Government Collaboration: Developing Tools to Facilitate Community Air Monitoring." *Environmental Justice* 10 (2): 51–61.

Keahey, Jennifer. 2021. "Sustainable Development and Participatory Action Research: A Systematic Review." *Systemic Practice and Action Research* 34 (3): 291–306.

Keck, Margaret E. 1995. "Social Equity and Environmental Politics in Brazil: Lessons from the Rubber Tappers of Acre." *Comparative Politics* 27 (4): 409–24.

Keene, Danya E., and Mark B. Padilla. 2014. "Spatial Stigma and Health Inequality." *Critical Public Health* 24 (4): 392–404.

Kelley, Todd R., J. Geoffery Knowles, Jeffrey D. Holland, and Jung Han. 2020. "Increasing High School Teachers Self-Efficacy for Integrated STEM Instruction Through a Collaborative Community of Practice." *International Journal of STEM Education* 7: 14.

Kendi, Ibram X. 2019. *How to Be an Antiracist*. New York: One World.

Kennedy, Heather, Jonah DeChants, Kimberly Bender, and Yolanda Anyon. 2019. "More than Data Collectors: A Systematic Review of the Environmental Outcomes of Youth Inquiry Approaches in the United States." *American Journal of Community Psychology* 63 (1–2): 208–26.

Kent-Stoll, Peter. 2020. "The Racial and Colonial Dimensions of Gentrification." *Sociology Compass* 14 (12).

Kentuckians for the Commonwealth. 2013. "Appalachia's Bright Future: Working Together to Shape a Just Transition." https://archive.kftc.org/sites/default/files/docs/resources/soar4pager.pdf.

Kerstetter, Katie. 2012. "Insider, Outsider, or Somewhere Between: The Impact of Researchers' Identities on the Community-Based Research Process." *Journal of Rural Social Sciences* 27 (2): 99–117.

Khodyakov, Dmitry, Terrance D. Savitsky, and Siddhartha Dalal. 2016. "Collaborative Learning Framework for Online Stakeholder Engagement." *Health Expectations* 19 (4): 868–82.

Kidd, Dorothy. 2019. "Extra-Activism: Counter-Mapping and Data Justice." *Information, Communication & Society* 22 (7): 954–70.

Killean, Rachel, and Lauren Dempster. 2021. "'Greening' Transitional Justice?" In *Beyond Transitional Justice: Transformative Justice and the State of the Field (or Non-Field)*, edited by Matthew Evans. New York: Routledge.

Kim, Annice E., Alicea J. Lieberman, and Daniel Dench. 2015. "Crowdsourcing Data Collection of the Retail Tobacco Environment: Case Study Comparing Data from Crowdsourced Workers to Trained Data Collectors." *Tobacco Control* 24 (e1): e6–9.

Kindon, Sara, Rachel Pain, and Mike Kesby, eds. 2007. *Participatory Action Research Approaches and Methods: Connecting People, Participation and Place*. New York: Routledge.

Kingdon, John W. 2011. *Agendas, Alternatives, and Public Policies*. 2nd ed. Harlow: Pearson Longman.

Kinney, Patrick, Maneesha Aggarwal, Mary E. Northridge, Nicole Janssen, and Peggy Shepard. 2000. "Airborne Concentrations of PM 2.5 and Diesel Exhaust Particles on Harlem Sidewalks: A Community-Based Pilot Study." *Environmental Health Perspectives* 108 (3): 213–18.

Kirk, Gwyn, and Margo Okazawa-Rey. 2019. *Gendered Lives: Intersectional Perspectives*. 7th ed. New York: Oxford University Press.

Kitzman-Ulrich, Heather, and Cheryl L. Holt. 2017. "Community-Based Participatory Research Studies in Faith-Based Settings." In *Handbook of Community-Based Participatory Research*, edited by Steven S. Coughlin, Selina A. Smith, and Maria E. Fernández, 71–80. Oxford: Oxford University Press.

Kline, Nolan, and Rachel Newcomb. 2013. "The Forgotten Farmworkers of Apopka, Florida: Prospects for Collaborative Research and Activism to Assist African American Former Farmworkers." *Anthropology & Humanism* 38 (2): 160–76.

Koh, Howard K., Amy Bantham, Alan C. Geller, Mark A. Rukavina, Karen M. Emmons, Pamela Yatsko, and Robert Restuccia. 2020. "Anchor Institutions: Best Practices to Address Social Needs and Social Determinants of Health." *American Journal of Public Health* 110 (3): 309–16.

Konisky, David M. 2015. *Failed Promises: Evaluating the Federal Government's Response to Environmental Justice*. Cambridge: MIT Press.

Korn, Edward L., and Barry I. Graubard. 1991. "Epidemiologic Studies Utilizing Surveys: Accounting for the Sampling Design." *American Journal of Public Health* 81 (9): 1166–73.

Kothari, Ashish, Federico Demaria, and Alberto Acosta. 2014. "Buen Vivir, Degrowth and Ecological Swaraj: Alternatives to Sustainable Development and the Green Economy." *Development* 57 (3–4): 362–75.

Kraft, Michael E., and Denise Scheberle. 1995. "Environmental Justice and the Allocation of Risk: The Case of Lead and Public Health." *Policy Studies Journal* 23 (1): 113–22.

Kresge, Lisa, and Chelsea Eastman. 2010. "Increasing Food Security Among Agricultural Workers in California's Salinas Valley." Davis: California Institute for Rural Studies. https://cirsinc.org/wp-content/uploads/2021/06/Increasing-Food-Security-among-Agricultural-Workers-in-Californias-Salinas-Valley.pdf.

Kresge Foundation. 2019. "Climate Resilience and Urban Opportunity Initiative." https://kresge.org/initiative/climate-resilience-and-urban-opportunity-cruo/.

Krieger, Nancy. 2001. "Theories for Social Epidemiology in the 21st Century: An Ecosocial Perspective." *International Journal of Epidemiology* 30 (4): 668–77.

———. 2005. "Embodiment: A Conceptual Glossary for Epidemiology." *Journal of Epidemiology & Community Health* 59 (5): 350–55.

———. 2020. "Measures of Racism, Sexism, Heterosexism, and Gender Binarism for Health Equity Research: From Structural Injustice to Embodied Harm—An Ecosocial Analysis." *Annual Review of Public Health* 41: 37–62.

Krumholz, Norman. 1982. "A Retrospective View of Equity Planning Cleveland 1969–1979." *Journal of the American Planning Association* 48 (2): 163–74.

Kuan, Da-Wei. 2021. "From Territorial Claim to Land-Use Plan: The Experience of Dialoging Indigenous Ecological Knowledge and State Management Regime in Taiwan." In *Indigenous Peoples, Heritage and Landscape in the Asia Pacific: Knowledge Co-Production and Empowerment*, edited by Stephen Acabado and Da-Wei Kuan, 88–107. New York: Routledge.

Kuehn, Robert R. 2000. "A Taxonomy of Environmental Justice." *Environmental Law Reporter* 30: 10681–703.

Kummitha, Rama Krishna Reddy. 2017. *Social Entrepreneurship and Social Inclusion: Processes, Practices, and Prospects*. Singapore: Palgrave Macmillan.

Kwan, Mei-Po. 2013. "Beyond Space (As We Knew It): Toward Temporally Integrated Geographies of Segregation, Health, and Accessibility." *Annals of the Association of American Geographers* 103 (5): 1078–86.

Labor Network for Sustainability and Strategic Practice Grassroots Policy Project. 2017. "'Just Transition': Just What Is It? An Analysis of Language, Strategies, and Projects." https://labor4sustainability.org/files/Just_Transition_Just_What_Is_It.pdf.

LaDuke, Winona. 1999. *All Our Relations: Native Struggles for Land and Life*. Boston: South End Press.

LaDuke, Winona, and Deborah Cowen. 2020. "Beyond Wiindigo Infrastructure." *South Atlantic Quarterly* 119 (2): 243–68.

Lake, Frank K., and Jonathan W. Long. 2014. *Fire and Tribal Cultural Resources*. Albany: U.S. Department of Agriculture, Forest Service. https://www.fs.usda.gov/treesearch/pubs/46960.

Le Dantec, Christopher A., Mariam Asad, Aditi Misra, and Kari E. Watkins. 2015. "Planning with Crowdsourced Data: Rhetoric and Representation in Transportation Planning." In *Proceedings of the 18th ACM Conference on Computer Supported Cooperative Work & Social Computing*, 1717–27. Vancouver: ACM.

Lebaron, Genevieve. 2018. "The Global Business of Forced Labour: Report of Findings." Sheffield: SPERI and University of Sheffield. http://globalbusinessofforcedlabour.ac.uk/wp-content/uploads/2018/05/Report-of-Findings-Global-Business-of-Forced-Labour.pdf.

LeClair, Amy, Jean J. Lim, and Carolyn Rubin. 2018. "Lessons Learned from Developing and Sustaining a Community-Research Collaborative Through Translational Research." *Journal of Clinical and Translational Science* 2 (2): 79–85.

Ledogar, Robert J., Luis Garden Acosta, and Analia Penchaszadeh. 1999. "Building International Public Health Vision Through Local Community Research: The El Puente-CIET Partnership." *American Journal of Public Health* 89 (12): 1795–97.

Legal Services Corporation. 2015. "LSC Agricultural Worker Population Estimate Update." https://www.lsc.gov/sites/default/files/LSC_Report_AgWrkr_Update_Jan_30_2015.pdf.

Lepczyk, Christopher A., Owen D. Boyle, and Timothy L. V. Vargo, eds. 2020. *Handbook of Citizen Science in Ecology and Conservation*. Berkeley: University of California Press.

Lerner, Josh A. 2014. *Making Democracy Fun: How Game Design Can Empower Citizens and Transform Politics*. Cambridge: MIT Press.

Lett, Elle, Emmanuella Ngozi Asabor, Theodore Corbin, and Dowin Boatright. 2021. "Racial Inequity in Fatal US Police Shootings, 2015–2020." *Journal of Epidemiological Community Health* 75 (4): 394–97.

Leung, May May, Alen Agaronov, Tara Entwistle, Lorene Harry, Julie Sharkey-Buckley, and Nicholas Freudenberg. 2017. "Voices Through Cameras: Using Photovoice to Explore Food Justice Issues with Minority Youth in East Harlem, New York." *Health Promotion Practice* 18 (2): 211–20.

Levenda, Anthony M., and Eliot Tretter. 2020. "The Environmentalization of Urban Entrepreneurialism: From Technopolis to Start-Up City." *EPA: Economy and Space* 52 (3): 490–509.

Levidow, Les, and Theo Papaioannou. 2018. "Which Inclusive Innovation? Competing Normative Assumptions Around Social Justice." *Innovation and Development* 8 (2): 209–26.

Lewin, Kurt. 1946. "Action Research and Minority Problems." *Journal of Social Issues* 2 (4): 34–46.

———. 1948. *Resolving Social Conflicts: Selected Papers on Group Dynamics*. New York: Harper and Row.

Li, Tania Murray. 2000. "Articulating Indigenous Identity in Indonesia: Resource Politics and the Tribal Slot." *Comparative Studies in Society and History* 42 (1): 149–79.

Liboiron, Max, Manuel Tironi, and Nerea Calvillo. 2018. "Toxic Politics: Acting in a Permanently Polluted World." *Social Studies of Science* 48 (3): 331–49.

Linder, Marc. 1986. "Farm Workers and the Fair Labor Standards Act: Racial Discrimination in the New Deal." *Texas Law Review* 65: 1335–93.

Lipperman-Kreda, Sharon, Christopher Morrison, Joel W. Grube, and Andrew Gaidus. 2015. "Youth Activity Spaces and Daily Exposure to Tobacco Outlets." *Health & Place* 34: 30–33.

Liss, Gary, Ruth Abbe, Neil Seldman, and Brenda Platt. 2020. "Baltimore's Fair Development Plan for Zero Waste." Baltimore: The Fair Development Roundtable. https://ilsr.org/wp-content/uploads/2020/02/BaltimoreZeroWastePlan2020.pdf.

Liu, Yvonne Yen. 2010. *Translating Green into Navajo: Alternatives to Coal Mining and the Campaign for a Navajo Green Economy*. Oakland: Applied Research Center.

Lockie, Stewart. 2018. "Privilege and Responsibility in Environmental Justice Research." *Environmental Sociology* 4 (2): 175–80.

Loh, Penn. 2016. "Community-University Collaborations for Environmental Justice: Toward a Transformative Co-Learning Model." *New Solutions* 26 (3): 412–28.

Lombardi, Kristen, Talia Buford, and Ronnie Greene. 2015. "Environmental Racism Persists, and the EPA Is One Reason Why." *Center for Public Integrity*. August 3, 2015. https://publicintegrity.org/environment/environmental-racism-persists-and-the-epa-is-one-reason-why/#note.

London, Jonathan K., Julie Sze, and Mary L. Cadenasso. 2018. "Facilitating Transdisciplinary Conversations in Environmental Justice Studies." In *The Routledge Handbook of Environmental Justice*, edited by Ryan Holifield, Jayajit Chakraborty, and Gordon Walker, 252–63. New York: Routledge.

Long, Jonathan W., Frank K. Lake, and Ron W. Goode. 2021. "The Importance of Indigenous Cultural Burning in Forested Regions of the Pacific West, USA." *Forest Ecology and Management* 500: 119597.

Los Angeles Collaborative for Environmental Health and Justice. 2010. "Hidden Hazards: A Call to Action for Healthy, Livable Communities." Santa Monica: Liberty Hill Foundation. https://dornsife.usc.edu/assets/sites/242/docs/hidden-hazards-low-res-version.pdf.

Low, Margaret, and Karena Shaw. 2011. "Indigenous Rights and Environmental Governance: Lessons from the Great Bear Rainforest." *BC Studies* 172: 9–33.

Lucero, Julie E., B. Boursaw, Milton "Mickey" Eder, Ella Greene-Moton, Nina Wallerstein, and John G. Oetzel. 2020. "Engage for Equity: The Role of Trust and Synergy in Community-Based Participatory Research." *Health Education & Behavior* 47 (3): 372–79.

Lucero, Julie E., Amber D. Emerson, David Beurle, and Yvette Roubideaux. 2020. "The Holding Space: A Guide for Partners in Tribal Research." *Progress in Community Health Partnerships* 14 (1): 101–7.

Lucero, Julie E., and Nina B. Wallerstein. 2013. "Trust in Community-Academic Research Partnerships: Increasing the Consciousness of Conflict and Trust Development." In *The SAGE Handbook of Conflict Communication: Integrating Theory, Research, and Practice*, edited by John G. Oetzel and Stella Ting-Toomey, 2nd ed., 537–61. Thousand Oaks: SAGE Publications.

Lucero, Julie, Nina Wallerstein, Bonnie Duran, Margarita Alegria, Ella Greene-Moton, Barbara Israel, Sarah Kastelic, et al. 2018. "Development of a Mixed Methods Investigation of Process and Outcomes of Community-Based Participatory Research." *Journal of Mixed Methods Research* 12 (1): 55–74.

Ludwig, Jason. 2021. "The Anthropocene Blues: Notes from Mississippi." *The Anthropocene Review* 8 (3): 230–40.

Ma, Jing, Yinhua Tao, Mei-Po Kwan, and Yanwei Chai. 2020. "Assessing Mobility-Based Real-Time Air Pollution Exposure in Space and Time Using Smart Sensors and GPS Trajectories in Beijing." *Annals of the American Association of Geographers* 110 (2): 434–48.

Madrigal, Daniel, Alicia Salvatore, Gardenia Casillas, Crystal Casillas, Irene Vera, Brenda Eskenazi, and Meredith Minkler. 2014. "Health in My Community: Conducting and Evaluating Photovoice as a Tool to Promote Environmental Health and Leadership Among Latino/a Youth." *Progress in Community Health Partnerships* 8 (3): 317–29.

Mah, Alice. 2017. "Environmental Justice in the Age of Big Data: Challenging Toxic Blind Spots of Voice, Speed, and Expertise." *Environmental Sociology* 3 (2): 122–33.

Maharawal, Manissa M., and Erin McElroy. 2018. "The Anti-Eviction Mapping Project: Counter Mapping and Oral History Toward Bay Area Housing Justice." *Annals of the American Association of Geographers* 108 (2): 380–89.

Malin, Stephanie A., and Stacia S. Ryder. 2018. "Developing Deeply Intersectional Environmental Justice Scholarship." *Environmental Sociology* 4 (1): 1–7.

Mamtora, Jayshree, Claire Ovaska, and Bronwyn Mathiesen. 2021. "Reconciliation in Australia: The Academic Library Empowering the Indigenous Community." *IFLA Journal* 47 (3): 351–60.

Manning, Beth Rose Middleton, and Kaitlin Reed. 2019. "Returning the Yurok Forest to the Yurok Tribe: California's First Tribal Carbon Credit Project." *Stanford Environmental Law Journal* 39: 71–124.

Mansuri, Ghazala, and Vijayendra Rao. 2004. "Community-Based and -Driven Development: A Critical Review." *World Bank Research Observer* 19 (1): 1–39.

Mares, Teresa Marie. 2019. *Life on the Other Border: Farmworkers and Food Justice in Vermont*. Berkeley: University of California Press.

Marks-Block, Tony, Frank K. Lake, and Lisa M. Curran. 2019. "Effects of Understory Fire Management Treatments on California Hazelnut, an Ecocultural Resource of the Karuk and Yurok Indians in the Pacific Northwest." *Forest Ecology and Management* 450: 117517.

Marley, Tennille L. 2019. "Indigenous Data Sovereignty: University Institutional Review Board Policies and Guidelines and Research with American Indian and Alaska Native Communities." *American Behavioral Scientist* 63 (6): 722–42.

Marquez, Emily, Kristin Schafer, Gabrielle Aldern, and Kristin VanderMolen. 2016. "Kids on the Frontline: How Pesticides are Undermining the Health of Rural Children." Oakland, CA: Pesticide Action Network North America. www.panna.org/resources/kids-frontline.

Marquez, Erika, Stephanie Smith, Alisa Howard, and David Perez. 2022. "Approaches to Vaccine Equity." Nevada Vaccine Equity Collaborative. https://nmhec.org/event/approaches-to-vaccine-equity/.

Martinez-Alier, Joan, Leah Temper, Daniela Del Bene, and Arnim Scheidel. 2016. "Is There a Global Environmental Justice Movement?" *The Journal of Peasant Studies* 43 (3): 731–55.

Martinez-Alier, Joan. 2002. *The Environmentalism of the Poor: A Study of Ecological Conflicts and Valuation*. Cheltenham: Edward Elgar.

Martini, Natalia. 2020. "Using GPS and GIS to Enrich the Walk-along Method." *Field Methods* 32 (2): 180–92.

Mascarenhas-Swan, Michelle. 2017. "The Case for a Just Transition." In *Energy Democracy: Advancing Equity in Clean Energy Solutions*, edited by Denise Fairchild and Al Weinrub, 37–56. Washington, DC: Island Press.

Mason, Maryann, Benjamin Rucker, Monique Reed, Darby Morhardt, William Healy, Gina Curry, Jen Kauper-Brown, and Chirstine Dunford. 2013. "'I Know What CBPR Is, Now What Do I Do?': Community Perspectives on CBPR Capacity Building." *Progress in Community Health Partnerships* 7 (3): 235–41.

Matson, Laura, Gene-Hua Ng, Michael Dockry, Madeline Nyblade, Hannah Jo King, Mark Bellcourt, Jeremy Bloomquist, et al. 2021. "Transforming Research and Relationships Through Collaborative Tribal-University Partnerships on Manoomin (Wild Rice)." *Environmental Science & Policy* 115: 108–15.

Matsuoka, Martha. 2017. "Democratic Development for Thriving Communities: Framing the Issues, Solutions, and Funding Strategies to Address Gentrification and Displacement." Los Angeles: Occidental College Urban and Environmental Policy Institute and

Neighborhood Funders Group. https://d3n8a8pro7vhmx.cloudfront.net/nfg/pages/476/attachments/original/1501798462/Democratic_Development_Report.pdf?1501798462.

Matsuoka, Martha, Andrea Hricko, Robert Gottlieb, and Juan DeLara. 2011. *Global Trade Impacts: Addressing the Health, Social and Environmental Consequences of Moving International Freight Through Our Communities*. Los Angeles: Occidental College and University of Southern California. https://scholar.oxy.edu/handle/20.500.12711/12377.

Matsuoka, Martha, and John Urquiza. 2021. "Building Community Knowledge, Resilience and Resistance Through Research." *GeoJournal* 87 (2): 249–66.

Matthews, Stephen A., James E. Detwiler, and Linda M. Burton. 2005. "Geo-Ethnography: Coupling Geographic Information Analysis Techniques with Ethnographic Methods in Urban Research." *Cartographica* 40 (4): 75–90.

Mayan, Maria J., and Christine H. Daum. 2016. "Worth the Risk? Muddled Relationships in Community-Based Participatory Research." *Qualitative Health Research* 26 (1): 69–76.

McCarthy, Deborah. 2004. "Environmental Justice Grantmaking: Elites and Activists Collaborate to Transform Philanthropy." *Sociological Inquiry* 74 (2): 250–70.

McCarthy, James. 2019. "Authoritarianism, Populism, and the Environment: Comparative Experiences, Insights, and Perspectives." *Annals of the American Association of Geographers* 109 (2): 301–13.

McCartney, Gerry, Mel Bartley, Ruth Dundas, Srinivasa Vittal Katikireddi, Rich Mitchell, Frank Popham, David Walsh, and Welcome Wami. 2019. "Theorising Social Class and Its Application to the Study of Health Inequalities." *SSM—Population Health* 7: 100315.

McCauley, Darren, and Raphael Heffron. 2018. "Just Transition: Integrating Climate, Energy and Environmental Justice." *Energy Policy* 119: 1–7.

McCutcheon, Priscilla. 2013. "'Returning Home to Our Rightful Place': The Nation of Islam and Muhammad Farms." *Geoforum* 49: 61–70.

———. 2021. "Growing Black Food on Sacred Land: Using Black Liberation Theology to Imagine an Alternative Black Agrarian Future." *Environment and Planning, Society and Space*, 39 (5): 887–905.

McElfish, Pearl A., Britni L. Ayers, Holly C. Felix, Christopher R. Long, Zoran Bursac, Joseph Keawe'aimoku Kaholokula, Sheldon Riklon, Williamina Bing, Anita Iban, and Karen Hye-Cheon Kim Yeary. 2019. "How Stakeholder Engagement Influenced a Randomized Comparative Effectiveness Trial Testing Two Diabetes Prevention Program Interventions in a Marshallese Pacific Islander Community." *Journal of Translational Medicine* 17: 42.

McGreavy, Bridie, Darren Ranco, John Daigle, Suzanne Greenlaw, Nolan Altvater, Tyler Quiring, Natalie Michelle, et al. 2021. "Science in Indigenous Homelands: Addressing Power and Justice in Sustainability Science from/with/in the Penobscot River." *Sustainability Science* 16 (3): 937–47.

McGregor, Deborah, Jean-Paul Restoule, and Rochelle Johnston, eds. 2018. *Indigenous Research: Theories, Practices, and Relationships*. Toronto: Canadian Scholars.

McGuire-Adams, Tricia. 2021. "Settler Allies Are Made, Not Self-Proclaimed: Unsettling Conversations for Non-Indigenous Researchers and Educators Involved in Indigenous Health." *Health Education Journal* 80 (7): 761–72.

McKittrick, Katherine. 2011 "On Plantations, Prisons, and a Black Sense of Place." *Social & Cultural Geography* 12 (8): 947–63.

Meehan, James. 2014. "Reinventing Real Estate: The Community Land Trust as a Social Invention in Affordable Housing." *Journal of Applied Social Science* 8 (2): 113–33.

Mello, Michelle M., and Leslie E. Wolf. 2010. "The Havasupai Indian Tribe Case—Lessons for Research Involving Stored Biologic Samples." *New England Journal of Medicine* 363 (3): 204–7.

Mendez, Michael. 2020. *Climate Change from the Streets: How Conflict and Collaboration Strengthen the Environmental Justice Movement.* New Haven: Yale University Press.

Mennis, Jeremy, and Megan Heckert. 2018. "Application of Spatial Statistical Techniques." In *The Routledge Handbook of Environmental Justice*, edited by Ryan Holifield, Jayajit Chakraborty, and Gordon Walker, 207–21. New York: Routledge.

Mennis, Jeremy, Michael Mason, and Andreea Ambrus. 2018. "Urban Greenspace Is Associated with Reduced Psychological Stress Among Adolescents: A Geographic Ecological Momentary Assessment (GEMA) Analysis of Activity Space." *Landscape and Urban Planning* 174: 1–9.

Merino, Roger. 2020. "Rethinking Indigenous Politics: The Unnoticed Struggle for Self-Determination in Peru." *Bulletin of Latin American Research* 39 (4): 513–28.

Meyer, Manulani Aluli. 2008. "Indigenous and Authentic: Hawaiian Epistemology and the Triangulation of Meaning." In *Handbook of Critical and Indigenous Methodologies*, edited by Norman Denzin, Yvonna Lincoln, and Linda Smith, 217–32. Thousand Oaks: SAGE Publications.

Michener, Lloyd, Jennifer Cook, Syed M. Ahmed, Michael A. Yonas, Tamera Coyne-Beasley, and Sergio Aguilar-Gaxiola. 2012. "Aligning the Goals of Community-Engaged Research: Why and How Academic Health Centers Can Successfully Engage with Communities to Improve Health." *Academic Medicine* 87 (3): 285–91.

Middleton, Beth Rose, Sabine Talaugon, Thomas M. Young, Luann Wong, Suzanne Fluharty, Kaitlin Reed, Christine Cosby, and Richard Myers. 2019. "Bi-Directional Learning: Identifying Contaminants on the Yurok Indian Reservation." *International Journal of Environmental Research and Public Health* 16 (19): 3513.

Migrant Justice and Milk with Dignity Standards Council. 2020. "Milk with Dignity: First Biennial Report, 2018–2019." Burlington: Migrant Justice and Milk with Dignity Standards Council. https://migrantjustice.net/sites/default/files/2020MDReport.pdf.

Miller, Ethan. 2009. "Solidarity Economy: Key Concepts and Issues." In *Solidarity Economy 1: Building Alternatives for People and Planet*, edited by Emily Kawano, Thomas Neal Masterson, and Jonathan Teller-Elsberg, 25–42. Amherst: Center for Popular Economics.

Minkler, Meredith, and Andrea Corage Baden. 2008. "Impacts of CBPR on Academic Researchers, Research Quality and Methodology, and Power Relations." In *Community-Based Participatory Research for Health: From Process to Outcomes*, edited by Meredith Minkler and Nina Wallerstein, 2nd ed., 243–62. San Francisco: Jossey-Bass.

Minkler, Meredith, Analilia P. Garcia, Joy Williams, Tony LoPresti, and Jane Lilly. 2010. "Sí Se Puede: Using Participatory Research to Promote Environmental Justice in a Latino Community in San Diego, California." *Journal of Urban Health* 87 (5): 796–812.

Minkler, Meredith, Cheri Pies, and Cheryl A. Hyde. 2012. "Ethical Issues in Community Organizing and Capacity Building." In *Community Organizing and Community Building for Health and Welfare*, edited by Meredith Minkler, 3rd ed., 110–29. New Brunswick: Rutgers University Press.

Minkler, Meredith, Alicia L. Salvatore, and Charlotte Chang. 2018. "Participatory Approaches for Study Design and Analysis in Dissemination and Implementation Research." In *Dissemination and Implementation Research in Health: Translating Science to Practice*, edited by Ross C. Brownson, Graham A. Colditz, and Enola Knisley Proctor, 175–90. New York: Oxford University Press.

Minkler, Meredith, Victoria Breckwich Vásquez, Charlotte Chang, Jenesse Miller, Victor Rubin, Angela Glover Blackwell, Rebecca Thompson, Rebecca Flournoy, and Judith Bell. 2008. *Promoting Healthy Public Policy through Community-Based Participatory Research: Ten Case Studies*. Oakland: PolicyLink. https://www.policylink.org/resources-tools/promoting-healthy-public-policy-through-community-based-participatory-research-ten-case-studies.

Minkler, Meredith, Victoria Breckwich Vásquez, and Peggy Shepard. 2006. "Promoting Environmental Health Policy Through Community Based Participatory Research: A Case Study from Harlem, New York." *Journal of Urban Health* 83 (1): 101–10.

Minkler, Meredith, and Patricia Wakimoto, eds. 2022. *Community Organizing and Community Building for Health and Social Equity*. New Brunswick: Rutgers University Press.

Miraftab, Faranak. 2012. "Planning and Citizenship." In *The Oxford Handbook of Urban Planning*, edited by Randall Crane and Rachel Weber, 786–802. Oxford: Oxford University Press.

Mitchell, Felicia M. 2018a. "Engaging in Indigenous CBPR Within Academia: A Critical Narrative." *Affilia* 33 (3): 379–94.

———. 2018b. "'Water Is Life': Using Photovoice to Document American Indian Perspectives on Water and Health." *Social Work Research* 42 (4): 277–89.

Mohammed, Selina A., Karina L. Walters, June LaMarr, Teresa Evans-Campbell, and Sheryl Fryberg. 2012. "Finding Middle Ground: Negotiating University and Tribal Community Interests in Community-Based Participatory Research." *Nursing Inquiry* 19 (2): 116–27.

Montenegro de Wit, Maywa, Annie Shattuck, Alastair Iles, Garrett Graddy-Lovelace, Antonio Roman-Alcalá, and M. Jahi Chappell. 2021. "Operating Principles for Collective Scholar-Activism: Early Insights from the Agroecology Research-Action Collective." *Journal of Agriculture, Food Systems, and Community Development* 10 (2): 319–37.

Moore, Jason W., and Raj Patel. 2017. "Unearthing the Capitalocene: Towards a Reparations Ecology." *ROAR Magazine*. https://roarmag.org/magazine/moore-patel-seven-cheap-things-capitalocene/.

Moore, Michele-Lee, Darcy Riddell, and Dana Vocisano. 2015. "Scaling Out, Scaling Up, Scaling Deep Strategies of Non-Profits in Advancing Systemic Social Innovation." *Journal of Corporate Citizenship* 58: 67–84.

Moore de Peralta, Arelis, Julie Smithwick, and Myriam E. Torres. 2020. "Perceptions and Determinants of Partnership Trust in the Context of Community-Based Participatory Research." *Participatory Research* 13 (1): 4.

Morello-Frosch, Rachel, Phil Brown, and Julia G. Brody. 2017. "Democratizing Ethical Oversight of Research Through CBPR." In *Community-Based Participatory Research for Health: Advancing Social and Health Equity*, edited by Nina Wallerstein, Bonnie Duran, John G. Oetzel, and Meredith Minkler, 3rd ed., 215–26. San Francisco: Jossey-Bass.

Morello-Frosch, Rachel, Phil Brown, Julia Green Brody, Rebecca Gasior Altman, Ruthann A. Rudel, Ami Zota, and Carla Perez. 2011. "Experts, Ethics, and Environmental

Justice: Communicating and Contesting Results from Personal Exposure Science." In *Technoscience and Environmental Justice: Expert Cultures in a Grassroots Movement*, edited by Gwen Ottinger and Benjamin Cohen, 93–118. Cambridge: MIT Press.

Morello-Frosch, Rachel, and Bill M. Jesdale. 2006. "Separate and Unequal: Residential Segregation and Estimated Cancer Risks Associated with Ambient Air Toxics in U.S. Metropolitan Areas." *Environmental Health Perspectives* 114 (3): 386–93.

Morello-Frosch, Rachel, Manuel Pastor, James Sadd, and Madeline Wander. 2015a. "Environmental Justice Screening Method." *Equity Research Institute*. https://dornsife.usc.edu/assets/sites/242/docs/EJSM_Convening_Presentations_Parts1and2_Jan_2015_for_website.pdf.

Morello-Frosch, Rachel, Julia Varshavsky, Max Liboiron, Phil Brown, and Julia G. Brody. 2015b. "Communicating Results in Post-Belmont Era Biomonitoring Studies: Lessons from Genetics and Neuroimaging Research." *Environmental Research* 136: 363–72.

Morello-Frosch, Rachel, Miriam Zuk, Michael Jerrett, Bhavna Shamasunder, and Amy D. Kyle. 2011. "Understanding the Cumulative Impacts of Inequalities in Environmental Health: Implications for Policy." *Health Affairs* 30 (5): 879–87.

Morgan-Trimmer, Sarah. 2014. "Policy Is Political; Our Ideas About Knowledge Translation Must Be Too." *Journal of Epidemiological Community Health* 68 (11): 1010–11.

Morris, Aldon. 2017. *The Scholar Denied: W. E. B. Du Bois and the Birth of Modern Sociology*. Berkeley: University of California Press.

Moss, Melissa A. 2020. "The Escambia Project: An Experiment in Community-Led Legal Design." *Design Issues* 36 (3): 45–60.

Movement Generation Justice and Ecology Project. n.d. "From Banks and Tanks to Cooperation and Caring: A Strategic Framework for a Just Transition." Berkeley: Movement Generation Justice and Ecology Project. https://movementgeneration.org/wp-content/uploads/2016/11/JT_booklet_English_SPREADs_web.pdf.

Mueller, Michael P., and Deborah J. Tippins, eds. 2015. *EcoJustice, Citizen Science and Youth Activism: Situated Tensions for Science Education*. New York: Springer.

Muhammad, Michael, Catalina Garzón, Angela Reyes, and West Oakland Environmental Indicators Project. 2017. "Understanding Contemporary Racism, Power and Privilege and Their Impacts on CBPR." In *Community-Based Participatory Research for Health: Advancing Social and Health Equity*, edited by Nina Wallerstein, Bonnie Duran, John Oetzel, and Meredith Minkler, 3rd ed., 47–59. San Francisco: Jossey-Bass.

Muhammad, Michael, Nina Wallerstein, Andrew L. Sussman, Magdalena Avila, Lorenda Belone, and Bonnie Duran. 2015. "Reflections on Researcher Identity and Power: The Impact of Positionality on Community Based Participatory Research (CBPR) Processes and Outcomes." *Critical Sociology* 41 (7–8): 1045–63.

Mulrennan, Monica E., Rodney Mark, and Colin H. Scott. 2012. "Revamping Community-Based Conservation Through Participatory Research." *The Canadian Geographer* 56 (2): 243–59.

Munck, Ronaldo. 2014. "Community-Based Research: Genealogy and Prospects." In *Higher Education and Community-Based Research*, edited by Ronaldo Munck, Lorraine McIlrath, Budd Hall, and Rajesh Tandon, 11–26. New York: Palgrave Macmillan.

Munck, Ronaldo, Lorraine McIlrath, Budd Hall, and Rajesh Tandon, eds. 2014. *Higher Education and Community-Based Research: Creating a Global Vision*. London: Palgrave Macmillan.

Muntaner, Carles, Edwin Ng, Haejoo Chung, and Seth J. Prins. 2015. "Two Decades of Neo-Marxist Class Analysis and Health Inequalities: A Critical Reconstruction." *Social Theory & Health* 13 (3): 267–87.

Murphy, Fred, Andrea Hinojosa, and Sirad Osman. 2013. "A View from the Community: African-American, Hispanic, and African Immigrant Perspectives." In *Community-Based Participatory Health Research: Issues, Methods, and Translation to Practice*, edited by Daniel S. Blumenthal, Ralph J. DiClemente, Ronald Braithwaite, and Selina Smith, 2nd ed., 51–78. New York: Springer.

Murphy, Shannon, Shankar Prasad, John Faust, and George Alexeeff. 2018. "Community-Based Cumulative Impact Assessment: California's Approach to Integrating Nonchemical Stressors into Environmental Assessment Practices." In *Chemical Mixtures and Combined Chemical and Nonchemical Stressors: Exposure, Toxicity, Analysis, and Risk*, edited by Cynthia V. Rider and Jane Ellen Simmons, 515–44. Cham: Springer.

Myers, Rodd, and Mumu Muhajir. 2015. "Killing Us Without Blood: In Search of Recognition Justice in Bukit Baka Bukit Raya National Park, Indonesia." In *Proceedings of the Conference of the International Association for the Study of the Commons*. Edmonton: IASC. https://hdl.handle.net/10535/9849.

Nabatchi, Tina, and Matt Leighninger. 2015. *Public Participation for 21st Century Democracy*. Hoboken: John Wiley & Sons.

Nadasdy, Paul. 1999. "The Politics of TEK: Power and the 'Integration' of Knowledge." *Arctic Anthropology* 36 (1–2): 1–18.

Nagy, Rosemary. 2022. "Transformative Justice in a Settler Colonial Transition: Implementing the UN Declaration on the Rights of Indigenous Peoples in Canada." *The International Journal of Human Rights* 26 (2): 191–216.

Naik, Yannish, Peter Baker, Sharif A. Ismail, Taavi Tillmann, Kristin Bash, Darryl Quantz, Frances Hillier-Brown, et al. 2019. "Going Upstream: An Umbrella Review of the Macroeconomic Determinants of Health and Health Inequalities." *BMC Public Health* 19: 1678.

National Association of Climate Resilience Planners. n.d. "National Association of Climate Resilience Planners: Home." Accessed December 21, 2021. https://www.nacrp.org.

National Center for Cultural Competence. n.d. "Foundations." Accessed November 15, 2021. https://nccc.georgetown.edu/foundations/framework.php.

National Center for Healthy Housing. 2016. "A Systematic Review of Health Impact Assessments on Housing Decisions and Guidance for Future Practice." March. https://nchh.org/resource-library/report_a-systematic-review-of-health-impact-assessments-on-housing-decisions-and-guidance-for-future-practice.pdf.

National Environmental Justice Advisory Council. 2013. "Model Guidelines for Public Participation: An Update to the 1996 NEJAC Model Plan for Public Participation." Washington, DC: U.S. Environmental Protection Agency. https://www.epa.gov/sites/default/files/2015-02/documents/recommendations-model-guide-pp-2013.pdf.

National Research Council. 2011. *Improving Health in the United States: The Role of Health Impact Assessment*. Washington, DC: National Academies of Sciences, Engineering, and Medicine.

Neale, Timothy, Rodney Carter, Trent Nelson, and Mick Bourke. 2019. "Walking Together: A Decolonising Experiment in Bushfire Management on Dja Dja Wurrung Country." *Cultural Geographies* 26 (3): 341–59.

Neale, Timothy, and Will Smith. 2019. "Indigenous People in the Natural Hazards Management Sector: Examining Employment Data." *Australian Journal of Emergency Management* 34 (3): 15–20.

Neely, Brooke, and Michelle Samura. 2011. "Social Geographies of Race: Connecting Race and Space." *Ethnic and Racial Studies* 34 (11): 1933–52.

Nelson, Melissa K., and Daniel Shilling, eds. 2018. *Traditional Ecological Knowledge: Learning from Indigenous Practices for Environmental Sustainability*. Cambridge: Cambridge University Press.

Neubauer, Leah C., and Melvin E. Hall. 2020. "Is Inciting Social Change Something Evaluators Can Do? Should Do?" *New Directions for Evaluation* 166: 129–35.

Neubauer, Leah C., Dominica McBride, Andrea D. Guajardo, Wanda D. Casillas, and Melvin E. Hall. 2020. "Examining Issues Facing Communities of Color Today: The Role of Evaluation to Incite Change." *New Directions for Evaluation* 166: 7–11.

New York City Council. 2021. "Council Votes to Pass the 'Renewable Rikers' Act." February 11, 2021. https://council.nyc.gov/press/2021/02/11/2069/.

New York City Environmental Justice Alliance. n.d. "New York City Climate Justice Agenda 2020: A Critical Decade for Climate, Health, and Equity." Accessed December 21, 2021. https://nyc-eja.org/campaigns/climate-justice-community-resiliency/.

Newell, Peter, and Dustin Mulvaney. 2013. "The Political Economy of the 'Just Transition.'" *The Geographical Journal* 179 (2): 132–40.

Newman, Richard. 2012. "Darker Shades of Green: Love Canal, Toxic Autobiography, and American Environmental Writing." In *Histories of the Dustheap: Waste, Material Cultures, Social Justice*, edited by Stephanie Foote and Elizabeth Mazzolini, 21–48. Cambridge: MIT Press.

Newman, Susan D., Jeannette O. Andrews, Gayenell S. Magwood, Carolyn Jenkins, Melissa J. Cox, and Deborah C. Williamson. 2011. "Community Advisory Boards in Community-Based Participatory Research: A Synthesis of Best Processes." *Preventing Chronic Disease* 8 (3): A70.

Nicole, Wendee. 2013. "CAFOs and Environmental Justice: The Case of North Carolina." *Environmental Health Perspectives* 121 (6): 182–89.

Nicolosi, Emily, Jim French, and Richard Medina. 2020. "Add to the Map! Evaluating Digitally Mediated Participatory Mapping for Grassroots Sustainabilities." *The Geographical Journal* 186 (2): 142–55.

Nixon, Rob. 2011. *Slow Violence and the Environmentalism of the Poor*. Cambridge: Harvard University Press.

Njoh, Ambe J. 2010. "Europeans, Modern Urban Planning and the Acculturation of 'Racial Others.'" *Planning Theory* 9 (4): 369–78.

Nevada Minority Health and Equity Coalition. 2021. "NMHEC: Our Story." Accessed May 17, 2021. https://nmhec.org/our-story/.

Nolan, James E. S., Eric S. Coker, Bailey R. Ward, Yahna A. Williamson, and Kim G. Harley. 2021. "Freedom to Breathe": Youth Participatory Action Research (YPAR) to Investigate Air Pollution Inequities in Richmond, CA." *International Journal of Environmental Research and Public Health* 18 (2): 554.

Norgaard, Kari Marie. 2004. "The Effects of Altered Diet on the Health of the Karuk People: A Preliminary Report." Orleans: Karuk Tribe Department of Natural Resources. https://

www.waterboards.ca.gov/waterrights/water_issues/programs/bay_delta/california_waterfix/exhibits/docs/PCFFA&IGFR/part2/pcffa_195.pdf.

———. 2014. "Karuk Traditional Ecological Knowledge and the Need for Knowledge Sovereignty: Social, Cultural and Economic Impacts of Denied Access to Traditional Management." Orleans: Karuk Tribe Department of Natural Resources. https://karuktribeclimatechangeprojects.files.wordpress.com/2016/05/final-pt-1-karuk-tek-and-the-need-for-knowledge-sovereignty.pdf.

———. 2019. *Salmon and Acorns Feed Our People: Colonialism, Nature, and Social Action*. New Brunswick: Rutgers University Press.

North, Peter. 2010a. "Eco-Localisation as a Progressive Response to Peak Oil and Climate Change—A Sympathetic Critique." *Geoforum* 41 (4): 585–94.

———. 2010b. *Local Money: How toMake It Happen in Your Community*. Cambridge: UIT Cambridge.

Nussbaum, Martha. 2011. *Creating Capabilities: The Human Development Approach*. Cambridge: Harvard University Press.

Nyseth Brehm, Hollie, and David N. Pellow. 2014. "Environmental Justice: Pollution, Poverty, and Marginalized Communities." In *Routledge Handbook of Global Environmental Politics*, edited by Paul G. Harris, 308–20. New York: Routledge.

O'Brien, Mary. 2000. *Making Better Environmental Decisions: An Alternative to Risk Assessment*. Cambridge: MIT Press.

O'Campo, Patricia. 2012. "Are We Producing the Right Kind of Actionable Evidence for the Social Determinants of Health?" *Journal of Urban Health* 89 (6): 881–93.

Occidental Magazine. 2019. "Springing into Action." *Occidental Magazine*. Spring 2019. https://www.oxy.edu/magazine/issues/spring-2019/springing-action.

O'Meara, KerryAnn, Dawn Culpepper, Joya Misra, and Audrey Jaeger. 2021. "Equity-Minded Faculty Workloads: What We Can and Should Do Now." Washington, DC: American Council on Education. https://www.acenet.edu/Documents/Equity-Minded-Faculty-Workloads.pdf.

O'Meara, KerryAnn, and R. Eugene Rice. 2005. "Introduction." In *Faculty Priorities Reconsidered: Rewarding Multiple Forms of Scholarship*, edited by KerryAnn O'Meara and R. Eugene Rice, 1–15. San Francisco: Jossey-Bass.

O'Meara, KerryAnn, Lorilee R. Sandmann, John Saltmarsh, and Dwight E. Giles. 2011. "Studying the Professional Lives and Work of Faculty Involved in Community Engagement." *Innovative Higher Education* 36 (2): 83–96.

One Brooklyn Health System. 2019. "2019–2021 IRS Implementation Strategy & NYS Community Service Plan." https://onebrooklynhealth.org/wp-content/uploads/2020/09/OBHS-CHNA-CSP-2019-11-15-19.pdf.

O'Neil, Cathy. 2016. *Weapons of Math Destruction: How Big Data Increases Inequality and Threatens Democracy*. New York: Crown.

O'Reilly, Nicole L., Erin R. Hager, Donna Harrington, and Maureen M. Black. 2020. "Assessment of Risk for Food Insecurity Among African American Urban Households: Utilizing Cumulative Risk Indices and Latent Class Analysis to Examine Accumulation of Risk Factors." *Translational Behavioral Medicine* 10 (6): 1322–29.

Orihuela, José Carlos. 2020. "Embedded Countermovements: The Forging of Protected Areas and Native Communities in the Peruvian Amazon." *New Political Economy* 25 (1): 140–55.

Orpinas, Pamela, Rebecca A. Matthew, J. Maria Bermúdez, Luis R. Alvarez-Hernandez, and Alejandra Calva. 2020. "A Multistakeholder Evaluation of Lazos Hispanos: An Application of a Community-Based Participatory Research Conceptual Model." *Journal of Community Psychology* 48 (2): 464–81.

Osborne, Tracey. 2015. "Tradeoffs in Carbon Commodification: A Political Ecology of Common Property Forest Governance." *Geoforum* 67: 64–77.

Oscilowicz, Emilia, Emilia Lewartowska, Ariella Levitch, Jonathan Luger, Jonathan Hajtmarova, Ella O'Neill, Aina Planas Carbonell, Helen Cole, Carla Rivera Blanco, and Erin Monroe. 2021. "Policy and Planning Tools for Urban Green Justice: Fighting Displacement and Gentrification and Improving Accessibility and Inclusiveness to Green Amenities." Barcelona: Barcelona Lab for Urban Environmental Justice and Sustainability and ICLEI. http://www.bcnuej.org/wp-content/uploads/2021/04/Toolkit-Urban-Green-Justice.pdf.

Overdevest, Christine, and Brian Mayer. 2008. "Harnessing the Power of Information Through Community Monitoring: Insights from Social Science." *Texas Law Review* 86 (7): 1493–1526.

Ozer, Emily J., Amber Akemi Piatt, and Cathleen E. Willging. 2017. "Youth-Led Participatory Action Research (YPAR): Principles Applied to the US and Diverse Global Settings." In *Community-Based Participatory Research for Health: Advancing Social and Health Equity*, edited by Nina Wallerstein, Bonnie Duran, John Oetzel, and Meredith Minkler, 3rd ed., 95–106. San Francisco: Jossey-Bass.

Pabel, Anja, Josephine Pryce, and Allison Anderson, eds. 2021. *Research Paradigm Considerations for Emerging Scholars*. Berlin: Channel View Publications.

Pacheco, Christina M., Sean M. Daley, Travis Brown, K. Allen Greiner, Melissa Filippi, and Christine M. Daley. 2013. "Moving Forward: Breaking the Cycle of Mistrust Between American Indians and Researchers." *American Journal of Public Health* 103 (12): 2152–59.

Pandya, Rajul, and Kenne Ann Dibner, eds. 2018. *Learning Through Citizen Science: Enhancing Opportunities by Design*. Washington, DC: National Academies of Sciences, Engineering, and Medicine.

Pansera, Mario, and Soumodip Sarkar. 2016. "Crafting Sustainable Development Solutions: Frugal Innovations of Grassroots Entrepreneurs." *Sustainability* 8 (1): 51.

Paradies, Yin. 2016. "Colonisation, Racism and Indigenous Health." *Journal of Population Research* 33 (1): 83–96.

Park, Yoo Min, and Mei-Po Kwan. 2020. "Understanding Racial Disparities in Exposure to Traffic-Related Air Pollution: Considering the Spatiotemporal Dynamics of Population Distribution." *International Journal of Environmental Research and Public Health* 17 (3): 908.

Parker, Myra, Cynthia Pearson, Caitlin Donald, and Celia B. Fisher. 2019. "Beyond the Belmont Principles: A Community-Based Approach to Developing an Indigenous Ethics Model and Curriculum for Training Health Researchers Working with American Indian and Alaska Native Communities." *American Journal of Community Psychology* 64 (1–2): 9–20.

Paschal, Angelia M., Jermaine B. Mitchell, Wanda M. Burton, Jen Nickelson, Pillar Z. Murphy, and Frances Ford. 2020. "Using Community-Based Participatory Research to Explore Food Insecurity in African American Adults." *American Journal of Health Education* 51 (3): 186–95.

Pastor, Manuel, Chris Benner, and Martha Matsuoka. 2009. *This Could Be the Start of Something Big: How Social Movements for Regional Equity Are Reshaping Metropolitan America*. Ithaca: Cornell University Press.

Patel, Raj. 2009. "Food Sovereignty." *The Journal of Peasant Studies* 36 (3): 663–706.

Pateman, Rachel Mary, Alison Dyke, and Sarah Elizabeth West. 2021. "The Diversity of Participants in Environmental Citizen Science." *Citizen Science: Theory and Practice* 6 (1): 9.

Patler, Caitlin, and Whitney Laster Pirtle. 2018. "From Undocumented to Lawfully Present: Do Changes to Legal Status Impact Psychological Wellbeing Among Latino Immigrant Young Adults?" *Social Science & Medicine* 199: 39–48.

Paustenbach, Dennis, and David Galbraith. 2006. "Biomonitoring and Biomarkers: Exposure Assessment Will Never Be the Same." *Environmental Health Perspectives* 114 (8): 1143–49.

Pearce, Jone L., and Laura Huang. 2012. "Toward an Understanding of What Actionable Research Is." *Academy of Management Learning & Education* 11 (2): 300–1.

Pearsall, Hamil, and Isabelle Anguelovski. 2016. "Contesting and Resisting Environmental Gentrification: Responses to New Paradoxes and Challenges for Urban Environmental Justice." *Sociological Research Online* 21 (3): 121–27.

Pearson, Cynthia R., Myra Parker, Celia B. Fisher, and Claudia Moreno. 2014. "Capacity Building from the Inside Out: Development and Evaluation of a CITI Ethics Certification Training Module for American Indian and Alaska Native Community Researchers." *Journal of Empirical Research on Human Research Ethics* 9 (1): 46–57.

Pedler, Megan Louise, Royce Willis, and Johanna Elizabeth Nieuwoudt. 2022. "A Sense of Belonging at University: Student Retention, Motivation and Enjoyment." *Journal of Further and Higher Education* 46 (3): 397–408.

Pellow, David N. 2002. *Garbage Wars: The Struggle for Environmental Justice in Chicago*. Cambridge: MIT Press.

———. 2007. *Resisting Global Toxics: Transnational Movements for Environmental Justice*. Cambridge: MIT Press.

———. 2011. "Politics by Other Greens: The Importance of Transnational Environmental Justice Movement Networks." In *Environmental Inequalities Beyond Borders: Local Perspectives on Global Injustices*, edited by JoAnn Carmin and Julian Agyeman, 247–66. Cambridge: MIT Press.

———. 2018. *What Is Critical Environmental Justice?* Cambridge: Polity Press.

Pellow, David, Jasmine Vazin, Michaela Austin, and Ketia Johnson. 2019. *Capitalism in Practice: Free Market Influence on Environmental Injustice in America's Prisons*. Santa Barbara: Global Environmental Justice Project.

Peña, Devon G. 1997. *The Terror of the Machine: Technology, Work, Gender, and Ecology on the US-Mexico Border*. Austin: University of Texas Press.

Peña, Devon G., ed. 1998. *Chicano Culture, Ecology, Politics: Subversive Kin*. Tucson: University of Arizona Press.

Penniman, Leah. 2018. *Farming While Black: Soul Fire Farm's Practical Guide to Liberation on the Land.* White River Junction: Chelsea Green.

Pennington, Pamela M., Elizabeth Pellecer Rivera, Sandra De Urioste-Stone, Teresa M. Aguilar, and José Guillermo Juárez. 2020. "A Successful Community-Based Pilot Programme to Control Insect Vectors of Chagas Disease in Rural Guatemala." In *Area-Wide Integrated Pest Management: Development and Field Application*, edited by Jorge Hendrichs, Rui Pereira, and Marc J. B. Vreysen, 1st ed., 709–27. Boca Raton: CRC Press.

Peréa, Flavia C., Nina R. Sayles, Amanda J. Reich, Alyssa Koomas, Heather McMann, and Linda S. Sprague Martinez. 2019. "'Mejorando Nuestras Oportunidades': Engaging Urban Youth in Environmental Health Assessment and Advocacy to Improve Health and Outdoor Play Spaces." *International Journal of Environmental Research and Public Health* 16 (4): 571.

Perera, Frederica P., Susan M. Illman, Patrick L. Kinney, Robin M. Whyatt, Elizabeth A. Kelvin, Peggy Shepard, David Evans, et al. 2002. "The Challenge of Preventing Environmentally Related Disease in Young Children: Community-Based Research in New York City." *Environmental Health Perspectives* 110 (2): 197–204.

Peters, Scott J. 2010. *Democracy and Higher Education: Traditions and Stories of Civic Engagement*. Lansing: MSU Press.

Petteway, Ryan J. 2019. "Intergenerational Photovoice Perspectives of Place and Health in Public Housing: Participatory Coding, Theming, and Mapping in/of the 'Structure Struggle.'" *Health & Place* 60: 102229.

———. 2021. "Let's Re-Place the Health Opportunity Maps." *Shelterforce*. https://shelterforce.org/2021/01/20/re-placing-geographies-of-health-opportunity/.

———. 2022. "On epidemiology as racial-capitalist (re)colonization and epistemic violence." *Critical Public Health* 33 (1): 5–12.

———. 2022. *Representation, Re-Presentation, and Resistance: Participatory Geographies of Place, Health, and Embodiment*. Springer Nature.

Petteway, Ryan, and Shannon Cosgrove. 2020. "Health Impact Assessment and City Council Policy: Identifying Opportunities to Address Local Social Determinants of Health & Place-Health Relationships, 10 Years Later." *Chronicles of Health Impact Assessment* 5 (1).

Petteway, Ryan, Mahasin Mujahid, and Amani Allen. 2019. "Understanding Embodiment in Place-Health Research: Approaches, Limitations, and Opportunities." *Journal of Urban Health* 96 (2): 289–99.

Petteway, Ryan J., Mahasin Mujahid, Amani Allen, and Rachel Morello-Frosch. 2019a. "The Body Language of Place: A New Method for Mapping Intergenerational 'Geographies of Embodiment' in Place-Health Research." *Social Science & Medicine* 223: 51–63.

———. 2019b. "Towards a People's Social Epidemiology: Envisioning a More Inclusive and Equitable Future for Social Epi Research and Practice in the 21st Century." *International Journal of Environmental Research and Public Health* 16 (20): 3983.

Petteway, Ryan J., Payam Sheikhattari, and Fernando Wagner. 2019. "Toward an Intergenerational Model for Tobacco-Focused CBPR: Integrating Youth Perspectives Via Photovoice." *Health Promotion Practice* 20 (1): 67–77.

Petticrew, Mark, Margaret Whitehead, Sally J. Macintyre, Hilary Graham, and Matt Egan. 2004. "Evidence for Public Health Policy on Inequalities: The Reality According to Policymakers." *Journal of Epidemiology & Community Health* 58 (10): 811–16.

Pezzullo, Phaedra C. 2007. *Toxic Tourism: Rhetorics of Pollution, Travel, and Environmental Justice*. Tuscaloosa: University of Alabama Press.

Phelan, Jo C., and Bruce G. Link. 2015. "Is Racism a Fundamental Cause of Inequalities in Health?" *Annual Review of Sociology* 41 (1): 311–30.

Picaut, Judicaël, Nicolas Fortin, Erwan Bocher, Gwendall Petit, Pierre Aumond, and Gwenaël Guillaume. 2019. "An Open-Science Crowdsourcing Approach for Producing Community Noise Maps Using Smartphones." *Building and Environment* 148: 20–33.

Pine, Kathleen H., and Max Liboiron. 2015. "The Politics of Measurement and Action." In *Proceedings of the 33rd Annual ACM Conference on Human Factors in Computing Systems*, 3147–56. Seoul: ACM.

Pitman, Nigel, Adriana Bravo, Santiago Claramunt, Corine Vriesendorp, Diana Alvira Reyes, Ashwin Ravikumar, Álvaro del Campo, et al., eds. 2016. *Perú: Medio Putumayo-Algodón. Rapid Biological and Social Inventories 28*. Chicago: Field Museum of Natural History.

Pollin, Robert, and Brian Callaci. 2019. "The Economics of Just Transition: A Framework for Supporting Fossil Fuel–Dependent Workers and Communities in the United States." *Labor Studies Journal* 44 (2): 93–138.

Porter, Libby, Hirini Matunga, Leela Viswanathan, Lyana Patrick, Ryan Walker, Leonie Sandercock, Dana Moraes, et al. 2017. "Indigenous Planning: From Principles to Practice." *Planning Theory & Practice* 18 (4): 639–66.

Post, Margaret A., Elaine Ward, Nicholas V. Longo, and John Saltmarsh. 2016. *Publicly Engaged Scholars: Next-Generation Engagement and the Future of Higher Education*. Sterling: Stylus.

Powell, John A. 2007. "Structural Racism and Spatial Jim Crow." In *The Black Metropolis in the Twenty-First Century: Race, Power, and the Politics of Place*, edited by Robert D. Bullard. Lanham: Rowman & Littlefield.

Prado, Carolina. 2019. "Just Community Participation and Border Environmental Governance: A View from the Border 2020 Program." *Journal of Environmental Policy & Planning* 21 (6): 662–74.

Prado, Carolina, Colectivo Salud y Justicia Ambiental, and Red de Ciudadanos para el Mejoramiento de las Comunidades. 2021. "Border Environmental Justice PPGIS: Community-Based Mapping and Public Participation in Eastern Tijuana, México." *International Journal of Environmental Research and Public Health* 18 (3): 1349.

Praeli, Yvette Sierra. 2021. "Liz Chicaje: Defensora Del Parque Nacional Yaguas Gana El Premio Goldman." *Mongabay Latam*. June 15, 2021. https://es.mongabay.com/2021/06/peru-liz-chicaje-gana-el-premio-goldman-2021/.

Pratt, Bridget. 2021. "Sharing Power in Global Health Research: An Ethical Toolkit for Designing Priority-Setting Processes That Meaningfully Include Communities." *International Journal for Equity in Health* 20 (1): 127.

Privitera, Elisa, Marco Armiero, and Filippo Gravagno. 2021. "Seeking Justice in Risk Landscapes: Small Data and Toxic Autobiographies from an Italian Petrochemical Town (Gela, Sicily)." *Local Environment* 26 (7): 847–71.

Pulido, Laura. 1996. *Environmentalism and Economic Justice: Two Chicano Struggles in the Southwest*. Tucson: University of Arizona Press.

Pulido, Laura, and Juan De Lara. 2018. "Reimagining 'Justice' in Environmental Justice: Radical Ecologies, Decolonial Thought, and the Black Radical Tradition." *Environment and Planning* 1 (1–2): 76–98.

Pulido, Laura, Ellen Kohl, and Nicole-Marie Cotton. 2016. "State Regulation and Environmental Justice: The Need for Strategy Reassessment." *Capitalism Nature Socialism* 27 (2): 12–31.

Purucker, David. 2021. "Critical Environmental Justice and the State: A Critique of Pellow." *Environmental Sociology* 7 (3): 176–86.

PUSH Buffalo. 2017. "PUSH Buffalo's Green Development Zone: A Model for New Economic and Community Development." Buffalo: Partnership for the Public Good. https://www.pushbuffalo.org/wp-content/uploads/2019/06/PPG-PUSH-GDZ-Report.6.2017.pdf.

Pyles, Loretta. 2021. *Progressive Community Organizing: Transformative Practice in a Globalizing World*. 3rd ed. New York: Routledge.

Rainie, Stephanie Carroll, Desi Rodriguez-Lonebear, and Andrew Martinez. 2017. "Policy Brief: Indigenous Data Sovereignty in the United States." Tucson: Native Nations Institute, University of Arizona. https://nni.arizona.edu/application/files/1715/1579/8037/Policy_Brief_Indigenous_Data_Sovereignty_in_the_United_States.pdf.

Ramírez, Ivan J., Ana Baptista, Jieun Lee, Ana Traverso-Krejcarek, and Andreah Santos. 2019. "Fighting for Urban Environmental Health Justice in Southside (Los Sures) Williamsburg, Brooklyn: A Community-Engaged Pilot Study." In *Handbook of Global Urban Health*, edited by Igor Vojnovic, Amber L. Pearson, Gershim Asiki, Geoffrey DeVerteuil, and Adriana Allen, 584–613. New York: Routledge.

Ramirez-Andreotta, Monica D., Mark L. Brusseau, Janick Artiola, Raina M. Maier, and A. Jay Gandolfi. 2015. "Building a Co-Created Citizen Science Program with Gardeners Neighboring a Superfund Site: The Gardenroots Case Study." *International Public Health Journal* 7 (1): 13.

Ramirez-Gomez, Sara O. I., Greg Brown, and Annette Tjon Sie Fat. 2013. "Participatory Mapping with Indigenous Communities for Conservation: Challenges and Lessons from Suriname." *Electronic Journal of Information Systems in Developing Countries* 58 (1): 1–22.

Raphael, Chad. 2019a. "Engaged Communication Scholarship for Environmental Justice: A Research Agenda." *Environmental Communication* 13 (8): 1087–1107.

———. 2019b. "Engaged Scholarship for Environmental Justice: A Guide." Santa Clara: Santa Clara University. https://scholarcommons.scu.edu/comm/114.

Reason, Peter, and Hilary Bradbury, eds. 2008. *The SAGE Handbook of Action Research: Participative Inquiry and Practice*. 2nd ed. Thousand Oaks: SAGE Publications.

Redvers, Jennifer. 2020. "'The Land Is a Healer': Perspectives on Land-Based Healing from Indigenous Practitioners in Northern Canada." *International Journal of Indigenous Health* 15 (1): 90–107.

Reed, Mark S. 2008. "Stakeholder Participation for Environmental Management: A Literature Review." *Biological Conservation* 141 (10): 2417–31.

Reed, Maureen G., and Colleen George. 2011. "Where in the World Is Environmental Justice?" *Progress in Human Geography* 35 (6): 835–42.

———. 2018. "Just Conservation: The Evolving Relationship Between Society and Protected Areas." In *The Routledge Handbook of Environmental Justice*, edited by Ryan Holifield, Jayajit Chakraborty, and Gordon Walker, 463–76. New York: Routledge.

Reed, Ron, and Kari Marie Norgaard. 2010. "Salmon Feeds Our People: Challenging Dams on the Klamath River." In *Indigenous Peoples and Conservation: From Rights to Resource Management*, edited by Kristen Walker Painemilla, Anthony B. Rylands, and Alisa Woofter, 1–10. Arlington: Conservation International.

Reese, Ashanté M. 2019. *Black Food Geographies: Race, Self-Reliance, and Food Access in Washington, D.C.* Chapel Hill: University of North Carolina Press.

Reid, Aileen M., Ayesha S. Boyce, Adeyemo Adetogun, J. R. Moller, and Cherie Avent. 2020. "If Not Us, Then Who? Evaluators of Color and Social Change." *New Directions for Evaluation* 166: 23–36.

Reid, Andrea J., Lauren E. Eckert, John-Francis Lane, Nathan Young, Scott G. Hinch, Chris T. Darimont, Steven J. Cooke, Natalie C. Ban, and Albert Marshall. 2021. "'Two-Eyed Seeing': An Indigenous Framework to Transform Fisheries Research and Management." *Fish and Fisheries* 22 (2): 243–61.

Reitan, Ruth, and Shannon Gibson. 2012. "Climate Change or Social Change? Environmental and Leftist Praxis and Participatory Action Research." *Globalizations* 9 (3): 395–410.

Research Data Alliance International Indigenous Data Sovereignty Interest Group. 2019. "CARE Principles for Indigenous Data Governance." The Global Indigenous Data Alliance. https://www.gida-global.org/care.

Resilient Power Puerto Rico. n.d-a. "Our Work." Accessed January 25, 2022. https://resilientpowerpr.org/our-work.

———. n.d-b. "Resilient Power Puerto Rico." Accessed January 25, 2022. https://static1.squarespace.com/static/59ee55e1f09ca4c5c5f43004/t/60805b8208a0311cb00e6c89/1619024794769/rppr-lookbook.pdf.

Reyes, Diana Alvira, Adriana Bravo Ordoñez, Álvaro del Campo, Richard Chase, Santiago Claramunt, Sebastian Heilpern, Nigel Pitman, et al., eds. 2016. Perú, Medio Putumayo-Algodón. Chicago: Field Museum of Natural History.

Reynolds, Kristin. 2021. "Food, Agriculture, and Environmental Justice: Perspectives on Scholarship and Activism in the Field." In *Environmental Justice: Key Issues*, edited by Brendan Coolsaet, 176–92. New York: Routledge.

Reynolds, Kristin, Daniel Block, and Katharine Bradley. 2018. "Food Justice Scholar-Activism and Activist-Scholarship." *ACME: An International Journal for Critical Geographies* 17 (4): 988–98.

Rhodes, Sarah, K. D. Brown, Larry Cooper, Naeema Muhammad, and Devon Hall. 2020. "Environmental Injustice in North Carolina's Hog Industry: Lessons Learned from Community-Driven Participatory Research and the 'People's Professor.'" In *Toxic Truths: Environmental Justice and Citizen Science in a Post-Truth Age*, edited by Thom Davies and Alice Mah, 99–116. Manchester: Manchester University Press.

Rhodes, Scott D., Lilli Mann, Florence M. Simán, Jorge Alonzo, Aaron T. Vissman, Jennifer Nall, and Amanda E. Tanner. 2017. "Engaged for Change: An Innovative CBPR Strategy to Intervention Development." In *Community-Based Participatory Research for Health: Advancing Social and Health Equity*, edited by Nina Wallerstein, Bonnie Duran, John G. Oetzel, and Meredith Minkler, 3rd ed., 189–206. San Francisco: Jossey-Bass.

Rhodes, Scott D., Aaron T. Vissman, Jason Stowers, Cindy Miller, Thomas P. McCoy, Kenneth C. Hergenrather, Aimee M. Wilkin, et al. 2011. "A CBPR Partnership Increases HIV Testing Among Men Who Have Sex with Men (MSM): Outcome Findings from a Pilot Test of the 'CyBER'/Testing Internet Intervention." *Health Education & Behavior* 38 (3): 311–20.

Richardson, Dawn M., and Amani M. Nuru-Jeter. 2012. "Neighborhood Contexts Experienced by Young Mexican-American Women: Enhancing Our Understanding of Risk for Early Childbearing." *Journal of Urban Health* 89: 59–73.

Rickenbacker, Harold, Fred Brown, and Melissa Bilec. 2019. "Creating Environmental Consciousness in Underserved Communities: Implementation and Outcomes of Community-Based Environmental Justice and Air Pollution Research." *Sustainable Cities and Society* 47: 101473.

Rigolon, Alessandro, and Jeremy Németh. 2020. "Green Gentrification or 'Just Green Enough': Do Park Location, Size and Function Affect Whether a Place Gentrifies or Not?" *Urban Studies* 57 (2): 402–20.

Rinfret, Sara R., and Michelle C. Pautz. 2014. *US Environmental Policy in Action: Practice and Implementation*. New York: Palgrave Macmillan.

Rivera, Felix G., and John Erlich. 1998. *Community Organizing in a Diverse Society*. Boston: Allyn & Bacon.

Rocha, Elizabeth M. 1997. "A Ladder of Empowerment." *Journal of Planning Education and Research* 17 (1): 31–44.

Rocheleau, Dianne, Barbara Thomas-Slayter, and Esther Wangari, eds. 2013. *Feminist Political Ecology: Global Issues and Local Experience*. New York: Routledge.

Rodríguez, Iokiñe. 2021. "Latin American Decolonial Environmental Justice." In *Environmental Justice: Key Issues*, edited by Brendan Coolsaet, 78–93. New York: Routledge.

Rollins, Latrice, Tara Carey, Adrianne Proeller, Mary Anne Adams, Margaret Hooker, Rodney Lyn, Olayiwola Taylor, Kisha Holden, and Tabia Henry Akintobi. 2021. "Community-Based Participatory Approach to Increase African Americans' Access to Healthy Foods in Atlanta, GA." *Journal of Community Health* 46 (1): 41–50.

Romsdahl, Rebecca, Gwendolyn Blue, and Andrei Kirilenko. 2018. "Action on Climate Change Requires Deliberative Framing at Local Governance Level." *Climatic Change* 149 (3): 277–87.

Roos, Michelle, Kathy Pope, and Roseann Stevenson. 2018. *Climate Justice Report: California's Fourth Climate Change Assessment*. Sacramento: California Energy Commission. https://www.energy.ca.gov/data-reports/reports/californias-fourth-climate-change-assessment.

Rosenberg, Marshall B. 2015. *Nonviolent Communication: A Language of Life; Life-Changing Tools for Healthy Relationships*. 3rd ed. Encinitas: PuddleDancer Press.

Rowe, Gene, and Lynn J. Frewer. 2004. "Evaluating Public-Participation Exercises: A Research Agenda." *Science, Technology, & Human Values* 29 (4): 512–56.

Roy, Ananya. 2011. "Commentary: Placing Planning in the World—Transnationalism as Practice and Critique." *Journal of Planning Education and Research* 31 (4): 406–15.

Roy, Bunker, and Jesse Hartigan. 2008. "Empowering the Rural Poor to Develop Themselves: The Barefoot Approach." *Innovations: Technology, Governance, Globalization* 3 (2): 67–93.

Rubin, Carolyn Leung, Linda Sprague Martinez, Jocelyn Chu, Karen Hacker, Doug Brugge, Alex Pirie, Nathan Allukian, Angie Mae Rodday, and Laurel K. Leslie. 2012. "Community-Engaged Pedagogy: A Strengths-Based Approach to Involving Diverse Stakeholders in Research Partnerships." *Progress in Community Health Partnerships* 6 (4): 481–90.

Ryder, Stacia, Kathryn Powlen, Melinda Laituri, Stephanie A. Malin, Joshua Sbicca, and Dimitris Stevis, eds. 2021. *Environmental Justice in the Anthropocene: From (Un)Just Presents to Just Futures*. New York: Routledge.

Sabzwari, Sidra, and Dayna Nadine Scott. 2012. "The Quest for Environmental Justice on a Canadian Aboriginal Reserve." In *Poverty Alleviation and Environmental Law*, edited by Yves Le Bouthillier, Miriam Alfie Cohen, Jose Juan Gonzalez Marquez, Albert Mumma, and Susan Smith, 85–99. Cheltenham: Edward Elgar.

Sadd, James, Rachel Morello-Frosch, Manuel Pastor, Martha Matsuoka, Michele Prichard, and Vanessa Carter. 2014. "The Truth, the Whole Truth, and Nothing but the Ground-Truth: Methods to Advance Environmental Justice and Researcher–Community Partnerships." *Health Education & Behavior* 41 (3): 281–90.

Saez, Marc, and Guillem López-Casasnovas. 2019. "Assessing the Effects on Health Inequalities of Differential Exposure and Differential Susceptibility of Air Pollution and

Environmental Noise in Barcelona, 2007–2014." *International Journal of Environmental Research and Public Health* 16 (18): 3470.

Safo, Stella, Chinazo Cunningham, Alice Beckman, Lorlette Haughton, and Joanna L. Starrels. 2016. "'A Place at the Table': A Qualitative Analysis of Community Board Members' Experiences with Academic HIV/AIDS Research." *BMC Medical Research Methodology* 16 (1): 80.

Salimi, Yahya, Khandan Shahandeh, Hossein Malekafzali, Nina Loori, Azita Kheiltash, Ensiyeh Jamshidi, Ameneh S. Frouzan, and Reza Majdzadeh. 2012. "Is Community-Based Participatory Research (CBPR) Useful? A Systematic Review on Papers in a Decade." *International Journal of Preventive Medicine* 3 (6): 386–93.

Salsberg, Jon, David Parry, Pierre Pluye, Soultana Macridis, Carol P. Herbert, and Ann C. Macaulay. 2015. "Successful Strategies to Engage Research Partners for Translating Evidence into Action in Community Health: A Critical Review." *Journal of Environmental and Public Health* 2015: 191856.

Saltmarsh, John. 2008. "Why Dewey Matters." *The Good Society* 17(2): 63–68.

———. 2010. "Changing Pedagogies." In *Handbook of Engaged Scholarship: Contemporary Landscapes, Future Directions*, edited by Hiram E. Fitzgerald, Cathy Burack, and Sarena Seifer, 331–52. Lansing: Michigan State University Press.

Saltmarsh, John, and Matthew Hartley. 2011. "Democratic Engagement." In *"To Serve a Larger Purpose": Engagement for Democracy and the Transformation of Higher Education*, edited by John Saltmarsh and Matthew Hartley, 14–26. Philadelphia: Temple University Press.

Sampson, Natalie, Joan Nassauer, Amy Schulz, Kathleen Hurd, Cynthia Dorman, and Khalil Ligon. 2017. "Landscape Care of Urban Vacant Properties and Implications for Health and Safety: Lessons from Photovoice." *Health & Place* 46: 219–28.

Sampson, Natalie R., Amy J. Schulz, Edith A. Parker, and Barbara A. Israel. 2014. "Improving Public Participation to Achieve Environmental Justice: Applying Lessons from Freight's Frontline Communities." *Environmental Justice* 7 (2): 45–54.

Sanchez, Ruben Elias Canedo, and Meng L. So. 2015. "UC Berkeley's Undocumented Student Program: Holistic Strategies for Undocumented Student Equitable Success Across Higher Education." *Harvard Educational Review* 85 (3): 464–77.

Sandlos, John, and Arn Keeling. 2016. "Toxic Legacies, Slow Violence, and Environmental Injustice at Giant Mine, Northwest Territories." *Northern Review* 42: 7–21.

Sandoval, Jennifer A., Julie Lucero, John Oetzel, Magdalena Avila, Lorenda Belone, Marjorie Mau, and Nina Wallerstein. 2011. "Process and Outcome Constructs for Evaluating Community Based Participatory Research Projects: A Matrix of Existing Measures and Measurement Tools." *Health Education Research* 27 (4): 680–90.

Sandy Regional Assembly. 2013. "Recovery from the Ground Up: Strategies for Community-Based Resiliency in New York and New Jersey." New York: Sandy Regional Assembly. https://www.issuelab.org/resources/15926/15926.pdf.

Sansom, Garett, Philip Berke, Thomas McDonald, Eva Shipp, and Jennifer Horney. 2016. "Confirming the Environmental Concerns of Community Members Utilizing Participatory-Based Research in the Houston Neighborhood of Manchester." *International Journal of Environmental Research and Public Health* 13 (9): 839.

Saunders, Ruth P., Martin H. Evans, and Praphul Joshi. 2005. "Developing a Process-Evaluation Plan for Assessing Health Promotion Program Implementation: A How-To Guide." *Health Promotion Practice* 6 (2): 134–47.

Sbicca, Joshua. 2018. *Food Justice Now! Deepening the Roots of Social Struggle*. Minneapolis: University of Minnesota Press.

Schlosberg, David. 1999. "Networks and Mobile Arrangements: Organisational Innovation in the US Environmental Justice Movement." *Environmental Politics* 8 (1): 122–48.

———. 2009. *Defining Environmental Justice: Theories, Movements, and Nature*. New York: Oxford University Press.

Schlosberg, David, and Lisette B. Collins. 2014. "From Environmental to Climate Justice: Climate Change and the Discourse of Environmental Justice." *Wiley Interdisciplinary Reviews: Climate Change* 5 (3): 359–74.

Schlosberg, David, Lisette B. Collins, and Simon Niemeyer. 2017. "Adaptation Policy and Community Discourse: Risk, Vulnerability, and Just Transformation." *Environmental Politics* 26 (3): 413–37.

Schwartz, Norah Anita, Christine Alysse Von Glascoe, Victor Torres, Lorena Ramos, and Claudia Soria-Delgado. 2015. "'Where They (Live, Work and) Spray': Pesticide Exposure, Childhood Asthma and Environmental Justice Among Mexican-American Farmworkers." *Health & Place* 32: 83–92.

Schwartz, Sharon, Seth J. Prins, Ulka B. Campbell, and Nicolle M. Gatto. 2016. "Is the 'Well-Defined Intervention Assumption' Politically Conservative?" *Social Science & Medicine* 166: 254–57.

Scott, Shaunna L. 2012. "What Difference Did It Make? The Appalachian Land Ownership Study After Twenty-Five Years." In *Confronting Ecological Crisis in Appalachia and the South: University and Community Partnerships*, edited by Stephanie McSpirit, Lynne Faltraco, and Connor Bailey, 39–60. Lexington: University Press of Kentucky.

Scriven, Michael. 1998. "Minimalist Theory: The Least Theory That Practice Requires." *American Journal of Evaluation* 19 (1): 57–70.

Scurr, Ivy, and Vanessa Bowden. 2021. "'The Revolution's Never Done': The Role of 'Radical Imagination' Within Anti-Capitalist Environmental Justice Activism." *Environmental Sociology* 7 (4): 316–26.

Seeman, Teresa, Elissa Epel, Tara Gruenewald, Arun Karlamangla, and Bruce S. McEwen. 2010. "Socio-Economic Differentials in Peripheral Biology: Cumulative Allostatic Load." *Annals of the New York Academy of Sciences* 1186 (1): 223–39.

Sen, Amartya. 2010. *The Idea of Justice*. London: Penguin Books.

Serrant-Green, Laura. 2002. "Black on Black: Methodological Issues for Black Researchers Working in Minority Ethnic Communities." *Nurse Researcher* 9 (4): 30–44.

Shaffer, Timothy J. 2017. "The Politics of Knowledge: Challenges and Opportunities for Social Justice Work in Higher Education Institutions." *EJournal of Public Affairs* 6 (1): 2.

Shamasunder, Bhavna, and Rachel Morello-Frosch. 2016. "Scientific Contestations Over 'Toxic Trespass': Health and Regulatory Implications of Chemical Biomonitoring." *Journal of Environmental Studies and Sciences* 6: 556–68.

Sharpe, Patricia A., Mary L. Greaney, Peter R. Lee, and Sherer W. Royce. 2000. "Assets-Oriented Community Assessment." *Public Health Reports* 115 (2–3): 205–11.

Shepard, Peggy M., Alma Idehen, Joann Casado, Elmer Freeman, Carol Horowitz, Sarena Seifer, and Hal Strelnick. 2013. "Amplifying the Community Voice in Community–Academic Partnerships: A Summary of and Commentary on a Thematic Issue." *Progress in Community Health Partnerships* 7 (3): 231–33.

———, eds. 2013. "Maximizing Community Contributions, Benefits, and Outcomes in Clinical and Translational Research (Special Issue)." *Progress in Community Health Partnerships* 7 (3).

Shepard, Peggy M., Mary E. Northridge, Swati Prakash, and Gabriel Stover. 2002. "Advancing Environmental Justice Through Community-Based Participatory Research." *Environmental Health Perspectives* 110 (S2): 139–40.

Shiva, Vandana. 2016a. *Biopiracy: The Plunder of Nature and Knowledge*. Berkeley: North Atlantic Books.

———. 2016b. *Staying Alive: Women, Ecology, and Development*. Berkeley: North Atlantic Books.

———. 2016c. *Water Wars: Privatization, Pollution, and Profit*. Berkeley: North Atlantic Books.

Sicotte, Diane M., and Robert J. Brulle. 2018. "Social Movements for Environmental Justice Through the Lens of Social Movement Theory." In *The Routledge Handbook of Environmental Justice*, edited by Ryan Holifield, Jayajit Chakraborty, and Gordon Walker, 25–36. New York: Routledge.

Sikor, Thomas, and Peter Newell. 2014. "Globalizing Environmental Justice?" *Geoforum* 54: 151–57.

Simckes, Maayan, Dale Willits, Michael McFarland, Cheryl McFarland, Ali Rowhani-Rahbar, and Anjum Hajat. 2021. "The Adverse Effects of Policing on Population Health: A Conceptual Model." *Social Science & Medicine* 281: 114103.

Simmons, Vani Nath, Lynne B. Klasko, Khaliah Fleming, Alexis M. Koskan, Nia T. Jackson, Shalewa Noel-Thomas, John S. Luque, et al. 2015. "Participatory Evaluation of a Community-Academic Partnership to Inform Capacity-Building and Sustainability." *Evaluation and Program Planning* 52: 19–26.

Simpson, Leanne Betasamosake. 2017. *As We Have Always Done: Indigenous Freedom through Radical Resistance*. Minneapolis: University of Minnesota Press.

Sladek, Emily, ed. 2019. "The Transformative Power of Anchor Institutions." *Metropolitan Universities* 30 (S1). https://journals.iupui.edu/index.php/muj/issue/view/1264.

Smith, Kimberly, and Black Mesa Water Coalition. 2007. "Pollution of the Navajo Nation Lands." Paper presented at the United Nations International Expert Group Meeting on Indigenous Peoples and Protection of the Environment, Khabarovsk, Russian Federation, August.

Smith, Linda F. 2004. "Why Clinical Programs Should Embrace Civic Engagement, Service Learning and Community Based Research." *Clinical Law Review* 10 (2): 723–54.

Smith, Linda Tuhiwai. 2021. *Decolonizing Methodologies: Research and Indigenous Peoples*. 3rd ed. London: Zed Books.

Smith, Selina A., Benjamin E. Ansa, and Daniel S. Blumenthal. 2017. "Colorectal Cancer Disparities and Community-Based Participatory Research." In *Handbook of Community-Based Participatory Research*, edited by Steven S. Coughlin, Selina A. Smith, and Maria E. Fernández, 167–84. Oxford: Oxford University Press.

Smith, Ted, and Chad Raphael. 2015. "The future of activism for electronics workers." In *The Routledge Companion to Labor and Media*, edited by Rick Maxwell, 78–90. New York: Routledge.

Smith, Ted, David A. Sonnenfeld, and David N. Pellow, eds. 2006. *Challenging the Chip: Labor Rights and Environmental Justice in the Global Electronics Industry*. Philadelphia: Temple University Press.

Soergel, Allison Arteaga. 2021. "Research Partnership Will Highlight STEM Learning in Local Community Garden." *UC Santa Cruz News* (blog). June 9, 2021. https://news.ucsc.edu/2021/06/calabasas-garden.html.

Solis, Miriam, Will Davies, and Abby Randall. 2022. "Climate Justice Pedagogies in Green Building Curriculum." *Curriculum Inquiry* 52 (2): 235–49.

Solomon, Gina M., Rachel Morello-Frosch, Lauren Zeise, and John B. Faust. 2016. "Cumulative Environmental Impacts: Science and Policy to Protect Communities." *Annual Review of Public Health* 37: 83–96.

Southwest Organizing Project. 1990. "Letter to Big Ten Environmental Groups." Southwest Community Resources. March 16, 1990. http://www.ejnet.org/ej/swop.pdf.

Sowerwine, Jennifer, Megan Mucioki, Daniel Sarna-Wojcicki, and Lisa Hillman. 2019. "Reframing Food Security by and for Native American Communities: A Case Study Among Tribes in the Klamath River Basin of Oregon and California." *Food Security* 11 (3): 579–607.

Sowerwine, Jennifer, Daniel Sarna-Wojcicki, Megan Mucioki, Lisa Hillman, Frank Lake, and Edith Friedman. 2019. "Enhancing Indigenous Food Sovereignty: A Five-Year Collaborative Tribal-University Research and Extension Project in California and Oregon." *Journal of Agriculture, Food Systems, and Community Development* 9 (S2): 167–90.

Spence, Mark David. 1999. *Dispossessing the Wilderness: Indian Removal and the Making of the National Parks*. Oxford: Oxford University Press.

Spiegel, Samuel J., Sarah Thomas, Kevin O'Neill, Cassandra Brondgeest, Jen Thomas, Jiovanni Beltran, Terena Hunt, and Annalee Yassi. 2020. "Visual Storytelling, Intergenerational Environmental Justice and Indigenous Sovereignty: Exploring Images and Stories Amid a Contested Oil Pipeline Project." *International Journal of Environmental Research and Public Health* 17 (7): 2362.

Stack, Erin E., and Katherine McDonald. 2018. "We Are 'Both in Charge, the Academics and Self-Advocates': Empowerment in Community-Based Participatory Research." *Journal of Policy and Practice in Intellectual Disabilities* 15 (1): 80–89.

Standing Strong (Mash Koh Wee Kah Pooh Win) Task Force. 2021. "Standing Strong Task Force Report and Recommendations: Acknowledging the Past, Learning from the Present, Looking to the Future." Toronto: Ryerson University. https://www.ryerson.ca/content/dam/next-chapter/Report/SSTF-report-and-recommendations-Aug_24_FINAL.pdf.

Stanton, Timothy K. 2008. "New Times Demand New Scholarship: Opportunities and Challenges for Civic Engagement at Research Universities." *Education, Citizenship and Social Justice* 3 (1): 19–42.

Staples, Lee. 2016. *Roots to Power: A Manual for Grassroots Organizing*. 3rd ed. Santa Barbara: Praeger.

Stephenson Jr., Max, and Lisa A. Schweitzer. 2011. "Learning from the Quest for Environmental Justice in the Niger River Delta." In *Environmental Inequalities Beyond Borders: Local Perspectives on Global Injustices*, edited by JoAnn Carmin and Julian Agyeman, 45–66. Cambridge: MIT Press.

Sterling, Eleanor J., Christopher Filardi, Anne Toomey, Amanda Sigouin, Erin Betley, Nadav Gazit, Jennifer Newell, et al. 2017. "Biocultural Approaches to Well-Being and Sustainability Indicators Across Scales." *Nature Ecology & Evolution* 1 (12): 1798–1806.

Sterling, Eleanor, Tamara Ticktin, Tē Kipa Kepa Morgan, Georgina Cullman, Diana Alvira, Pelika Andrade, Nadia Bergamini, et al. 2017. "Culturally Grounded Indicators of Resilience in Social-Ecological Systems." *Environment and Society* 8 (1): 63–95.

Strauss, Ronald P., Sohini Sengupta, Sandra Crouse Quinn, Jean Goeppinger, Cora Spaulding, Susan M. Kegeles, and Greg Millett. 2001. "The Role of Community Advisory Boards: Involving Communities in the Informed Consent Process." *American Journal of Public Health* 91 (12): 1938–43.

Su, Celina, Maddy Fox, Anna Ortega-Williams, Catherine McBride, and Red Hook Initiative. 2018. "Towards New Ethics Protocols for Community-Based Research and Community Research Ethics in Red Hook." New York: URBAN Research Network. https://urbanresearchnetwork.org/wp-content/uploads/2018/06/community-protocols-urban.pdf.

Suiseeya, Kimberly R. Marion. 2021. "Procedural Justice Matters: Power, Representation, and Participation in Environmental Governance." In *Environmental Justice: Key Issues*, edited by Brendan Coolsaet, 37–51. New York: Routledge.

Suiter, Sarah V., Amie Thurber, and Clare Sullivan. 2016. "A Co-Learning Model for Community-Engaged Program Evaluation." *Progress in Community Health Partnerships* 10 (4): 551–58.

Sullivan, John, and Juan Parras. 2008. "Environmental Justice and Augusto Boal's Theatre of the Oppressed: A Unique Community Tool for Outreach, Communication, Education and Advocacy." *Theory in Action* 1 (2): 20–39.

Sultana, Farhana. 2007. "Reflexivity, Positionality and Participatory Ethics: Negotiating Fieldwork Dilemmas in International Research." *ACME: An International Journal for Critical Geographies* 6 (3): 374–85.

Suman, Anna and Sven Schade. 2021. "The Formosa Case: A Step Forward on the Acceptance of Citizen-Collected Evidence in Environmental Litigation?" *Citizen Science: Theory and Practice*, 6 (1).

Sun, Yeran, and Amin Mobasheri. 2017. "Utilizing Crowdsourced Data for Studies of Cycling and Air Pollution Exposure: A Case Study Using Strava Data." *International Journal of Environmental Research and Public Health* 14 (3): 274.

Swiftwolfe, Dakota. 2019. "Indigenous Ally Toolkit." Montreal: Montreal Urban Aboriginal Community Strategy Network. http://reseaumtlnetwork.com/wp-content/uploads/2019/04/Ally_March.pdf.

Symanski, Elaine, Heyreoun An Han, Loren Hopkins, Mary Ann Smith, Sheryl McCurdy, Inkyu Han, Maria Jimenez, et al. 2020. "Metal Air Pollution Partnership Solutions: Building an Academic-Government-Community-Industry Collaboration to Improve Air Quality and Health in Environmental Justice Communities in Houston." *Environmental Health* 19: 39.

Sze, Julie. 2004. "Asian American Activism for Environmental Justice." *Peace Review* 16 (2): 149–56.

———. 2007. *Noxious New York: The Racial Politics of Urban Health and Environmental Justice*. Cambridge: MIT.

———. 2020. *Environmental Justice in a Moment of Danger*. Oakland: University of California Press.

Sze, Julie, and Jonathan K. London. 2008. "Environmental Justice at the Crossroads." *Sociology Compass* 2 (4): 1331–54.

Tachine, Amanda R., and Nolan L. Cabrera. 2021. "'I'll Be Right Behind You': Native American Families, Land Debt, and College Affordability." *AERA Open* 7 (1). https://journals.sagepub.com/doi/full/10.1177/23328584211025522.

Tajik, Mansoureh. 2012. "Environmental Justice from the Roots: Tillery, North Carolina." In *Confronting Ecological Crisis in Appalachia and the South: University and Community Partnerships*, edited by Stephanie McSpirit, Lynne Faltraco, and Connor Bailey, 131–46. Lexington: University Press of Kentucky.

Tajik, Mansoureh, and Meredith Minkler. 2006. "Environmental Justice Research and Action: A Case Study in Political Economy and Community-Academic Collaboration." *International Quarterly of Community Health Education* 26 (3): 213–31.

Tandon, Rajesh. 1996. "The Historical Roots and Contemporary Tendencies in Participatory Research: Implications for Health Care." In *Participatory Research in Health: Issues and Experiences*, edited by Korrie De Koning and Marrion Martin, 19–26. London: Zed Books.

Tao, Yinhua, Lirong Kou, Yanwei Chai, and Mei-Po Kwan. 2021. "Associations of Co-Exposures to Air Pollution and Noise with Psychological Stress in Space and Time: A Case Study in Beijing, China." *Environmental Research* 196: 110399.

Tarus, Lyndsay, Mary Hufford, and Betsy Taylor. 2017. "A Green New Deal for Appalachia: Economic Transition, Coal Reclamation Costs, Bottom-Up Policymaking (Part 2)." *Journal of Appalachian Studies* 23 (2): 151–69.

Taylor, Dorceta E. 1997. "American Environmentalism: The Role of Race, Class and Gender in Shaping Activism 1820–1995." *Race, Gender & Class* 5 (1): 16–62.

———. 2000. "The Rise of the Environmental Justice Paradigm: Injustice Framing and the Social Construction of Environmental Discourses." *American Behavioral Scientist* 43 (4): 508–80.

———. 2009. *The Environment and the People in American Cities, 1600s-1900s: Disorder, Inequality, and Social Change*. Durham: Duke University Press.

———. 2016. *The Rise of the American Conservation Movement: Power, Privilege, and Environmental Protection*. Durham: Duke University Press.

———. 2018. "Black Farmers in the USA and Michigan: Longevity, Empowerment, and Food Sovereignty." *Journal of African American Studies* 22: 49–76.

Temper, Leah. 2018. "Globalizing Environmental Justice: Radical and Transformative Movements Past and Present." In *The Routledge Handbook of Environmental Justice*, edited by Ryan Holifield, Jayajit Chakraborty, and Gordon Walker, 490–503. New York: Routledge.

Tervalon, Melanie, and Jann Murray-García. 1998. "Cultural Humility Versus Cultural Competence: A Critical Distinction in Defining Physician Training Outcomes in Multicultural Education." *Journal of Health Care for the Poor and Underserved* 9 (2): 117–25.

Tickner, Joel, Christopher Weis, and Molly Jacobs. 2017. "Alternatives Assessment: New Ideas, Frameworks and Policies." *Journal of Epidemiology and Community Health* 71 (7): 655–56.

Todd, Zoe. 2014. "Fish Pluralities: Human-Animal Relations and Sites of Engagement in Paulatuuq, Arctic Canada." *Études Inuit / Inuit Studies* 38 (1–2): 217–38.

Town Charts. 2021. "Salinas, California Education Data." Accessed November 23, 2021. https://www.towncharts.com/California/Education/Salinas-city-CA-Education-data.html.

Townsend, Claire K. M., Adrienne Dillard, Kelsea K. Hosoda, Gregory G. Maskarinec, Alika K. Maunakea, Sheryl R. Yoshimura, Claire Hughes, Donna-Marie Palakiko, Bridget Puni Kehauoha, and Joseph Keawe'aimoku Kaholokula. 2016. "Community-Based Participatory Research Integrates Behavioral and Biological Research to Achieve Health Equity for Native Hawaiians." *International Journal of Environmental Research and Public Health* 13 (1): 4.

Tran, Emma, Kim Blankenship, Shannon Whittaker, Alana Rosenberg, Penelope Schlesinger, Trace Kershaw, and Danya Keene. 2020. "My Neighborhood Has a Good Reputation: Associations Between Spatial Stigma and Health." *Health & Place* 64: 102392.

Tremblay, Crystal, and Bruno de Oliveira Jayme. 2015. "Community Knowledge Co-Creation Through Participatory Video." *Action Research* 13 (3): 298–314.

Tuck, Eve. 2009. "Suspending Damage: A Letter to Communities." *Harvard Educational Review* 79 (3): 409–27.

Tuck, Eve, and Marcia McKenzie. 2014. *Place in Research: Theory, Methodology, and Methods*. New York: Routledge.

Tuck, Eve, and K. Wayne Yang. 2012. "Decolonization Is Not a Metaphor." *Decolonization: Indigeneity, Education & Society* 1 (1): 1–40.

Tucker, Carolyn M., Jaime L. Williams, Julia Roncoroni, and Martin Heesacker. 2017. "A Socially Just Leadership Approach to Community-Partnered Research for Reducing Health Disparities." *The Counseling Psychologist* 45 (6): 781–809.

Turney, Kristin, and Dylan B. Jackson. 2021. "Mothers' Health Following Youth Police Stops." *Preventive Medicine* 150: 106693.

UCDEHSC and UMLEEDC (University of California, Davis Environmental Health Sciences Center and University of Michigan Lifestage Environmental Exposures and Disease Center). 2018. "Building Equitable Partnerships for Environmental Justice: Curriculum." Davis and Ann Arbor: UCDEHSC and UMLEEDC. https://environmentalhealth.ucdavis.edu/sites/g/files/dgvnsk2556/files/inline-files/building-equitable-partnerships-for-environmental-justice.pdf.

UCLA Abolitionist Planning Group. 2018. "Abolitionist Planning for Resistance." Los Angeles: UCLA Institute on Inequality and Democracy. https://challengeinequality.luskin.ucla.edu/wp-content/uploads/sites/16/2017/05/AboPlan_Pub_FINAL_online-v2-1.pdf.

Ulrich, Mary Eileen. 2016. "Learning Relational Ways of Being: What Globally Engaged Scholars Have Learned About Global Engagement and Sustainable Community Development." PhD diss., Montana State University.

United Frontline Table. 2020. "A People's Orientation to a Regenerative Economy: Protect, Repair, Invest, and Transform." https://climatejusticealliance.org/wp-content/uploads/2020/06/ProtectRepairInvestTransformdoc24s.pdf.

United Nations Development Programme. 2018. "Human Development Indices and Indicators: 2018 Statistical Update." New York: United Nations Development Programme. http://hdr.undp.org/sites/default/files/2018_human_development_statistical_update.pdf.

United Nations Environment Programme. 2011. "Towards a Green Economy: Pathways to Sustainable Development and Poverty Eradication; A Synthesis for Policy Makers." Nairobi: United Nations Environment Programme. https://sustainabledevelopment.un.org/content/documents/126GER_synthesis_en.pdf.

UPROSE. n.d. "The GRID." Accessed December 20, 2021. https://www.uprose.org/the-grid.

UpstreamPublicHealth.2015."TobaccoRetailLicensingPolicy:AHealthEquityImpactAssessment." https://www.pewtrusts.org/-/media/assets/external-sites/health-impact-project/upstream-2015-tobacco-licensing-report.pdf.

Urban Planners for Liberation. 2021. "Urban Planners for Liberation Resource Guide and Framework." July. www.planforliberation.space.

Urkidi, Leire, and Mariana Walter. 2018. "Environmental Justice and Large-Scale Mining." In *The Routledge Handbook of Environmental Justice*, edited by Ryan Holifield, Jayajit Chakraborty, and Gordon Walker, 374–87. New York: Routledge.

U.S. Centers for Disease Control. n.d. "Original Essential Public Health Services Framework." Accessed February 4, 2022. https://www.cdc.gov/publichealthgateway/publichealthservices/originalessentialhealthservices.html.

U.S. Environmental Protection Agency. 2014. "The Road to Executive Order 12898 on Environmental Justice." YouTube video, 3:36. https://youtu.be/Sx93yKLxSyk.

———. 2015. "EPA Strategic Directions on Using Citizen Science for Environmental Protection." Washington, DC: U.S. Environmental Protection Agency. https://www.epa.gov/sites/default/files/2015-09/documents/nacept_charge_on_citizen_science_final.pdf.

———. 2016. "Environmental Justice Research Roadmap." Washington, DC: U.S. Environmental Protection Agency. https://www.epa.gov/sites/production/files/2017-01/documents/researchroadmap_environmentaljustice_508_compliant.pdf.

U.S. General Accounting Office. 1983. *Siting of Hazardous Waste Landfills and Their Correlation with Racial and Economic Status of Surrounding Communities*. Washington, DC: U.S. General Accounting Office.

Valley, Will, Molly Anderson, Nicole Tichenor Blackstone, Eleanor Sterling, Erin Betley, Sharon Akabas, Pamela Koch, Colin Dring, Joanne Burke, and Karen Spiller. 2020. "Towards an Equity Competency Model for Sustainable Food Systems Education Programs." *Elementa: Science of the Anthropocene* 8: 33.

Van Brakel, Martin L., Md. Nahiduzzaman, A. B. M. Mahfuzul Haque, Md. Golam Mustafa, Md. Jalilur Rahman, and Md. Abdul Wahab. 2018. "Reimagining Large-Scale Open-Water Fisheries Governance Through Adaptive Comanagement in Hilsa Shad Sanctuaries." *Ecology and Society* 23 (1): 26.

van Daalen, Kim, Laura Jung, Roopa Dhatt, and Alexandra L. Phelan. 2020. "Climate Change and Gender-Based Health Disparities." *The Lancet Planetary Health* 4 (2): 44–45.

Van der Arend, Sonja. 2018. "Really Imagined: Policy Novels as a Mode of Action Research." In *Action Research in Policy Analysis*, edited by Koen P. R. Bartels and Julia M. Wittmayer, 225–44. New York: Routledge.

Vanderwarker, Amy. 2012. "Water and Environmental Justice." In *A Twenty-First Century U.S. Water Policy*, edited by Juliet Christian-Smith, Peter H. Gleick, Heather Cooley, Lucy Allen, Amy Vanderwarker, and Kate A. Berry, 52–89. Oxford: Oxford University Press.

Varese, Stefano. 2006. *Witness to Sovereignty: Essays on the Indian Movement in Latin America*. Copenhagen: International Work Group for Indigenous Affairs.

Vaughn, Lisa M., and Farrah Jacquez. 2017. "Community-Based Participatory Research Studies Involving Immigrants." In *Handbook of Community-Based Participatory Research*, edited by Steven S. Coughlin, Selina A. Smith, and Maria E. Fernández, 115–30. Oxford: Oxford University Press.

Vaughn, Lisa M., Crystal Whetstone, Alicia Boards, Melida D. Busch, Maria Magnusson, and Sylvia Määttä. 2018. "Partnering with Insiders: A Review of Peer Models Across Community-Engaged Research, Education and Social Care." *Health & Social Care in the Community* 26 (6): 769–86.

Velasco, Gabriella, Caroline Faria, and Jayme Walenta. 2020. "Imagining Environmental Justice 'Across the Street': Zine-Making as Creative Feminist Geographic Method." *GeoHumanities* 6 (2): 347–70.

Vera, Lourdes A., Dawn Walker, Michelle Murphy, Becky Mansfield, Ladan Mohamed Siad, and Jessica Ogden. 2019. "When Data Justice and Environmental Justice Meet: Formulating a Response to Extractive Logic Through Environmental Data Justice." *Information, Communication & Society* 22 (7): 1012–28.

Verplanke, Jeroen, Michael K. McCall, Claudia Uberhuaga, Giacomo Rambaldi, and Muki Haklay. 2016. "A Shared Perspective for PGIS and VGI." *The Cartographic Journal* 53 (4): 308–17.

Vickery, Jamie, and Lori M. Hunter. 2016. "Native Americans: Where in Environmental Justice Research?" *Society & Natural Resources* 29 (1): 36–52.

Villarejo, Don, David Lighthall, Daniel Williams, Ann Souter, Richard Mines, Bonnie Bade, Steve Samuels, and Stephen A. McCurdy. 2000. "Suffering in Silence: A Report on the Health of California's Agricultural Workers." Davis, CA: California Institute for Rural Studies. https://donvillarejo.github.io/Fulltext/Suffering_in_Silence_report.pdf.

Vineis, Paolo, Cyrille Delpierre, Raphaële Castagné, Giovanni Fiorito, Cathal McCrory, Mika Kivimaki, Silvia Stringhini, Cristian Carmeli, and Michelle Kelly-Irving. 2020. "Health Inequalities: Embodied Evidence across Biological Layers." *Social Science & Medicine* 246: 112781.

Vogelgesang, Lori J., Nida Denson, and Uma M. Jayakumar. 2010. "What Determines Faculty-Engaged Scholarship?" *The Review of Higher Education* 33 (4): 437–72.

Voida, Amy, Ellie Harmon, Willa Weller, Aubrey Thornsbury, Ariana Casale, Samuel Vance, Forrest Adams, et al. 2017. "Competing Currencies: Designing for Politics in Units of Measurement." In *Proceedings of the 2017 ACM Conference on Computer Supported Cooperative Work and Social Computing*, 847–60. Portland: ACM.

Wali, Alaka, Diana Alvira, Paula S. Tallman, Ashwin Ravikumar, and Miguel O. Macedo. 2017. "A New Approach to Conservation: Using Community Empowerment for Sustainable Well-Being." *Ecology and Society* 22 (4): 6.

Walker, Gordon. 2012. *Environmental Justice: Concepts, Evidence and Politics*. New York: Routledge.

Wallerstein, Nina, and Bonnie Duran. 2017. "Theoretical, Historical, and Practice Roots of CBPR." In *Community-Based Participatory Research for Health: Advancing Social and Health Equity*, edited by Nina Wallerstein, Bonnie Duran, John G. Oetzel, and Meredith Minkler, 3rd ed., 17–30. San Francisco: Jossey-Bass.

Wallerstein, Nina, Bonnie Duran, John G. Oetzel, and Meredith Minkler, eds. 2017. *Community-Based Participatory Research for Health: Advancing Social and Health Equity*. 3rd ed. San Francisco: Jossey-Bass.

Wallerstein, Nina, Michael Muhammad, Shannon Sanchez-Youngman, Patricia Rodriguez Espinosa, Magdalena Avila, Elizabeth A. Baker, Steven Barnett, et al. 2019. "Power Dynamics in Community-Based Participatory Research: A Multiple–Case Study Analysis of Partnering Contexts, Histories, and Practices." *Health Education & Behavior* 46 (S1): 19–32.

Wallerstein, Nina B., Irene H. Yen, and S. Leonard Syme. 2011. "Integration of Social Epidemiology and Community-Engaged Interventions to Improve Health Equity." *American Journal of Public Health* 101 (5): 822–30.

Warner, Keith Douglass. 2008. "Agroecology as Participatory Science: Emerging Alternatives to Technology Transfer Extension Practice." *Science, Technology, & Human Values* 33 (6): 754–77.

Warren, Mark R. 2018. "Research Confronts Equity and Social Justice–Building the Emerging Field of Collaborative, Community Engaged Education Research: Introduction to the Special Issue." *Urban Education* 53 (4): 439–44.

Warren, Mark R., José Calderón, Luke Aubry Kupscznk, Gregory Squires, and Celina Su. 2018. "Is Collaborative, Community-Engaged Scholarship More Rigorous Than Traditional Scholarship? On Advocacy, Bias, and Social Science Research." *Urban Education* 53 (4): 445–72.

Watene, Krushil. 2016. "Valuing Nature: Māori Philosophy and the Capability Approach." *Oxford Development Studies* 44 (3): 287–96.

WE ACT for Environmental Justice. n.d. "Planning." Accessed December 20, 2021. https://www.weact.org/whatwedo/planning/.

Welch, Marshall. 2016. *Engaging Higher Education: Purpose, Platforms, and Programs for Community Engagement*. Sterling: Stylus.

Westley, Frances. 2013. "Social Innovation and Resilience: How One Enhances the Other." *Stanford Social Innovation Review* 11 (3): 28–39.

Wheeler, Stephen M., and Timothy Beatley, eds. 2014. *Sustainable Urban Development Reader*. 3rd ed. New York: Routledge.

White, Monica M. 2011. "D-Town Farm: African American Resistance to Food Insecurity and the Transformation of Detroit." *Environmental Practice* 13 (4): 406–17.

———. 2018. *Freedom Farmers: Agricultural Resistance and the Black Freedom Movement*. Chapel Hill: University of North Carolina Press.

Whyte, Kyle. 2018a. "The Recognition Paradigm of Environmental Injustice." In *The Routledge Handbook of Environmental Justice*, edited by Ryan Holifield, Jayajit Chakraborty, and Gordon Walker, 113–23. New York: Routledge.

———. 2018b. "What Do Indigenous Knowledges Do for Indigenous Peoples?" In *Traditional Ecological Knowledge: Learning from Indigenous Practices for Environmental Sustainability*, edited by Melissa K. Nelson and Dan Shilling, 57–81. Cambridge: Cambridge University Press.

———. 2020. "Too Late for Indigenous Climate Justice: Ecological and Relational Tipping Points." *Wiley Interdisciplinary Reviews: Climate Change* 11 (1): 603.

———. 2021. "Indigenous Environmental Justice: Anti-Colonial Action Through Kinship." In *Environmental Justice: Key Issues*, edited by Brendan Coolsaet, 266–78. New York: Routledge.

Widener, Michael J., Leia M. Minaker, Jessica L. Reid, Zachary Patterson, Tara Kamal Ahmadi, and David Hammond. 2018. "Activity Space-Based Measures of the Food Environment and Their Relationships to Food Purchasing Behaviours for Young Urban Adults in Canada." *Public Health Nutrition* 21 (11): 2103–16.

Wiggins, Noelle, Laura Chanchien Parajon, Chris M. Coombe, Aileen Alfonso Duldulao, Leticia Rodriguez Garcia, and Pei-Ru Wang. 2017. "Participatory Evaluation as a Process of Empowerment." In *Community-Based Participatory Research for Health: Advancing*

Social and Health Equity, edited by Nina Wallerstein, Bonnie Duran, John G. Oetzel, and Meredith Minkler, 251–64. San Francisco: Jossey-Bass.

Wild, Christopher P. 2005. "Complementing the Genome with an 'Exposome': The Outstanding Challenge of Environmental Exposure Measurement in Molecular Epidemiology." *Cancer Epidemiology, Biomarkers & Prevention* 14: 1847–50.

Wilkins, Karin Gwinn Wilkins, Peter Nsubuga, John Mendlein, David F. Mercer, and Marguerite Pappaioanou. 2008. "The Data for Decision Making Project: Assessment of Surveillance Systems in Developing Countries to Improve Access to Public Health Information." *Public Health* 122 (9): 914–22.

Wilkinson, Mark D., Michel Dumontier, IJsbrand Jan Aalbersberg, Gabrielle Appleton, Myles Axton, Arie Baak, Niklas Blomberg, et al. 2016. "The FAIR Guiding Principles for Scientific Data Management and Stewardship." *Scientific Data* 3: 160018.

Willett, Jennifer, Alonso Tamayo, and Alexi Rayo. 2021. "Making the Invisible Visible: Documenting Slow Violence Through Photovoice with Youth in Nevada." *Journal of Community Practice* 29 (2): 112–32.

Willow, Anna J. 2013. "Doing Sovereignty in Native North America: Anishinaabe Counter-Mapping and the Struggle for Land-Based Self-Determination." *Human Ecology* 41 (6): 871–84.

Wilson, Elena, Amanda Kenny, and Virginia Dickson-Swift. 2018. "Ethical Challenges in Community-Based Participatory Research: A Scoping Review." *Qualitative Health Research* 28 (2): 189–99.

Wilson, Nicole J., Edda Mutter, Jody Inkster, and Terre Satterfield. 2018. "Community-Based Monitoring as the Practice of Indigenous Governance: A Case Study of Indigenous-Led Water Quality Monitoring in the Yukon River Basin." *Journal of Environmental Management* 210: 290–98.

Wilson, Paul M., Mark Petticrew, Mike W. Calnan, and Irwin Nazareth. 2010. "Disseminating Research Findings: What Should Researchers Do? A Systematic Scoping Review of Conceptual Frameworks." *Implementation Science* 5: 91.

Wilson, Sacoby, Aaron Aber, Lindsey Wright, and Vivek Ravichandran. 2018. "A Review of Community-Engaged Research Approaches Used to Achieve Environmental Justice and Eliminate Disparities." In *The Routledge Handbook of Environmental Justice*, edited by Ryan Holifield, Jayajit Chakraborty, and Gordon Walker, 283–96. New York: Routledge.

Wilson, Shawn. 2008. *Research Is Ceremony: Indigenous Research Methods*. Winnipeg: Fernwood.

Windchief, Sweeney, and Timothy San Pedro, eds. 2019. *Applying Indigenous Research Methods: Storying with Peoples and Communities*. New York: Routledge.

Wing, Steve. 2002. "Social Responsibility and Research Ethics in Community-Driven Studies of Industrialized Hog Production." *Environmental Health Perspectives* 110 (5): 437–44.

Wing, Steve, Dana Cole, and Gary Grant. 2000. "Environmental Injustice in North Carolina's Hog Industry." *Environmental Health Perspectives* 108 (3): 225–31.

Wing, Steve, Rachel Avery Horton, Naeema Muhammad, Gary R. Grant, Mansoureh Tajik, and Kendall Thu. 2008. "Integrating Epidemiology, Education, and Organizing for Environmental Justice: Community Health Effects of Industrial Hog Operations." *American Journal of Public Health* 98 (8): 1390–97.

Wing, Steve, Rachel Avery Horton, and Kathryn M. Rose. 2013. "Air Pollution from Industrial Swine Operations and Blood Pressure of Neighboring Residents." *Environmental Health Perspectives* 121 (1): 92–96.

Winterbauer, Nancy L., Kathy C. Garrett, Samantha Hyde, Valerie Feinberg, Laureen Husband, Karen Landry, and James E. Sylvester. 2013. "A Communications Tool to Recruit Policymakers to a CBPR Partnership for Childhood Obesity Prevention." *Progress in Community Health Partnerships* 7 (4): 443–49.

Wolch, Jennifer R., Jason Byrne, and Joshua P. Newell. 2014. "Urban Green Space, Public Health, and Environmental Justice: The Challenge of Making Cities 'Just Green Enough.'" *Landscape and Urban Planning* 125: 234–44.

Wolferman, Nicholas, Trendha Hunter, Jennifer S. Hirsch, Shamus R. Khan, Leigh Reardon, and Claude A. Mellins. 2019. "The Advisory Board Perspective from a Campus Community-Based Participatory Research Project on Sexual Violence." *Progress in Community Health Partnerships* 13 (1): 115–19.

Wong, David W. S., and Shih-Lung Shaw. 2011. "Measuring Segregation: An Activity Space Approach." *Journal of Geographical Systems* 13 (2): 127–45.

Woodbury, R. Brian, Julie A. Beans, Vanessa Y. Hiratsuka, and Wylie Burke. 2019. "Data Management in Health-Related Research Involving Indigenous Communities in the United States and Canada: A Scoping Review." *Frontiers in Genetics* 10: 942.

Woodford, Michael R., Christopher L. Kolb, Gabrielle Durocher-Radeka, and Gabe Javier. 2014. "Lesbian, Gay, Bisexual, and Transgender Ally Training Programs on Campus: Current Variations and Future Directions." *Journal of College Student Development* 55 (3): 317–22.

Work, Courtney, Arnim Scheidel, Ida Theilade, Sen Sothea, and Danik Song. 2021. "Engaged Research Uncovers the Grey Areas and Trade-Offs in Climate Justice." In *Indigenous Peoples, Heritage and Landscape in the Asia Pacific: Knowledge Co-Production and Empowerment*, edited by Stephen Acabado and Da-Wei Kuan, 16–30. New York: Routledge.

World Commission on Environment and Development. 1987. *Our Common Future: Report of the World Commission on Environment and Development*. Oxford: Oxford University Press.

World Health Organization. 2014. *Implementation Research Toolkit: Workbook*. Geneva: World Health Organization Special Programme for Research and Training in Tropical Diseases.

Wright, Walter, Kathryn W. Hexter, and Nick Downer. 2016. "Cleveland's Greater University Circle Initiative: An Anchor-Based Strategy for Change." *Maxine Goodman Levin School of Urban Affairs*. https://engagedscholarship.csuohio.edu/urban_facpub/1360.

Wyeth, George, LeRoy C. Paddock, Alison Parker, Robert L. Glicksman, and Jecoliah Williams. 2019. "The Impact of Citizen Environmental Science in the United States." *Environmental Law Reporter* 49 (3): 10237–9.

Wyndham, Felice S. 2017. "The Trouble with TEK." *Ethnobiology Letters* 8 (1): 78–80.

Yonas, Michael, Robert Aronson, Nettie Coad, Eugenia Eng, Regina Petteway, Jennifer Schaal, and Lucille Webb. 2013. "Infrastructure for Equitable Decision-Making in Research." In *Methods for Community-Based Participatory Research for Health*, edited by Barbara A. Israel, Eugenia Eng, Amy J. Schulz, and Edith A. Parker, 2nd ed., 97–126. San Francisco: Jossey-Bass.

Yosso, Tara J. 2005. "Whose Culture Has Capital? A Critical Race Theory Discussion of Community Cultural Wealth." *Race Ethnicity and Education* 8 (1): 69–91.

Young, Iris Marion. 2000. *Inclusion and Democracy*. Oxford: Oxford University Press.

Zakocs, Ronda, Jessica A. Hill, Pamela Brown, Jocelyn Wheaton, and Kimberley E. Freire. 2015. "The Data-to-Action Framework: A Rapid Program Improvement Process." *Health Education & Behavior* 42 (4): 471–9.

Zúñiga, Ximena, Gretchen Lopez, and Kristie A. Ford, eds. 2014. *Intergroup Dialogue: Engaging Difference, Social Identities and Social Justice.* New York: Routledge.

LIST OF CONTRIBUTORS

ANA ISABEL BAPTISTA is Associate Professor in Environmental Policy and Sustainability Management, and Co-Director of the Tishman Environment and Design Center, at The New School.

FLORIDALMA BOJ LOPEZ is Assistant Professor of Chicana/o and Central American Studies at University of California, Los Angeles.

ZSEA BOWMANI is Assistant Professor of Law at University of Toledo College of Law.

CELESTINA CASTILLO is Executive Director of the Center for Community Based Learning at Occidental College and a Ph.D. student in Gender Studies with a concentration in American Indian Studies at University of California, Los Angeles.

VERA L. CHANG is a Ph.D. Candidate in Environmental Science, Policy, and Management at University of California, Berkeley.

SARAH COMMODORE is Assistant Professor in the School of Public Health at Indiana University Bloomington.

JEANYNA GARCIA earned the Bachelor's degree in Environmental Studies at Amherst College.

MALAYA JULES earned the Bachelor's degree in Political Science and Environmental Studies at Amherst College, and is Program Manager, Google Research.

JULIE E. LUCERO is Associate Professor of Health and Kinesiology, and Associate Dean of Equity, Diversity, and Inclusion, in the College of Health at the University of Utah.

TERESA MARES is Associate Professor of Anthropology and Gund Institute for Environment Fellow at University of Vermont.

ERIKA MARQUEZ is Assistant Professor in the School of Public Health at the University of Nevada, Las Vegas.

DENISS MARTINEZ is a Ph.D. Candidate in Ecology at University of California, Davis.

FELICIA M. MITCHELL is Associate Professor in the School of Social Work at Arizona State University.

RYAN PETTEWAY is Associate Professor in the OHSU-PSU School of Public Health at Portland State University.

CAROLINA PRADO is Assistant Professor of Environmental Studies at San Francisco State University.

ASHWIN J. RAVIKUMAR is Assistant Professor of Environmental Studies, and Latinx and Latin American Studies, at Amherst College.

R. DAVID REBANAL is Associate Professor of Public Health, and Affiliate Faculty in the Health Equity Institute, in the College of Health and Social Sciences at San Francisco State University.

MIRIAM SOLIS is Assistant Professor of Community and Regional Planning at the University of Texas at Austin.

INDEX

Note: *f* indicates a figure and *t* indicates a table.